Get the eBook FREE!

(PDF, ePub, Kindle, and liveBook all included)

We believe that once you buy a book from us, you should be able to read it in any format we have available. To get electronic versions of this book at no additional cost to you, purchase and then register this book at the Manning website.

Go to https://www.manning.com/freebook and follow the instructions to complete your pBook registration.

That's it!
Thanks from Manning!

Data Storytelling with Altair and AI

Data Storytelling
with Altair and AI

ANGELICA LO DUCA

MANNING
SHELTER ISLAND

For online information and ordering of this and other Manning books, please visit
www.manning.com. The publisher offers discounts on this book when ordered in quantity.
For more information, please contact

Special Sales Department
Manning Publications Co.
20 Baldwin Road
PO Box 761
Shelter Island, NY 11964
Email: orders@manning.com

Manning Publications Co.
20 Baldwin Road
PO Box 761
Shelter Island, NY 11964

Development editor:	Ian Hough
Technical editor:	Ninoslav Čerkez
Review editors:	Dunja Nikitović and Isidora Isakov
Production editor:	Keri Hales
Copy editor:	Christian Berk
Proofreader:	Katie Tennant
Typesetter:	Dennis Dalinnik
Cover designer:	Marija Tudor

ISBN: 9781633437920
Printed in the United States of America

To Andrea, my unique love and best friend of my life

brief contents

         10 ■ Common issues while using generative AI   285
         11 ■ Publishing the data story   300

contents

preface

Since ancient times, humans have told stories to transmit values and communicate with their fellow humans. The power of a story is enormous. It involves those who listen and excites those who tell. A very strong bond is established between the narrator and the listener, which goes beyond the content of the story itself. For this, every storyteller should know their audience and adapt stories according to their audience's values, traditions, and culture.

The power of a data-based story should be even stronger because it is anchored in the evidence provided by the data. It's impossible to disbelieve a story based on data.

Personally, I have always loved stories, although to tell the truth, when I was little, my mother didn't tell me many. It was my grandmother who told me ancient stories of imaginary characters—some even terrifying—who invited my childhood imagination to create a world entirely apart. I have always had a passion for stories, and I started writing them when I was very young. I remember my first story, written when I was seven. Over the years, I have continued to write short stories, poems, and short novels.

More than 15 years ago, I began my adventure in software development and data science applied to research and university education of students. I developed a great love for data and the people behind it, as well as a passion for software development. At the same time, I continued to write stories, poems, and short stories in the brief moments I found to dedicate to my hobbies and passions.

Then, a few years ago, I had a flash of genius, thanks to my boss, Andrea Marchetti. I will never tire of thanking him for helping me combine data with stories. From this discovery, my passion for data storytelling was born. In my own small way, I had the same experience as Steve Jobs, who, as a young man, had learned the art of calligraphy in a university course and then put it aside. Ten years later, it all returned to him: Jobs designed the first Macintosh and used the information he learned in calligraphy class to design the first computer with beautiful typography. Jobs said, "*You can't connect the dots looking forward; you can only connect them looking backwards. So you have to trust that the dots will somehow connect in your future.*" (Excerpt from Jobs's speech on June 12, 2005, to the recent graduates of Stanford University.) And so it happened to me, too: I combined my passion for data with my passion for stories.

My hope is that you, too, can connect your dots and that by reading this book, you can find ideas for tackling your daily work in a new and exciting way.

acknowledgments

First, I would like to thank my husband, Andrea. His patience and support have helped to bring this book to life. I am also grateful to my two children, Giulia and Antonio, for their smiles and love. A big thank you goes to my father, Angelo, for teaching me to never give up, even in the face of difficulties. I would also like to thank my sweet mother for her tenderness and encouraging cuddles. Mum, I miss you.

This book wouldn't have seen the light of day without the incredible work of the fantastic team at Manning: Andy Waldron, acquisitions editor; Ian Hough, development editor; Ninoslav Čerkez, technical editor; and the rest of the production team, who worked diligently to get this book to press.

Also, I thank all the reviewers: Alain Couniot, Ali Shakiba, Andre Weiner, Ankit Anchlia, Bin Hu, Daniel Paes, David Cronkite, George Carter, Greg Grimes, João Marcelo Borovina Josko, Jeremy Chen, Joel Kotarski, Jose San Leandro, Karan Gupta, Kaushik Kompella, Keerthivasan Santhanakrishnan, Krishnamurthy TV, Madiha Khalid, Manos Parzakonis, Maxim Volgin, Mikael Dautrey, Monica Guimaraes, Nadir Doctor, Oliver Korten, Peter Henstock, Radhakrishna MV, Richard Vaughan, Sarang S. Brahme, Scott Chaussee, Shaurya Khurana, Shiroshica Kulatilake, Sriram Macharla, Subhasis Ghosh, Thomas Joseph Heiman, Vidhya Vinay, and Xiangbo Mao. Without their suggestions and expertise, this book would not be what it is.

A final thank-you goes to you, reader, who, by purchasing this copy of the book, has placed your trust in me. I hope I don't disappoint your expectations.

Many thanks to all.

about this book

I wrote *Data Storytelling with Altair and AI* to help you improve your skills in communicating data exploration and analysis results. Although many tools exist to build data stories, such as Tableau and Power BI, there is a need for Python data scientists to build stories directly using their preferred programming language. In fact, if they perform all data exploration and analysis in Python, it should be natural to communicate data results using the same programming language. This book combines theoretical concepts, such as the data, information, knowledge, wisdom pyramid, and practice (Python, Altair, and generative AI) to make you aware of the potential of data storytelling while programming.

Who should read this book

Data Storytelling with Altair and AI is for Python data scientists or analysts and developers looking to learn the topic of data storytelling. Both beginners and experienced "Pythonists" will learn to use Python and generative AI for data storytelling. Although many blog posts and online resources exist, this book organizes concepts progressively so that the reader can start learning a new concept after they have assimilated the previous one.

How this book is organized: A road map

This book comprises three parts covering 11 chapters and 3 appendixes. Chapters 1–4 introduce the topic and explain the logic followed in the book: how to combine data storytelling, the DIKW pyramid, Altair, and generative AI. Chapter 5 dives deep into the DIKW pyramid, and chapters 6–9 analyze each step of the DIKW pyramid progressively. Chapter 10 contains some practical considerations for using generative AI correctly. Finally, in chapter 11, you'll see how to publish your story.

Part 1 introduces the topic of data storytelling and the tools used to implement it: Python Altair, generative AI, and the data, information, knowledge, wisdom (DIKW) pyramid:

- Chapter 1 introduces the concepts of data storytelling and the DIKW pyramid and why you should use them to communicate the insights extracted during data exploration and analysis.
- Chapter 2 describes the basic technical concepts, enabling you to write your first data story using the DIKW pyramid, Altair, and generative AI.
- Chapter 3 deepens the discussion on Altair and Vega as well as Vega-Lite, the data visualization grammars behind Altair.
- Chapter 4 discusses the basic concepts behind generative AI and how to use it for data storytelling.

Part 2 deepens the concepts learned in part 1 and uses the DIKW pyramid as a guideline to build your data story:

- Chapter 5 reviews the DIKW pyramid through a practical case study and sets the foundations for the following chapters.
- Chapter 6 focuses on the bottom of the DIKW pyramid and discusses how to turn data into information. It also shows how to build the most popular charts in Altair, such as bar charts, line charts, pie charts, and geographical maps.
- Chapter 7 illustrates the next step of the DIKW pyramid: how to transform information into knowledge. It describes how to use ChatGPT to build a textual context and insert it as an annotation or a commentary to the basic chart at the bottom of the pyramid.
- Chapter 8 continues the process of transforming information into knowledge. But in this case, the focus is on visual context. The chapter explains how to use DALL-E to build images to insert into the chart.
- Chapter 9 reaches the top of the DIKW pyramid by describing how to turn knowledge into wisdom. The chapter discusses the concept of next steps and how you can use ChatGPT to suggest possible next steps in your story.

Part 3 discusses what to do once your data story is ready, focusing on ethical aspects to consider when dealing with generative AI and describing different techniques to publish the final data story:

- Chapter 10 reflects on the problems encountered when using generative AI, such as bias and hallucinations. It also describes some possible solutions to problems when applied to data storytelling.
- Chapter 11 proposes different ways to publish a data story, based on Streamlit, a Python framework to build web applications; Tableau; Power BI; and Comet, an experimentation platform for machine learning.

The book also includes three appendixes:

- Appendix A describes the technical requirements to set up the environment, such as how to install Python and the libraries used in the book.
- Appendix B describes an overview of the pandas library, focusing on the Data-Frame and the functions used in the book.
- Appendix C describes some common data visualization charts and their implementation in Python Altair.

There are two possible strategies to use when reading this book:

- Read it from cover to cover. With this strategy, you'll learn concepts gradually, and you'll be guided to reinforce already learned concepts and learn new concepts chapter after chapter.
- Read part 1 and then pick what you need in part 2. Finally, extract the relevant content for your projects in part 3.

Throughout the book, you'll implement seven case studies, one in each chapter in part 1 (four case studies), two in chapter 5, and one throughout chapters 6–9. You'll also find exercises and challenges inviting you to put the concepts learned into practice. The exercises are optional, so feel free to skip them. Often, the exercises will ask you to change the implemented use cases, while the challenges will invite you to look at the scenarios differently. You can find all the solutions to the exercises in the book's GitHub repository.

About the code

This book contains many examples of source code both in numbered listings and in line with normal text. In both cases, source code is formatted in a `fixed-width font like this` to separate it from ordinary text. Sometimes code is also **in bold** to highlight code that has changed from previous steps in the chapter, such as when a new feature adds to an existing line of code. To set up the environment, refer to appendix A.

In many cases, the original source code has been reformatted; we've added line breaks and reworked indentation to accommodate the available page space in the book. In rare cases, even this was not enough, and listings include line-continuation markers (➡). Additionally, comments in the source code have often been removed from the listings when the code is described in the text. Code annotations accompany many of the listings, highlighting important concepts.

You can get executable snippets of code from the liveBook (online) version of this book at https://livebook.manning.com/book/data-storytelling-with-altair-and-ai. The complete code for the examples in the book is available for download from the Manning website at https://www.manning.com/books/data-storytelling-with-altair-and-ai, and from GitHub at https://github.com/alod83/Data-Storytelling-with-Altair-and-AI/tree/main.

liveBook discussion forum

Purchase of *Data Storytelling with Altair and AI* includes free access to liveBook, Manning's online reading platform. Using liveBook's exclusive discussion features, you can attach comments to the book globally or to specific sections or paragraphs. It's a snap to make notes for yourself, ask and answer technical questions, and receive help from the author and other users. To access the forum, go to https://livebook .manning.com/book/data-storytelling-with-altair-and-ai/discussion. You can also learn more about Manning's forums and the rules of conduct at https://livebook.manning .com/discussion.

Manning's commitment to our readers is to provide a venue where a meaningful dialogue between individual readers and between readers and the author can take place. It is not a commitment to any specific amount of participation on the part of the author, whose contribution to the forum remains voluntary (and unpaid). We suggest you try asking the author some challenging questions lest her interest stray! The forum and the archives of previous discussions will be accessible from the publisher's website as long as the book is in print.

about the author

ANGELICA LO DUCA is a researcher at the Institute of Informatics and Telematics of the National Research Council, Italy. She is also an adjunct professor of Data Journalism at the University of Pisa. Her research interests include data storytelling, data science, data journalism, data engineering, and web applications. In the past, she worked with topics such as network security, semantic web, linked data, and blockchain. She has published over 40 scientific papers at national and international conferences and journals. She has participated in different national and international projects and events. She is the author of the book *Comet for Data Science* (Packt Publishing, 2022) and coauthor of the book *Learning and Operating Presto* (O'Reilly Media, 2023).

about the cover illustration

The figure on the cover of *Data Storytelling with Altair and AI* is "Habitude d'une Morlaque de Sluin en Croatie," or "Habit of a Morlaque from Sluin in Croatia," taken from a collection by Jacques Grasset de Saint-Sauveur, published in 1788. Each illustration is finely drawn and colored by hand.

In those days, it was easy to identify where people lived and what their trade or station in life was just by their dress. Manning celebrates the inventiveness and initiative of the computer business with book covers based on the rich diversity of regional culture centuries ago, brought back to life by pictures from collections such as this one.

Part 1

Introducing Altair and generative AI to data storytelling

What is data storytelling? How can you implement data-driven stories using Python Altair? What benefits would generative AI introduce to building data stories? You'll find the answers to each of these questions in part 1 of this book. This part introduces the use of generative AI in a progressive way: first, we will look at GitHub Copilot, and later, in chapter 4, ChatGPT and DALL-E. I chose to follow this learning strategy because it is better to first lay the theoretical foundations to understand the main concepts and then automate them, using the various tools generative AI makes available. Using this approach, you will be the full master of generative AI tools, rather than their servant.

Before we enter the wonderful world of data storytelling combined with generative AI, I want to warn you of one thing: generative AI is an ever-evolving field, so the code you read in this book may be obsolete by the time you attempt to run it (although it was updated to the latest version at the time of writing). However, the principles described always remain valid, and you can check the official documentation of generative AI tools to update the code. Indeed, the code described in this book is located on a GitHub repository, so you might even think about opening issues on broken code to keep the GitHub repository continuously updated. It would be a fantastic way to collaborate together, and I would be really grateful if you did.

In chapter 1, you'll learn what data storytelling is and why you should use it to communicate the insights extracted during data exploration and analysis. To build a data story, you'll also be introduced to the data, information, knowledge, wisdom (DIKW) pyramid, the model you'll use throughout the book. Although other models exist to build data stories (e.g., the storytelling arc), I've chosen the DIKW pyramid as a reference for this book because I believe it's more straightforward and effective. It acts like matryoshka dolls, where the most external piece contains all the previous ones, although every single piece can exist as a standalone object. Like the matryoshka pieces (aka nesting dolls), each step of the DIKW pyramid can live as a standalone (partial) story; however, only the last step, wisdom, contains a complete story. In the last part of this chapter, you'll see a practical use case that applies the DIKW pyramid, which will unveil the potential of this model. I hope this simple example will pique your interest. However, if it does not meet your expectations, I ask you to be patient. Throughout the book, you will encounter many other practical case studies and examples, which I hope you can use as references in your own real-world scenarios.

In chapter 2, you'll get your hands dirty by writing your first data story using Altair and generative AI. For now, you'll use only Copilot, but be patient. You'll learn how to use ChatGPT and DALL-E later in the book. In this chapter, you will only see Copilot at work. You will not simply implement examples, but you will learn the basic techniques to use Copilot as a working tool, not only with Altair, but to produce any type of code.

Chapter 3 will review the basic concepts of Altair as well as Vega and Vega-Lite, the data visualization grammars behind Altair. You'll implement practical exercises and a final case study using the DIKW pyramid. Compared to the case studies implemented in the previous chapters, this one is slightly complex. At the end of this chapter, you will have matured enough Altair skills to be ready to use ChatGPT and DALL-E.

In chapter 4, you'll learn how to structure a prompt for ChatGPT and DALL-E and how to use these tools for data storytelling. Before using them, you'll review some general concepts related to artificial intelligence, machine learning, deep learning, and generative AI. This will help you to set the context of generative AI tools. In the last part of the chapter, you'll implement a practical use case, showing the potential of generative AI in data storytelling.

As outlined in the following list, at the end of each chapter, you'll implement a practical case study, each with a different purpose:

- *Chapter 1*—This case study is only theoretical (without any code) and focuses on some statistics related to an *advertising campaign* about pets. You'll find the code for this example in the GitHub repository for the book.
- *Chapter 2*—This case study describes a simplified *decision-making process* related to the opportunity for a hotel to build a new swimming pool. You'll implement two versions: one using and one without using GitHub Copilot.

- *Chapter 3*—This case study focuses on a *data-journalism*-like example, studying population growth in North America. The example is more advanced than those implemented in the previous chapters.
- *Chapter 4*—This case study involves again a *decision-making process*. This one adds generative AI to the preceding case studies.

Introducing data storytelling

This chapter covers

- What data storytelling is
- The importance of data storytelling
- Why you should use Python Altair and generative AI tools for data storytelling
- When Altair and generative AI tools are not useful for data storytelling
- How to read this book
- The data, information, knowledge, wisdom (DIKW) pyramid

By purchasing this book, you have decided to embark on the stimulating path of combining data storytelling, Python, and generative AI. Likely, you already have some knowledge of data visualization and want to learn new techniques to improve your charts and how to use generative AI in the chart creation process. Simply put, by reading this book, you would like to learn how to do data storytelling using generative AI. I'll tell you right away that the book isn't only that—it's more than just background on data storytelling and generative AI. In fact, throughout the book, you will also learn a methodology that will allow you to create your own data-based

5

stories systematically. So it won't be another book of code and examples to implement, but here and there, I will give you the theoretical basis for implementing your own data-driven stories. This book is a mix of theory and practice, which will allow you to apply the techniques learned in other contexts.

The examples you will encounter throughout the book—and in this chapter—will be essential, as they aim to be as simple as possible to understand a working method. At the end of the book, you will realize that rather than having implemented examples, you will have learned a working methodology. Therefore, I ask you not to be disappointed by the simplicity of the examples but to look at the methodology behind them and how you can apply it in your work, which is certainly more complex than the examples described in this book. So let's start our journey!

1.1   *The art of data storytelling*

Data storytelling is a powerful way to share data insights by transforming them into narrative stories. It is an art you can use for any industry, such as government, education, finance, entertainment, and healthcare. Data storytelling is not only for data scientists and analysts; it's for anyone who has ever wanted to tell a story with data.

To move from data visualization to data storytelling, you need to change perspective. Instead of looking at the data from your point of view, you will have to look at it from the point of view of the people you will tell it to—in other words, the *audience*. Figure 1.1 shows the journey from data to the intended audience, as viewed through the eyes of both the skilled data scientist (represented on the left) and the eagerly

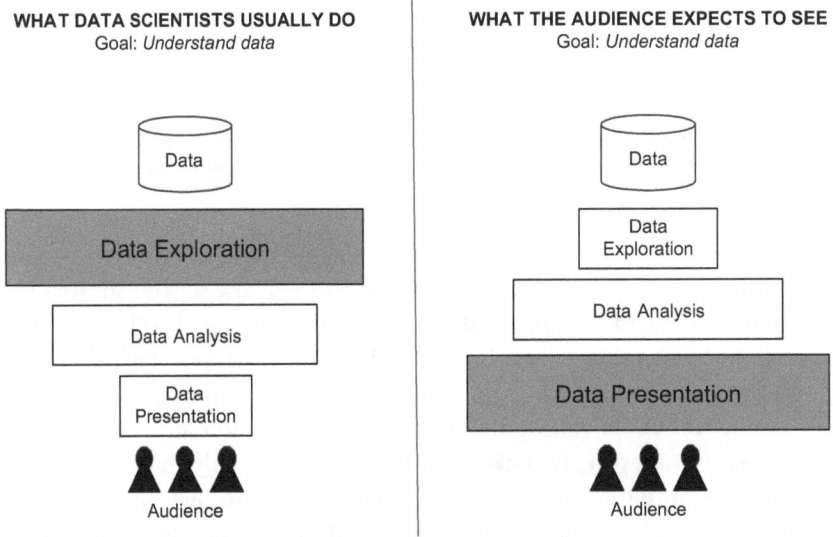

Figure 1.1   The data science flow from the data scientist's perspective (on the left) and the audience's perspective (on the right)

awaiting audience (represented on the right). The flow consists of three major phases: data exploration, analysis, and presentation. The size of each box reflects the amount of time devoted to its respective phase. While data scientists and audiences share a common goal (that is, to grasp the essence of the data truly), how they achieve said goal varies. Data scientists understand data during the data exploration phase, whereas audiences take center stage during data presentation.

Data storytelling can help data scientists and analysts to present and communicate data to an audience. You can think of data storytelling as the grand finale of the data science life cycle. It entails taking the results of the previous phases and transforming them into a narrative that effectively communicates the results of data analysis to the audience. Rather than relying on dull graphs and charts, data storytelling enables you to bring your data to life and communicate insights compellingly and persuasively. Data storytelling gives the audience an opportunity to feel emotions and experience wins and setbacks while pursuing a goal.

More formally, data storytelling builds compelling stories, supported by data, allowing analysts and data scientists to present and share their insights interestingly and interactively. The ultimate goal of data storytelling is to engage the audience and inspire them to make decisions. In some cases, including business scenarios, data is not the first step; to begin with, you have in mind a narrative or a hypothesis. Then, you search for data that confirms or negates it. In this case, you can still have data storytelling, but you must pay attention not to alter your data to support your hypothesis. Brent Dykes, a well-known consultant in storytelling training, suggests the following approach: "Whenever you start with the narrative and not the data, it requires discipline and open-mindedness. In these scenarios, one source of risk will be confirmation bias. You will be tempted to cherry-pick data that confirms your viewpoint and ignore conflicting data that doesn't." (Dykes, 2023) Remember to build your data stories on accurate and unbiased data analysis. In addition, always consider the data you are analyzing.

Some time ago, I had the opportunity to work on a cultural heritage project on which I had to automatically analyze entities from the transcripts of a registry of names dating back to around 1700–1800. The goal was to calculate some statistics about the people in the register, such as the most frequently appearing name, the number of births by year, and so on. Sitting at my computer, I calculated and visualized data statistics. The project also involved linking these people to their graves to build an interactive cemetery map. At some point in the project, I had the opportunity to visit the cemetery. As I walked through it, the rows upon rows of headstones made me stop. It hit me like a ton of bricks: every name etched into those stones represented a life. Suddenly, the numbers and statistics I had been poring over in my datasets became more than just data points—they were the stories of real people. It was a powerful realization that changed the way I approached my work. That's when I discovered the true power of data storytelling. It's not just about creating fancy graphs and charts—it's about bringing the people behind the data to life. We have a mission to give these people a voice, to ensure that their stories are heard. And that's exactly what data storytelling does; it gives a voice to people often buried deep within the numbers. Our mission as data storytellers is to bring these stories to the forefront and ensure they are heard loud and clear.

In this book, you'll learn two technologies to transform data into stories: Python Vega-Altair (or simply Altair) (https://altair-viz.github.io/) and generative AI tools. Python Altair is a Python library for data visualization. Unlike the most known Python libraries, such as Matplotlib and Seaborn, Altair is a declarative library where you specify only what you want to see in your visualization. This aspect is beneficial for quickly building data stories without caring about how to build a visualization. Altair also supports chart interactivity, so users can explore data and interact with it directly.

Generative AI is the second technology you'll use to build data stories in this book. We will focus on ChatGPT to generate text, DALL-E to generate images, and GitHub Copilot to generate the Altair code automatically. I chose to use GitHub Copilot to generate code, rather than ChatGPT, because Copilot was trained with domain-specific texts, including GitHub and Stack Overflow codes. Instead, ChatGPT is more general purpose. At the time of writing, generative AI is a very recent technology, still in progress, which translates a description of specifications or actions into text.

Data storytelling is about more than communicating data; it's about inspiring your audience and inviting them to take action. Good data storytelling requires a mix of art and science. The art is in finding the right story, while the science is in understanding how to use data to support that story. When done well, data storytelling can be a potent tool for change. In the remainder of this section, we'll briefly cover three fundamental questions about data storytelling.

1.1.1 Why should you use data storytelling?

Data storytelling enables you to convey the results of your data analysis process, using narratives that can be easily understood by an audience. Throughout this book, we will see many examples and case studies. For example, you will see how to transform the raw chart in figure 1.2 into the data story shown in figure 1.3.

Data storytelling allows you to fill the gap between simply visualizing data and communicating it to an audience. Data storytelling improves your communication skills and standardizes and simplifies the process of communicating results, making it easier for people to understand and remember information. Data storytelling also helps you learn to communicate more effectively with others, improving personal and professional relationships.

Use data storytelling if you want to do any of the following:

- Focus on the message you want to communicate and make data more understandable and relatable.
- Communicate your findings to others in a way that is clear and convincing.
- Connect with your audience on an emotional level, which makes them more likely to take action.
- Inspire the audience to make better decisions by helping them understand your data more deeply.

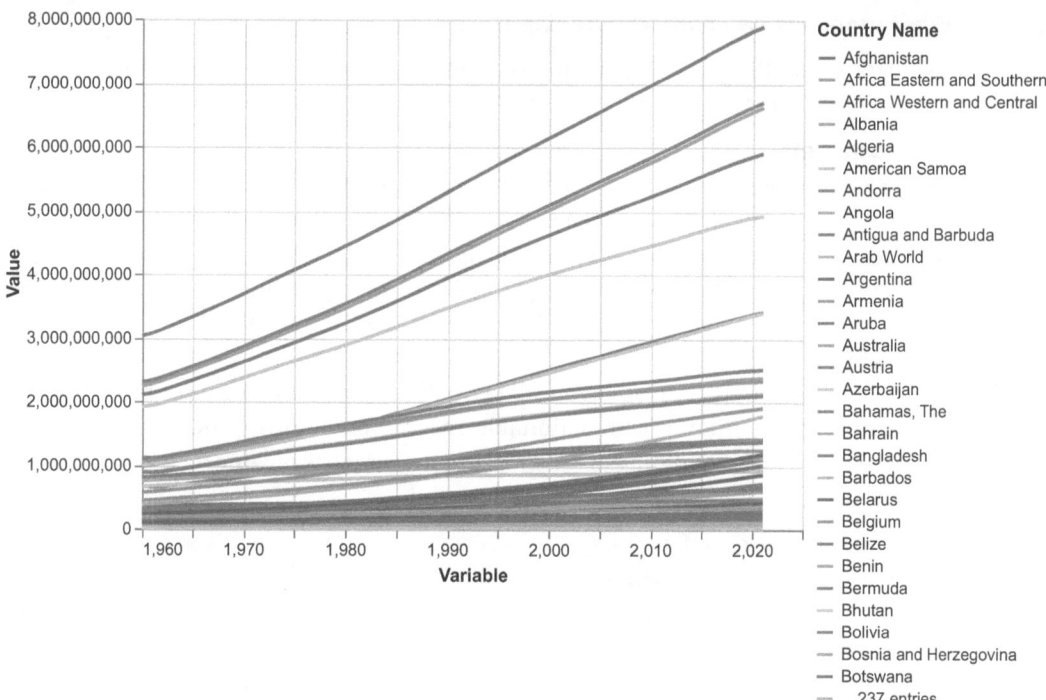

Figure 1.2 An example of a raw chart

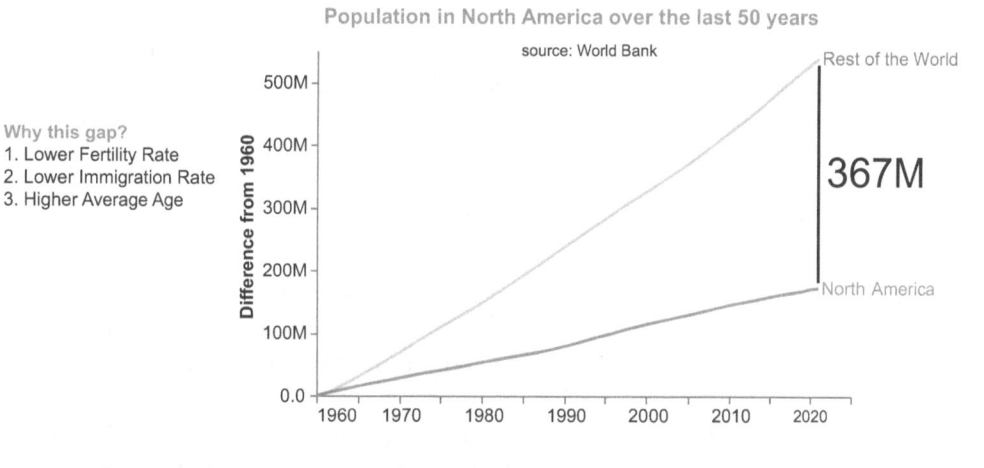

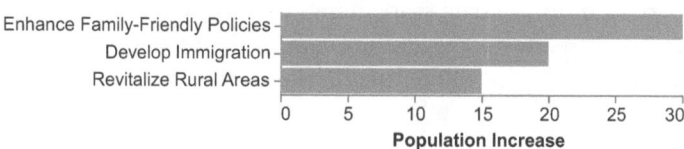

Figure 1.3 The raw chart from figure 1.2 transformed into a data story

1.1.2   *What problems can data storytelling solve?*

Use data storytelling if you want to communicate something to an audience in the form of writing reports, doing presentations, or building dashboards.

WRITING REPORTS

Imagine you must write a sales report for a retail company. Instead of presenting raw numbers and figures, you can weave a data story about the performance of different product categories. Start by identifying the most crucial aspects of the data, such as the top-selling products, emerging trends, or seasonal fluctuations. Then, use a combination of visualizations, anecdotes, and a logical narrative flow to present the information.

You can build a story around your data, such as by introducing a problem, building suspense, and concluding with actionable recommendations. When writing reports, use data storytelling to highlight the most important parts of your data and make your reports more engaging and easier to understand.

DOING PRESENTATIONS

Consider a marketing presentation in which you must demonstrate various marketing campaigns' effectiveness. Instead of bombarding the audience with numerous charts and statistics, focus on creating a compelling narrative that guides them through how the campaigns unfolded and their impact on the target audience. Finally, show potential next steps to follow. In presentations, use data storytelling to engage your audience and help them understand your message better.

BUILDING DASHBOARDS

Let's imagine you are developing a sales performance dashboard for a retail company. Instead of presenting a cluttered interface with overwhelming data, focus on guiding users through a narrative, highlighting key insights. When building dashboards, use data storytelling to build more user-friendly and informative dashboards.

1.1.3   *What are the challenges of data storytelling?*

Crafting compelling data stories is no easy feat. It demands *time* to ensure engaging narratives packed with valuable information. Moreover, it is a *team effort* because it involves bringing together individuals from diverse backgrounds, each with their own expertise and perspectives, to work collaboratively. This collaboration can be challenging, but it's essential for weaving together a cohesive and impactful data story.

Creating data stories involves two key challenges: time and teamwork. Investing in these areas is crucial to captivate audiences and effectively communicate insights.

Now that we've covered when to use data storytelling, what problems it can solve, and what makes it unique, we're ready to consider questions relating to our two tools: Python Altair and generative AI tools. We will do that in the next section.

1.2 Why should you use Python Altair and generative AI for data storytelling?

Python provides you with many libraries for data visualization. Many of them, including Matplotlib and Seaborn, are *imperative libraries*, meaning you must define exactly how you want to build a visualization. Python Altair, instead, is a *declarative library*, meaning you specify only what to visualize. Using Python Altair for data storytelling instead of other imperative libraries allows you to quickly build your visualizations.

For example, to plot a line chart using Matplotlib, you must specify explicitly the x and y coordinates, set the plot title and labels, and customize the appearance.

Listing 1.1   Imperative library

```python
import matplotlib.pyplot as plt

x = [1, 2, 3, 4, 5]
y = [1, 4, 9, 16, 25]

plt.plot(x, y)
plt.title('Square Numbers')
plt.xlabel('X')
plt.ylabel('Y')

plt.show()
```

> **NOTE**   The chart builds a line chart in Matplotlib. You must define the single steps to build the chart: (1) set the title, (2) set the x-axis, (3) set the y-axis.

Declarative visualization libraries, like Altair, enable you to define the desired outcome, without specifying the exact steps to achieve it. For instance, using Altair, you can simply define the data, define the *x* and *y* variables, and let the library handle the rest, including axes, labels, and styling, resulting in a more concise and intuitive code.

Listing 1.2   Declarative library

```python
import altair as alt
import pandas as pd

df = pd.DataFrame({'x': [1, 2, 3, 4, 5], 'y': [1, 4, 9, 16, 25]})

chart = alt.Chart(df).mark_line().encode(
    x='x',
    y='y'
).properties(
    title='Square Numbers'
)

chart.save('chart.png')
```

> **NOTE**   The chart builds a line chart in Altair. You must define the chart type
> (`mark_line`), the variables, and the title.

In imperative Matplotlib, you had to define the axis in a certain way and let the library handle it; however, in the declarative library, you define the code with *x* and *y* variables with axis labels and styling so that, as a creator, you can be more specific and directional in the chart you intend to create and expand.

Generative AI is a subset of artificial intelligence techniques involving creating new, original content based on patterns and examples from existing data. It enables computers to generate realistic and meaningful outputs, such as text, images, or even code. In this book, we will focus on ChatGPT to generate text, DALL-E to generate images, and GitHub Copilot to assist you while coding:

- *ChatGPT*—An advanced language model developed by OpenAI. Powered by the GPT-3.5 or GPT-4 model, it is designed to engage in human-like conversations and provide intelligent responses.
- *DALL-E*—A generative AI model created by OpenAI. It combines the power of GPT-3 with image generation capabilities, allowing it to create unique and realistic images from textual descriptions.
- *GitHub Copilot*—A new tool powered by OpenAI Codex that assists you while writing your code. In GitHub Copilot, you describe the sequence of actions that your software must run, and GitHub Copilot transforms it into a runnable code in your preferred programming language. The ability to use GitHub Copilot consists of learning how to describe the sequence of actions. GitHub Copilot is a for-fee tool, but you can apply the concepts described in this book to other popular AI code assistants as well.

Combining Python Altair and generative AI tools will enable you to write compelling data stories more quickly directly in Python. For example, we can use Copilot to assist us in generating the necessary code snippets, such as importing the required libraries, setting up the plot, and labeling the axes. In addition, Copilot's contextual understanding helps it propose relevant customization options, such as adding a legend or changing the color scheme, saving time and effort in searching for documentation or examples.

While generative AI tools are still in their early stages, some promising statistics reveal they increase workers' productivity. A study by Duolingo, one of the largest language learning apps, reveals that the use of Copilot in their company has increased developers' speed by 25% (Duolingo and GitHub Enterprise, 2022). Compass UOL, a digital media and technology company, ran another study, asking experienced developers to measure the time it took to complete a use case task (analysis, design, implementation, testing, and deployment) during three distinct periods: without using AI, before its availability was widespread; utilizing AI tools available through 2022; and employing modern generative AI tools, such as ChatGPT. Results demonstrated that developers completed the tasks in 78 hours before AI, 56 hours with the AI used through 2022, and 36 hours with the new generative AI. Compared to the pre-AI era, there is an increase in speed of 53.85% with the new generative AI (figure 1.4).

Increase in developers, speed compared to pre-AI era

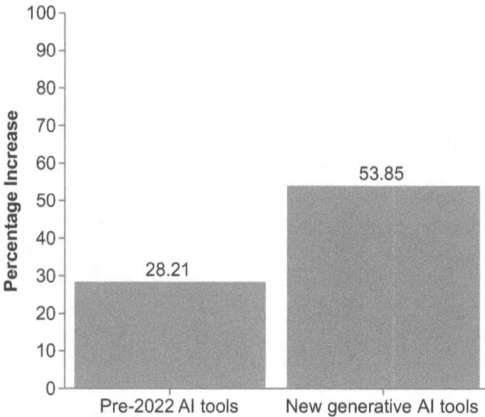

Figure 1.4   The results of the tests conducted by Compass UOL

1.2.1   The benefits of using Python in all the steps of the data science project life cycle

Many data scientists and analysts use Python to analyze their data. Thus, it should be natural to build the final report on the analyzed data in Python. However, data scientists and analysts often use Python only during the central phases of the data science project life cycle. Then, they move to other tools, such as Tableau and Power BI, to build the final report, as shown in figure 1.5. This requires adding other work, which includes exporting data from Python and importing them into the external application. This export/import operation, in itself, is not expensive, but if, while building the report, you realize that you have made a mistake, you need to modify the data in Python and then export the data again. If this process is repeated many times, there is a risk of significantly increasing the overhead until it becomes unmanageable.

This book enables data scientists and analysts to run each step of the data science project life cycle in Python, filling the gap of exporting data to an external tool or framework in the last phase of the project life cycle, as shown in figure 1.6. The advantage of using only Python is that programmers can build their reports even during the intermediate stages of their experiments, without wasting time transferring data to other tools, such as Tableau or Power BI.

1.2.2   The benefits of using generative AI for data storytelling

In general, you can use generative AI as an aid throughout the entire life cycle of a data science project. However, in this book, we will focus only on generative AI in the data presentation phase, which corresponds to the data storytelling phase.

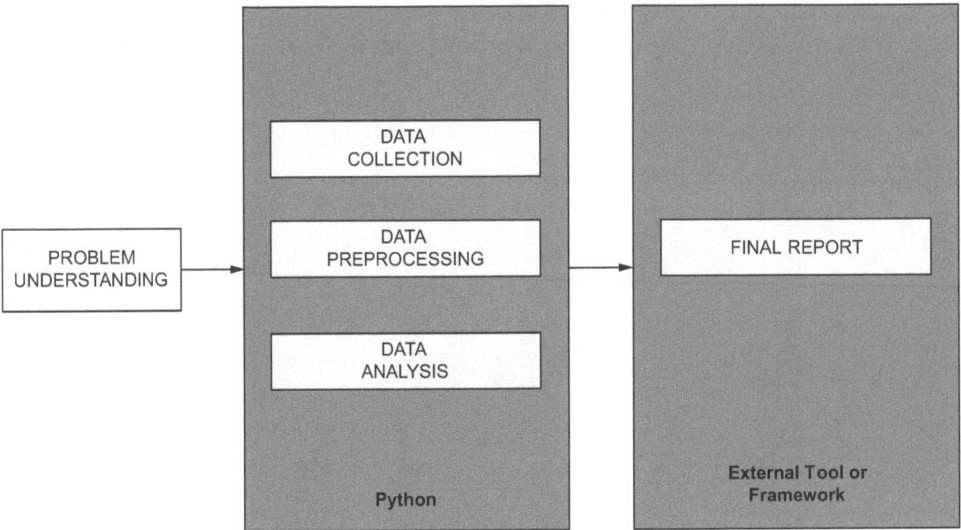

Figure 1.5   In the traditional approach, data scientists use different technologies during different phases of the data science project life cycle.

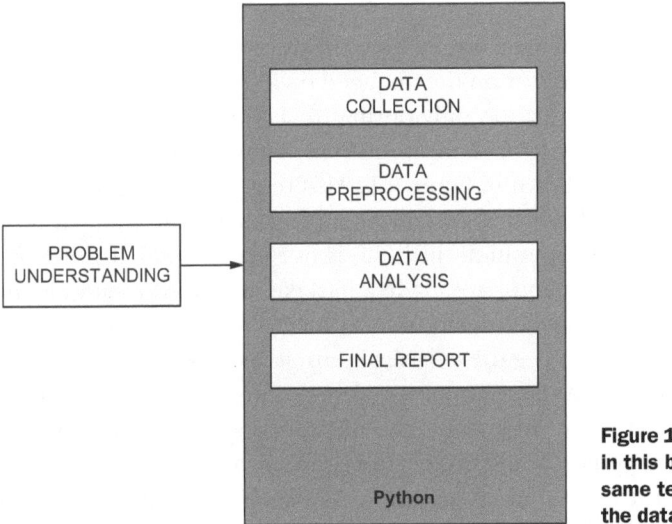

Figure 1.6   In the approach proposed in this book, data scientists use the same technology during all phases of the data science project life cycle.

The introduction of generative AI tools to aid the data presentation phase allows you to devote the effort and time you saved to the data presentation phase, obtaining better results. Thanks to generative AI tools, you can make the audience understand your data (figure 1.7). Now that we've discussed the benefits of the tools of choice in this book, we will next briefly discuss contexts in which these tools are not as effective.

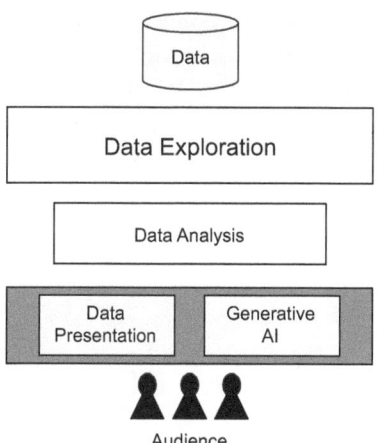

WHAT DATA SCIENTISTS DO IS WHAT THE AUDIENCE EXPECTS TO SEE
Goal: *Understand data*

Figure 1.7 Introducing generative AI in the data presentation phase helps you build better charts in less time, enabling the audience to understand your message.

1.3 When Altair and generative AI tools are not useful for data storytelling

While Python Altair and generative AI tools are handy for building data stories very quickly, they are not useful when analyzing big data, such as gigabytes of data. You should not use them for the following tasks:

- *Complex exploratory data analysis*—Exploratory data analysis helps data analysts summarize a dataset's main features, identify relationships between variables, and detect outliers. This approach is often used when working with large datasets or datasets with many variables.
- *Big data analytics*—Big data analytics analyzes large datasets to uncover patterns and trends. To be effective, big data analytics requires access to large amounts of data, powerful computers for processing that data, and specialized software for analyzing it.
- *Complex reports that summarize big data*—Generating detailed reports requires robust data processing capabilities and advanced reporting tools, especially when dealing with big data.

Altair enables you to build charts using datasets of up to 5,000 rows quickly. If the number of rows exceeds 5,000, Altair still builds the chart, but it's slower. For complex data analytics, use more sophisticated analytics platforms, such as Tableau and Power BI. Though we say this combo is not ideal for big data, if it is feasible to bring your data to below 5,000 rows using data preprocessing, then Altair and generative AI can be a good combo for your storytelling. In addition, consider that you should pay a fee to use generative AI tools, so avoid them if you do not have a sufficient budget.

1.4   Using the data, information, knowledge, wisdom pyramid for data storytelling

The primary focus of this book is a significant concept known as the *data, information, knowledge, wisdom* (DIKW) pyramid (figure 1.8), which we believe gives data scientists and analysts the macro steps to build data stories. We'll cover the DIKW pyramid more extensively in chapter 5. We introduce the DIKW pyramid here because using it to build data stories is a fundamental concept for this text.

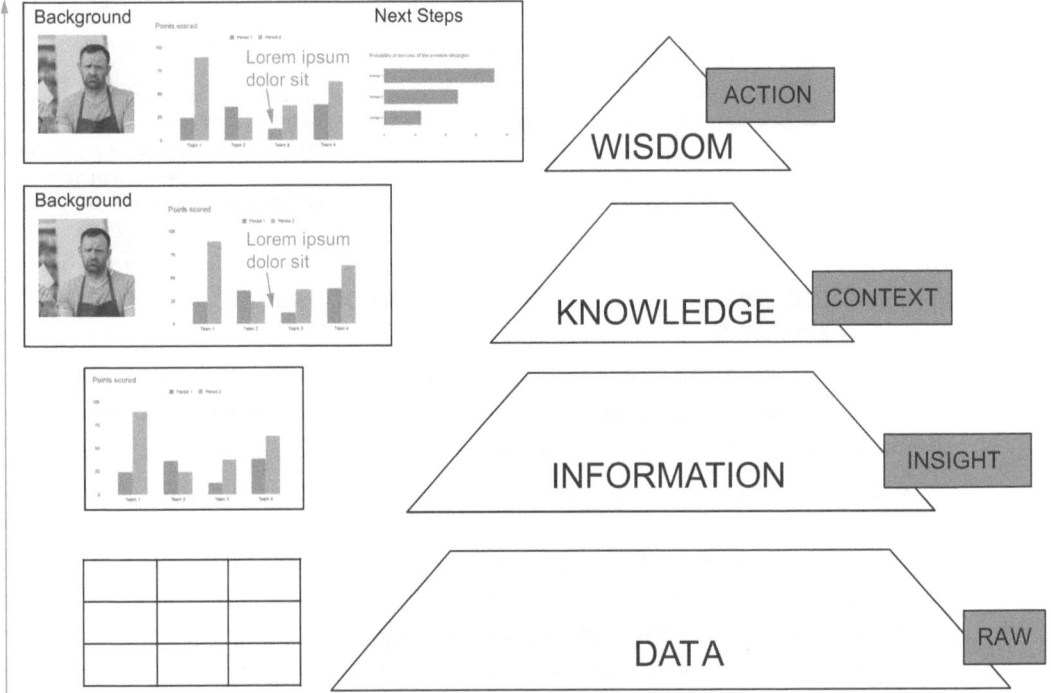

Figure 1.8   The DIKW pyramid

The DIKW pyramid provides macro steps to transform data into wisdom, following other intermediate steps, which include information and knowledge. It is composed of the following elements:

- *Data*—The building block at the bottom of the pyramid. Usually, we start from a significant amount of data, more or less cleaned.
- *Information*—Involves extracting insights from data. The information represents organized and processed data that are easy to understand.
- *Knowledge*—Information interpreted and understood through a context that defines the data background.

■ *Wisdom*—The knowledge enriched with specific ethics that invites you to act in some way. Wisdom also proposes the next steps after we understand the data.

This book describes how to use the elements of the DIKW pyramid as progressive steps to transform your data into compelling data stories. This idea is not new in data storytelling; Berengueres and Sandell proposed this approach (2019). The novelty of this book lies in the combination of the use of the DIKW pyramid, Python Altair, and generative AI. In this section, we will introduce the fundamentals of DIKW that will be applied throughout this book, and we'll explore how to climb each level of the pyramid.

1.4.1 From data to information

To turn data into information, extract insights from data. Consider the following scenario: the organizers of an event dedicated to pets are collecting the type of pets that will participate. For each pet category, the organizers advertise the event on specific websites dedicated to that category. The organizers ask you to build a quick report about the current situation. Table 1.1 shows the number of participants and the number of advertised websites, divided by type. First, you focus on the number of participants and build the following bar chart (figure 1.9).

Table 1.1 Data related to pets involved in the event

Pet	Number of Participants	Number of Advertised Websites
Cat	1138	150
Dog	130	28
Other	17	147

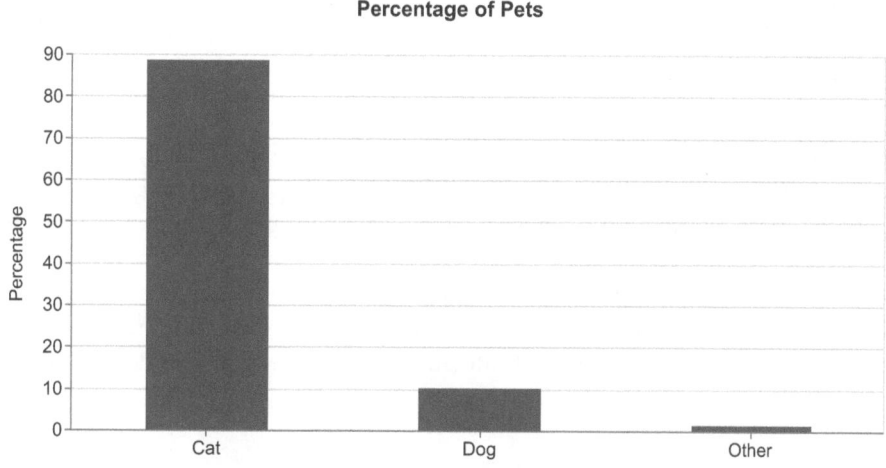

Figure 1.9 A bar chart showing the number of pets participating in the event

Figure 1.9 does not add any insight into table 1.1. It's simply a visual representation of the table. For some of the data, the table is even clearer than the bar chart.

To turn data into information, start by understanding data. Ask the following questions:

- Who is the audience of my story?
- What information do they want?
- Are all the data relevant to answer the previous questions?

Let's answer those questions. The organizers of the event are the audience for our story. They want to know the current situation of animals involved in the event. We could suppose they want to elaborate a promotion plan to increase the number of pets participating in the event.

Looking at the raw data in the table, you notice that the Cat category has the highest number of participants by some margin, followed by a relatively low number of dogs. It seems that owners of other pets, judging by the almost total lack of participants, are not interested in the event. You can use this insight to focus on one of the following options:

- Remove the Other category because it is irrelevant.
- Focus on the Other category, and propose a strategy to increase their participation.

Let's focus on the first possible course of action: removing the Other category. Start by calculating the percentage of participants for each category, as shown in table 1.2.

Table 1.2 Data related to pets involved in the event, with a focus on the percentage of the number of participants

Pet	Number of Participants	Percentage
Cat	1,138	88.56
Dog	130	10.12

Notice that 88.56% of pets are cats and 10.12% are dogs. Usually, the final audience will not be interested in the finer details, so you can approximate cats at 90% and dogs at 10%. The extracted information—your data's insight—is that 1 pet out of 10 is a dog, and 9 out of 10 are cats.

Consider the following stacked bar chart describing the situation (figure 1.10). The figure is self-explanatory because the audience can understand the situation immediately. You have extracted information (and, thus, an insight) from the data.

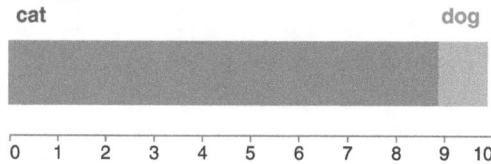

Figure 1.10   A stacked bar chart showing the current percentage of pets participating in the event

1.4.2   From information to knowledge

To turn information into knowledge, add context. Context involves all circumstances surrounding data, such as events, situations, and other details. Adding context helps the audience to understand data better.

Consider the previous example of dogs and cats. You already know that the percentage of cats is greater than that of dogs. Adding context here could involve, for example, describing the events and situations in which motivated dog owners do not participate. Let's focus on the third column of the dataset, described back in table 1.1 and recalled in table 1.3: the number of advertised websites. For cats, this is 150, and for dogs, it is 28. This information is considered context because it helps the audience understand why the number of cats is higher than the number of dogs.

Calculate the ratio between the number of participants and the number of advertised websites to understand the participants' rates. Table 1.3 shows the calculated values.

Table 1.3   Data related to pets involved in the event, with a focus on the participants' rates

Pet	Number of Participants	Percentage	Number of Advertised Websites	Participants-to-website rate (# of participants/# of websites)
Cat	1138	88.56	150	7.59
Dog	130	10.12	28	4.64

Cats' participants-to-website rate is 7.59, which is nearly 8. Dogs' participation rate is 4.64, which is almost 5. The participant rate helps the audience understand the context related to the number of participants in the event.

Figure 1.11 shows the stacked bar chart of figure 1.10, enriched with context. The context involves the following elements:

- *The title*—Summarizes the chart content
- *The header*—Describes the number of advertised websites
- *Two images*—One for cats and the other for dogs, to facilitate reading

The content described by the chart in figure 1.11 is easier to read and understand than that shown in the chart in figure 1.9. Adding context to the chart has enabled us to turn information into knowledge.

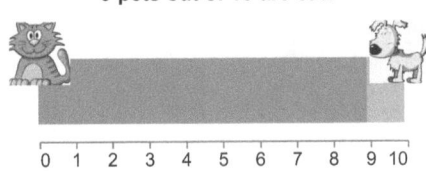

The cat participation rate
is almost 8%
over 150 sites advertised,
while the dog participation rate
is almost 5%
over 30 sites advertised.

Figure 1.11   A stacked bar chart, showing the current percentage of pets participating in the event, enriched with context

1.4.3   *From knowledge to wisdom*

Turning knowledge into wisdom means adding a call to action, which invites the audience to do something with the discovered knowledge. A call to action is a very effective way to help drive conversions. Here are some examples:

- What changes can we make?
- What opportunities do we have?
- What advantages does our program offer?
- What scenarios can we outline?
- What are some examples of scenarios?

Alternatively, you may propose a possible solution to the questions, or you may invite the audience to discuss and listen to their proposals and answers to these questions. It's here that the discussion takes place.

Your audience often needs the opportunity to voice their opinion and make suggestions. Sometimes, they may even have one or more questions.

Consider again the cats and dogs example. To add a call to action, change the title from *9 pets out of 10 are cats* to *Increase the advertising campaign on dog-related websites!* Figure 1.12 shows the resulting chart, which adds the call to action to the title.

**Increase the advertising campaign
on dog-related websites!**

The cat participation rate
is almost 8%
over 150 sites advertised,
while the dog participation rate
is almost 5%
over 30 sites advertised.

```
0  1  2  3  4  5  6  7  8  9 10
```

Figure 1.12   A stacked bar chart showing the current percentage of pets participating in the event, enriched with a call to action

The pets scenario demonstrated how to turn data into wisdom. First, we took data that contained the raw number of participants in the event. Then, we extracted information that told us that 1 participant out of 10 is a dog and the remaining 9 are cats. Next, we added a context that explained why dog owners were not interested in the event. The motivation was that the advertising campaign for dogs was poor. Finally, we added a call to action that invited the audience to increase the dog advertising campaign.

In this chapter, you have learned the basic concepts behind data storytelling and how to transform data into stories, using the DIKW pyramid. In the next chapter, you'll see how to use Python Altair and generative AI to implement the DIKW pyramid approach.

Summary

- Data storytelling is a powerful tool that helps you communicate your data more effectively. When used correctly, data storytelling makes complex data more relatable and easier to understand.
- Data storytelling requires changing your perspective from your point of view to that of the audience.
- Python Altair and generative AI are great tools for creating stunning data stories.
- Don't use Python Altair and generative AI if you want to perform big data analytics or write complex reports that summarize big data.
- Use the DIKW pyramid to turn your data into wisdom. Start from raw data, and then extract information by adding meaning to data. Next, add context to define knowledge. Finally, include action to provide wisdom.

References

- Berengueres, J. and Sandell, M. (2019). *Introduction to Data Visualization & Storytelling: A Guide for The Data Scientist.* Self-published.
- Compass UOL (2023). *Generative AI Speeds Up Software Development: Compass UOL Study.* Compass. https://blog.compass.uol/noticias/generative-ai-speeds-up-software-development-compass-uol-study/.
- Duolingo and GitHub Enterprise (2022). *Duolingo Empowers its Engineers to be Force Multipliers for Expertise with GitHub Copilot.* Codespaces. https://github.com/customer-stories/duolingo.
- Dykes, B. (2023). LinkedIn post discussing if a data story can begin with the narrative instead of the data: https://www.linkedin.com/feed/update/urn:li:activity:7061798908565336065/.

Running your
first data story in Altair
and GitHub Copilot

2

This chapter covers

- Introducing Altair
- A relevant use case: Describing the scenario
- Using Altair
- Using Copilot

One of my peculiarities is getting my hands dirty right away. If I don't feel something right away, I'm not happy. To start using software, or anything in general, I immediately go to the "Getting Started" section. As I need more details, I consult the documentation. This chapter was born with the same idea: to see how things work immediately. We'll look at a rough but complete sketch of what you'll learn in the book. In this chapter, we will look at the basic concepts behind Altair, and then we will implement a practical use case, which will allow us to transform a raw dataset into a story. We will progressively apply the data, information, knowledge, wisdom (DIKW) pyramid principles in Altair and see the results achieved step by step. In the second part of the chapter, we will use Copilot to automate some steps of the story creation process. We will focus only on Copilot as a generative AI tool to keep the chapter simple and the flow understandable. In the following chapters, we will introduce ChatGPT and DALL-E to the DIKW pyramid.

2.1 Introducing Altair

The Vega-Altair library (Altair, for short) is a declarative Python library for statistical visualization based on the Vega and Vega-Lite visualization grammars. Vega is a visualization grammar for creating expressive and interactive data visualizations. Vega-Lite is a high-level declarative language built on top of Vega, designed to simplify the creation of common data visualizations with concise and intuitive syntax. We will discuss Vega and Vega-Lite in the next chapter.

Altair is a declarative library, meaning you can describe the intended chart outcome rather than needing to manually program every step, defining a concise and intuitive syntax for creating interactive visualizations. We use declarative libraries to specify what we want to see in a chart. We can specify the data and the type of visualization we want, and the library creates the visualization for us automatically. Declarative libraries stand in contrast to imperative libraries, which instead focus on building a visualization manually (e.g., specifying the desired axis, size, legend, and labels). Matplotlib is an example of an imperative library.

Every Altair chart has at least three main elements: a chart, a mark, and encodings. We will cover each of these briefly in the following pages. Refer to appendix A for instructions on how to install Altair and the other Python libraries described in this chapter. Refer to appendix B for more details on the pandas DataFrame methods used in this chapter.

2.1.1 Chart

In Altair, a chart is an object that acts as a starting point for constructing and customizing interactive visualizations in Altair. Use the `alt.Chart()` method to input the dataset you would like to represent visually, as shown in listing 2.1 You can find this example in the GitHub repository for the book under 02/bar-chart.py. Start by cloning the GitHub repository for the book: https://mng.bz/PZVP. To get started with GitHub, follow the procedure described in appendix A of the book or the GitHub official documentation (https://mng.bz/1Geg).

Listing 2.1 Charts in Altair

```
import altair as alt
import pandas as pd

data = {'Name': ['Alice', 'Bob', 'Charlie'],
'Age': [25, 30, 35],
'City': ['New York', 'Paris', 'London']     Creates data for
}                                            the DataFrame

                                  Creates the DataFrame
                                  from the data
df = pd.DataFrame(data)

print(df)          Prints the DataFrame

chart = alt.Chart(df)       Creates the chart
```

NOTE   First, import the required libraries, then build the DataFrame containing your data. Finally, pass the DataFrame to `alt.Chart()`. The example does not create any visualization because we have not specified what we want to represent. Note that the code is still incomplete, so if you try to run it, it will not work. Be patient. In the next section, you'll add other pieces to run it!

2.1.2 Mark

A mark defines how to represent data. Examples of marks include bar charts, line charts, area charts, and many others. To specify a mark, append it to the Altair chart. For example, in the following listing, we will specify that we want to use a bar chart.

Listing 2.2   Mark in Altair

```
chart = alt.Chart(df).mark_bar()
```

NOTE   Use `mark_bar()` to create a bar chart in Altair.

The chart is not ready yet because you need other pieces before showing it. However, to get an idea, figure 2.1 shows how the final chart will look.

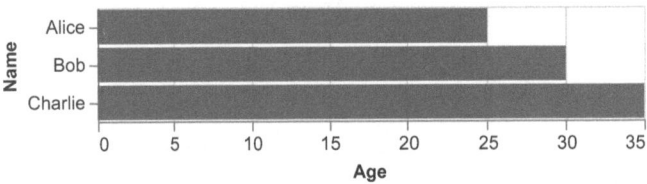

Figure 2.1   An example of a bar chart

There are other examples of marks, such as `mark_line()`, to plot a line chart (figure 2.2). Additionally, `mark_circle()` plots a scatter plot (figure 2.3).

2.1.3 Encodings

Encodings specify the mapping between the DataFrame columns and their visual representation in the Altair chart. In practice, encodings define where to represent data in the chart, such as their position, size, and color. In practice, encodings define how data attributes are mapped to visual properties in a chart. For example, you can encode a dataset's *x* values as positions on the *x*-axis and *y* values as positions on the y-axis. A *channel* in Altair refers to a visual property, like x, y, color, size, and shape, which can be used to represent data attributes in a chart.

Each channel is associated with a data type, which describes the type of data an attribute contains, such as *quantitative* for numerical data, *ordinal* for ordered categorical data, or *nominal* for categorical data with no inherent order. For example, in the

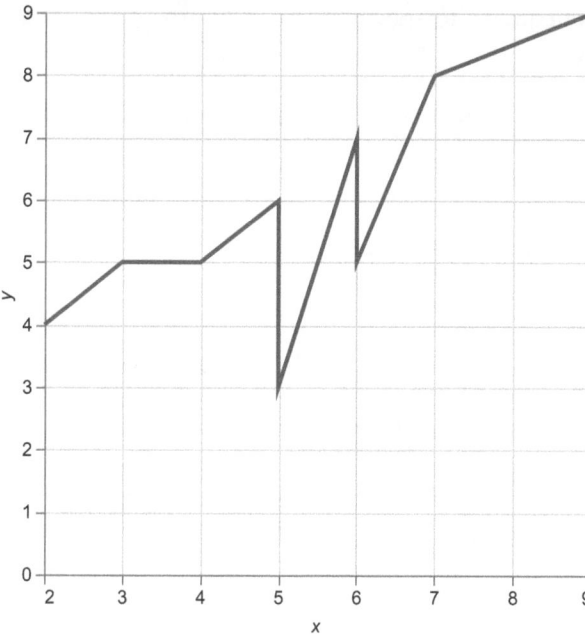

Figure 2.2 An example of a line chart

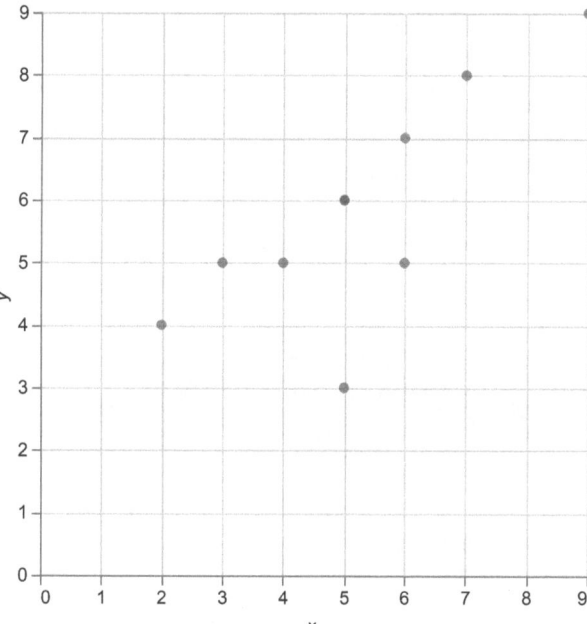

Figure 2.3 An example of a scatter plot

following listing, we use encodings to specify which columns of the DataFrame we must use in the x- and y-axes.

Listing 2.3   Encodings in Altair

```
import pandas as pd
import altair as alt

data = {'Name': ['Alice', 'Bob', 'Charlie'],
        'Age': [25, 30, 35],
        'City': ['New York', 'Paris', 'London']
}                                                    ◁─┐  Creates data for
                                                         the DataFrame
df = pd.DataFrame(data)              ◁─┐
                                        Creates the DataFrame
# Create the chart                      from the data
chart = alt.Chart(df).mark_bar(
).encode(
    x = 'Age:Q',
    y = 'Name:N'
)

chart.save('bar-chart.html')
```

NOTE   Use encode() to define encodings in Altair. Place the Age variable of df on the x-axis and the Name variable of df on the y-axis. Interpret Age as a quantity (Q) and Name as a nominal value (N). To show the chart, save it as an HTML file using the save() method. Alternatively, if you are using a Jupyter Notebook, simply write the name of the chart variable to show the chart inline.

Figure 2.4 shows the chart produced by the code contained in listing 2.4. Now that you have learned the basic concepts behind Altair, we will review the prerequisites for implementing a practical scenario in Altair and Copilot.

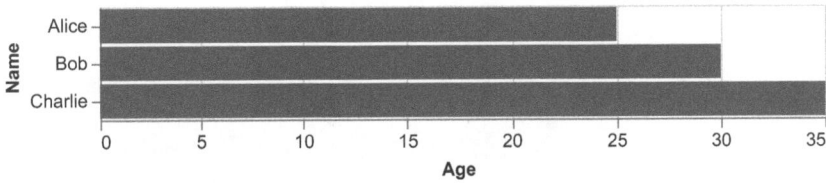

Figure 2.4   The chart produced by listing 2.3

EXERCISE 1

Draw the charts in figures 2.2 (line chart) and 2.3 (scatter plot) using the DataFrame in listing 2.4.

Listing 2.4

```
df = pd.DataFrame({
    'X' : [2,3,5,6,7,9,4,5,5,6],
    'Y' : [4,5,6,7,8,9,5,6,3,5]
})
```

The solutions to the exercises are in the GitHub repository for the book under 02/line-chart.py and 02/scatter-plot.py.

2.2 *Use case: Describing the scenario*

Suppose we are at the end of 2019, before the COVID-19 pandemic. You work as a data analyst at a Portuguese hotel. Your boss has experienced increased tourist arrivals at their hotel and wants to invest money in building a new swimming pool. Before investing money, they ask you to study the phenomenon to understand whether their investment will be successful. The objective of your task is to see whether Portugal is seeing an increased number of tourist arrivals in recent years relative to global trends.

You start your searches and find a dataset about arrivals at tourist accommodation establishments (https://mng.bz/2KOX), released as open data by Eurostat. For simplicity, in this chapter, you consider only one dataset, but in a real use case scenario, you should consider more data, including the popularity of swimming pools among tourists visiting Portugal, the cost of building a new swimming pool at the hotel, and the potential revenue that the new swimming pool could generate for the hotel.

The dataset contains the number of arrivals at tourist accommodation establishments since 2019 in all European countries. For simplicity, focus only on the following countries: Portugal, France, Italy, Germany, Spain, and the United Kingdom. The code described in this chapter can be found under CaseStudies/tourist-arrivals in the GitHub repository and the dataset under CaseStudies/tourist-arrivals/source.

2.2.1 *The dataset*

Table 2.1 shows a sample of the dataset we will analyze. The dataset contains the tourist arrivals at establishments from 1994 to 2019 for Italy (IT), France (FR), Germany (DE), Portugal (PT), Spain (ES), and the United Kingdom (UK).

Table 2.1 Arrivals at tourist arrivals establishments since 1994

Date	IT	FR	DE	PT	ES	UK
1990-01-01	2,543,920		3,185,877	325,138	1723,786	1,776,000
1990-02-01	2,871,632		3,588,879	381,539	1,885,718	2,250,000
1990-03-01	3,774,702		4,272,437	493,957	2,337,847	2,662,000
...	...	...	...	...	...	...
2019-08-01	11,649,500	13,692,822	14,570,339	2,531,809	12,893,366	8,889,049

Table 2.1   Arrivals at tourist arrivals establishments since 1994 *(continued)*

Date	IT	FR	DE	PT	ES	UK
2019-09-01	9,888,817	11,684,845	14,373,815	2,263,748		5,858,984
2019-10-01	7,692,388	10,401,793	13,780,441	1,995,942		7,455,781

The table contains some missing values for some countries and some years, so we should pay attention if, for example, we want to combine data by year, because aggregating incomplete years could lead to wrong results. Load the dataset as a pandas DataFrame, as described in the file from-data-to-information/raw-chart.py of the GitHub repository for the book and the following listing.

Listing 2.5   Loading the dataset as a pandas DataFrame

```
import pandas as pd

df = pd.read_csv('../source/tourist_arrivals_countries.csv', parse_dates=['Date'])
```

> **NOTE** Use pandas to load the dataset as a DataFrame. Use the `parse_dates` parameter to load the `Date` field as a date.

Now, you are ready to perform basic data exploration.

2.2.2 *Data exploration*

EDA is a mandatory phase to extract insights from your data, but you cannot use its output to build raw data stories. In fact, EDA aims to analyze and visualize data to extract insights and understand its underlying patterns, distributions, and relationships. You may even use generative AI to perform EDA. Performing a complete data exploration is outside of the scope of this book. You can refer to the bibliography at the end of this chapter to learn how to perform EDA. However, you can use an existing EDA library for quick data exploration, such as ydata-profiling (https://pypi .org/project/ydata-profiling/) or sweetviz (https://pypi.org/project/sweetviz/). In this chapter, we'll use ydata-profiling. Import the library, create a `ProfileReport()` object, and pass the pandas DataFrame as an input argument. Finally, build the report by invoking the `to_file()` method.

Listing 2.6   Building a summary report

```
from ydata_profiling import ProfileReport
import pandas as pd

df = pd.read_csv('../source/tourist_arrivals_countries.csv', parse_dates=['Date'])

eda = ProfileReport(df)
eda.to_file(output_file='eda.html')
```

NOTE   Use ydata-profiling to build a summary report of the dataset.

As a result, the `to_file()` method produces an HTML file that contains the report. Figure 2.5 shows a snapshot of the produced report.

Pandas Profiling Report   Overview   Variables   Interactions   Correlations   Missing values   Sample

Overview

Overview   Alerts (12)   Reproduction

Dataset statistics

Number of variables	7
Number of observations	358
Missing cells	50
Missing cells (%)	2.0%
Duplicate rows	0
Duplicate rows (%)	0.0%
Total size in memory	19.7 KiB
Average record size in memory	56.4 B

Variable types

DateTime	1
Numeric	6

Figure 2.5   A snapshot of the report produced by ydata-profiling

The report contains many sections, as shown in the top-right menu of figure 2.5. The report may help you do the following:

- *Understand the data.* The report provides an overview of the dataset, including the number of observations, variables, and missing values. Use this information to understand the data quality.
- *Identify data types.* The report identifies the data types of each variable and provides a summary of their distribution, including measures such as mean, median, standard deviation, and range. Use this information to identify variables that may require further preprocessing.
- *Identify correlations.* The report provides a correlation matrix, heatmap, and scatter plot matrix, which can help identify variables that are highly correlated or related.
- *Identify distributions.* The report provides distribution plots, including histograms and density plots, to help identify each variable's distribution shape and spread. Use this information to understand the underlying patterns, trends, and data quality problems in the data.

In your case, the Missing Values section describes the presence of missing values for some countries (France, Spain, and the United Kingdom), as shown in figure 2.6.

Missing values

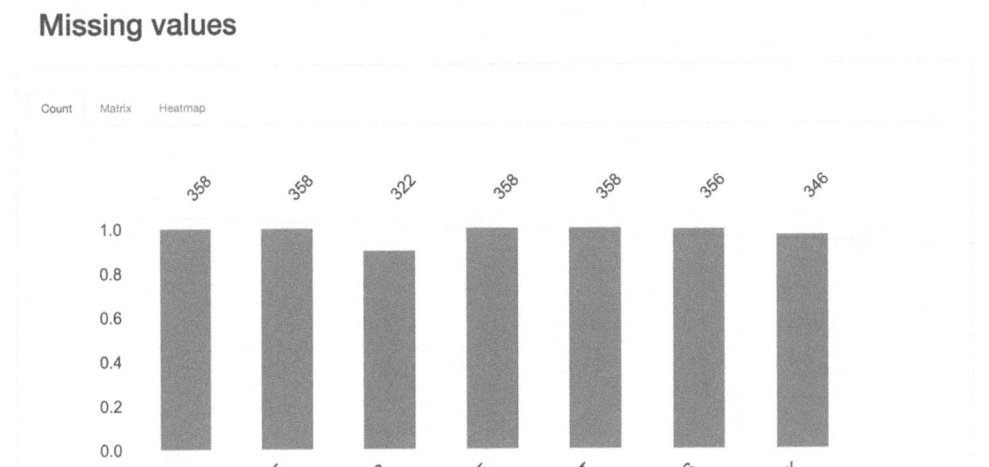

Figure 2.6   **The report highlights missing values for France, Spain, and the United Kingdom.**

Now that we have quickly explored the dataset, you are ready to build a data story from that dataset. Let's start with the first approach: using Altair.

2.3   *First approach: Altair*

To transform data into a story, we will proceed step by step. First, we'll build a basic chart, and then we'll gradually enrich it with the elements necessary to transform the chart into a story in its own right. We'll use the principles of the DIKW pyramid to implement this progressive enrichment:

- From data to information
- From information to knowledge
- From knowledge to wisdom

Let's start with the first step of the DIKW pyramid: turning data into information. In this section, we'll use the code under CaseStudies/tourist-arrivals in the book's GitHub repository.

2.3.1   *From data to information*

Turning data into information means extracting some insights from data. Start by drawing the raw chart, shown in the following code, and the file from-data-to-information/raw-chart.py.

Listing 2.7   **Building the raw chart in Altair**

```
import altair as alt
import pandas as pd
```

```
from ydata_profiling import ProfileReport

df = pd.read_csv('../source/tourist_arrivals_countries.csv',
    parse_dates=['Date'])

df = pd.melt(df, id_vars='Date', value_name='Tourist Arrivals',
    var_name='Country')

chart = alt.Chart(df).mark_line().encode(
    x = 'Date:T',
    y = 'Tourist Arrivals:Q',
    color=alt.Color('Country:N')
)

chart.save('raw-chart.html')
```

Uses the notation
:T to specify that the
data type is temporal

Uses the notation :Q to specify
that the data type is quantitative

Uses the notation :N to specify
that the data type is nominal

NOTE   Use the `mark_line()` property to build the raw chart in Altair. Draw
the number of tourist arrivals on the y-axis and the date on the x-axis.

Use the `melt()` function to unpivot the dataset (i.e., transform data rows into columns). For more details about the `melt()` function, refer to appendix B. Then, build
the chart. The chart uses three channels: x and y to describe the axes, and `color` to
group by countries. For each channel, specify the column in the dataset (e.g., `Date` for
the x channel) and the data type (`T` for temporal data, `Q` for quantitative data, and `N`
for nominal data). Figure 2.7 shows the produced chart.

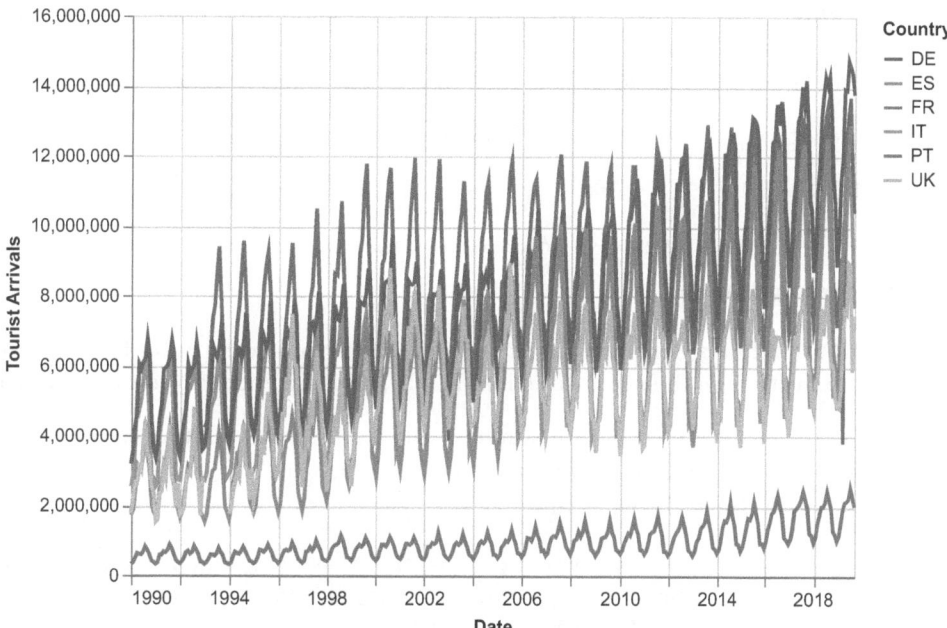

Figure 2.7   The raw chart produced in Altair without any manipulation

The chart is difficult to read because all countries overlap. However, you can extract the following highlight: all the trend lines have increased since 1990. This is a positive message to communicate to your boss. From 1990 until 2019, there has been a progressive increase in the arrival of tourists. Let's focus more on this information.

In all cases, you are not interested in knowing the intermediate data, as there are no outliers, only the starting point (1990) and the final point of the time series (2019). However, we note that for some countries, there is no data for 1990, and 2019 is incomplete (the data goes up to October). In this case, the question is, what should we do in the presence of missing values? The answer is, it depends. In your case, you are dealing with a time series that grows progressively (seasonality excluded), so you can restrict the analysis range to only the dates for which you have data. Thus, you narrow your search to the range 1994–2018.

Let's implement this strategy in Altair. Start by filtering out missing years from the dataset, as described in the script from-data-to-information/grouped-chart.py and in the following listing.

Listing 2.8   Removing missing years from the dataset

```
df.loc[:, 'Year'] = df['Date'].dt.year           ←⎯ Extract the year from the date.
df = df[(df['Year'] >= 1994) & (df['Year'] <= 2018)]     ←⎯
                                           Filter out years before
                                           1994 and after 2018.
```

NOTE   First, extract the year from the date column, and then filter out all the years before 1994 and after 2018. Use `loc` to access a group of rows and columns by labels and `iloc` to access group rows and columns by numerical indices.

Then, group data by year and calculate the sum.

Listing 2.9   Grouping by year and calculating the sum

```
df = df.groupby(['Year', 'Country'])['Tourist Arrivals'].sum().reset_index()
```
Group by year and country.

NOTE   Use `group_by()` to group data by year and country, and calculate the sum through the `sum()` function. Finally, reset the index to retrieve the year.

Use the `reset_index()` function to create a new index column starting from 0, and move the current index values to a new column.

Finally, build the chart in Altair.

Listing 2.10   Building the chart in Altair

```
chart = alt.Chart(df).mark_line().encode(
    x = 'Year:O',
    y = 'Tourist Arrivals:Q',
```

```
        color=alt.Color('Country:N')
)

chart.save('chart.html')
```

NOTE Use the `mark_line()` property to build a line chart in Altair.

Figure 2.8 shows the produced chart.

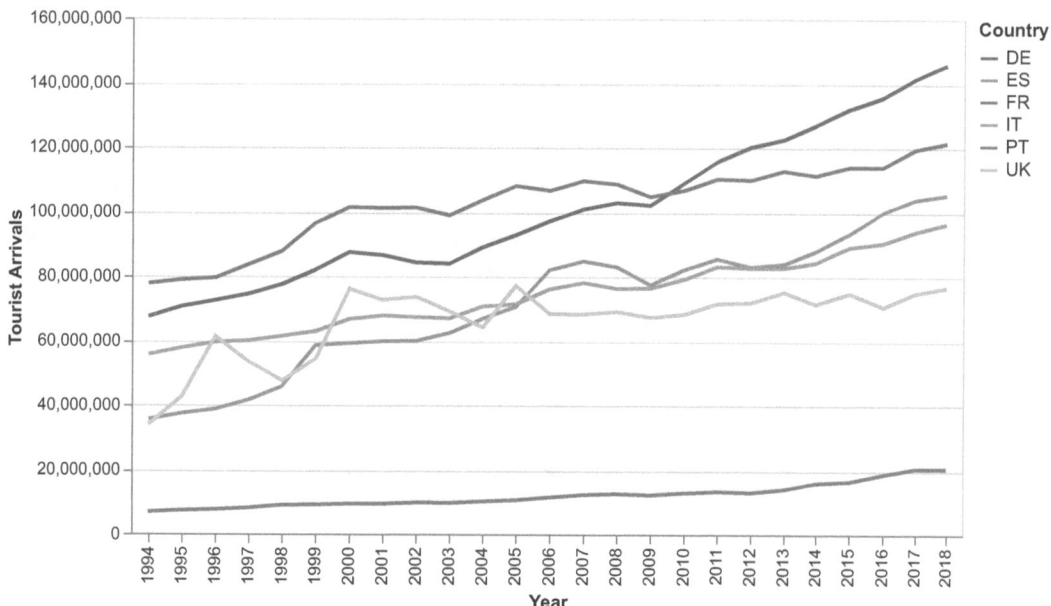

Figure 2.8   The produced chart after calculating the sum for each year and filtering out the boundary years

The chart is clearer than that in figure 2.7; you can easily distinguish the trend line for each country. However, the presence of too many colors and countries does not transmit the message. Your objective is to focus on Portugal, which is the country about which your boss has asked for information.

The increase in the number of tourists over time in the other countries is greater than that of Portugal. However, you are not interested in knowing the absolute values but rather the percentage increase for each country over time. If we use absolute values, instead, we can't answer the following questions immediately:

- For each country, what is the percentage increase since 1994?
- Which of the two nations experiences the greater percentage increase?

If you use the percentage increases, we can put the two countries on a more comparable level, and you can answer the previous questions. Let's calculate the percentage increase since 1994 for each country and then plot the chart of percentage increases instead of that of absolute values.

Calculate the percentage increase using the following formula:

Percentage Increase = (Final Value – Starting Value) / Starting Value * 100

In your case, calculate the percentage increase for each country as shown in the following listing and in the script from-data-to-information/chart.py.

Listing 2.11   Calculating the percentage increase for each country

```
for country in df['Country'].unique():
    current = df[df['Country'] == country]['Tourist Arrivals']
    base = df[(df['Country'] == country) & (df['Year'] == 1994)]['Tourist
    Arrivals'].values[0]
    df.loc[df['Country'] == country, 'PI'] = (current - base)/ base*100
```

**Add a new column containing the difference for each country
between the number of tourist arrivals in the current year and 1994.**

NOTE   For each country, calculate the difference between the number of tourist arrivals in each year and that in 1994.

Now, build the chart again.

Listing 2.12   Drawing the chart of percentage increases

```
base = alt.Chart(df).encode(
    x = alt.X('Year:O', title=''),
    y = alt.Y('PI:Q', title='Percentage Increase since 1994'),
    color=alt.Color('Country:N',
                    scale=alt.Scale(scheme='set2'),
                    legend=None),
    strokeWidth=alt.condition(alt.datum.Country == 'PT', alt.value(7),
     alt.value(0.5))
).properties(
    title='Tourist Arrivals in Portugal (1994-2018)'
)

chart = base.mark_line()

chart.save('chart.html')
```

NOTE   First, create a new DataFrame with the required values. Then, draw the chart. Use `title` to modify the title of the y-axis. Use the `scale` property to select the range of colors. In your case, you used a default scheme (set2). For the complete list of schemes, you can refer to the D3 documentation (https://observablehq.com/@d3/color-schemes). The chart also uses the `strokeWidth` channel, which enables us to configure the stroke width of each line. We use a conditional statement to set the stroke width. To access a column value directly within a conditional statement, use `alt.datum.<column>`. To set the value to a specific string or number, use `alt.value()`.

To encode the x channel, you used the `alt.X()` object—rather than a simple string, like we did before—the same as you did for the y and `color` channels. When you want to specify the channel details, you can use the channel object instead of a simple string. For example, you can use `alt.Y()` for the y channel, `alt.Color()` for the `color` channel, and so on. Within a channel object, you can set different properties that depend on the channel. For example, for the x and y channels, you can set the axis property. You can define a value or an additional object for each property within a channel. In practice, the same strategy of channels applies to the channel properties. For example, `axis` is a property of the x and y channels. You can use `alt.Axis()` to set additional properties of the axis. In the example, you have set the `legend` property of the `color` channel.

To build the chart, you have also used the `properties()` function, which configures the chart's general properties, such as the title. Figure 2.9 shows the resulting chart.

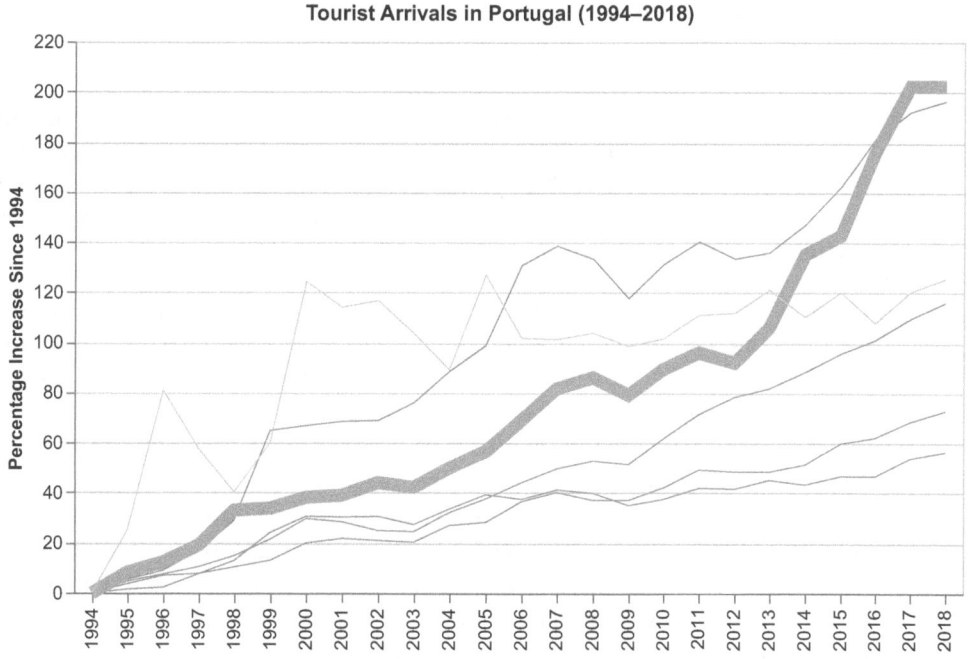

Figure 2.9   The chart produced after calculating percentage increases and highlighting Portugal

You removed the legend voluntarily because you will add it as an annotation following each line. The chart shows that Portugal has experienced a larger percentage increase in tourist arrivals than the other countries. This is an extraordinary discovery, and it answers your boss's question. You have extracted information from the data. To high-light your discovery, add an annotation to the chart that precisely describes the per-centage increase value in correspondence of 2018.

Listing 2.13   Adding an annotation to the chart

```
annotation = base.mark_text(         Sets the offset in pixel
    dx=10,                           from the x coordinate
    align='left',          ◄──────── Sets the text alignment
    fontSize=12            ◄───────
).encode(                            Sets the text font size

    text='label:N'
).transform_filter(
    alt.datum.Year == 2018
).transform_calculate(
    label='datum.Country + "(" + format(datum.PI, ".2f") + "%)"'   ◄──────
)
                                             Formats the text to show
chart = (chart + annotation                  only 2 decimal places and
)                                            adds a percentage sign

chart.save('chart.html')
```

NOTE   Use the `mark_text()` property to draw the text. Next, combine the pre-vious chart and the text annotation through the + operator. Use `transform _filter()` to select only 2018 (the position near which you want to place the annotation) and `transform_calculate()` to add a new column to the dataset containing the formatted percentage.

The `mark_text()` function (and the other `mark_*()` functions) can receive as input some parameters that define some static properties of the chart. For example, if you want to use a static color for all the texts in the chart, set the color in the `mark_*()` function. If, instead, the color of the text depends on a column in the DataFrame, set it in the `encode()` function. Figure 2.10 shows the resulting chart at the end of the first step, turning data into information.

The chart is clear and easy to understand. Adding information means extracting only the relevant data and representing it in a way that is easy for the audience to understand without further calculation. Now that you have turned data into informa-tion, it's time to move to the next step of the DIKW pyramid: turning information into knowledge.

EXERCISE 2
Modify the chart from figure 2.10 as follows:

1   Build a slope graph, which is a line chart showing only the first and the last val-ues (1994 and 2018).
2   On the y-axis, show the number of tourist arrivals, not the percentage increase.

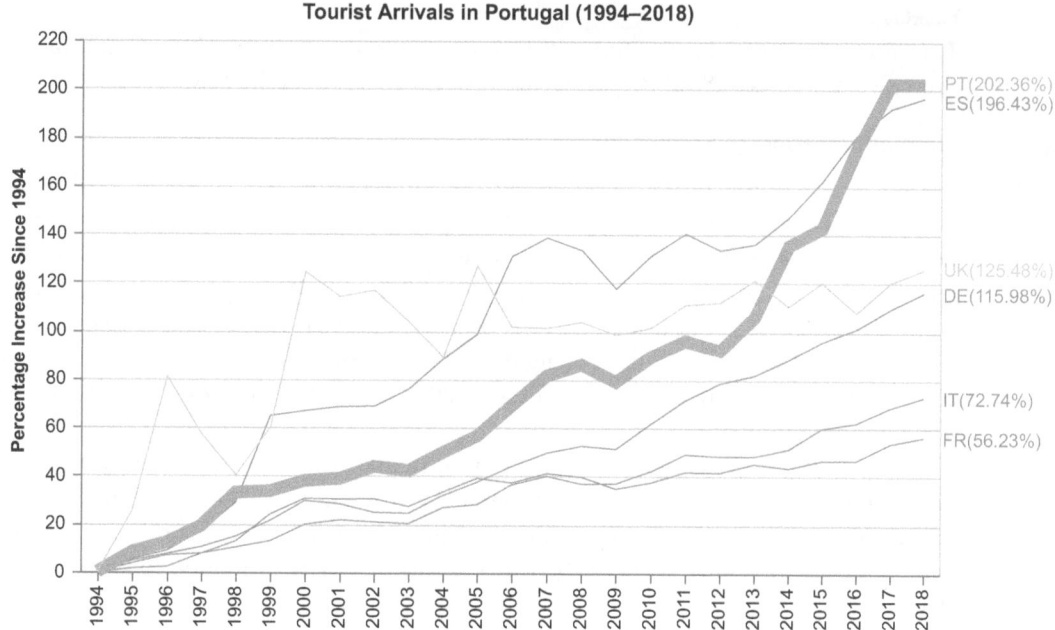

Figure 2.10   The chart produced at the end of the first step, turning data into information

As an output, you should produce the chart shown in figure 2.11. You can find the solution of the exercise under CaseStudies/tourist-arrivals/exercises/exercise1.py.

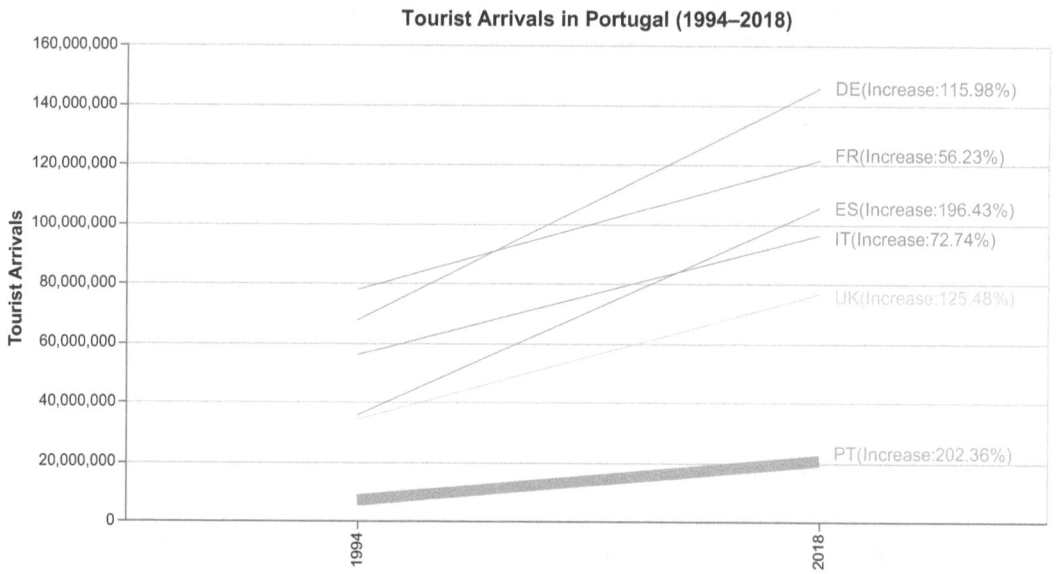

Figure 2.11   The output of exercise 1

EXERCISE 3

Transform the slope chart of figure 2.11 into a stacked bar chart. The solution of this exercise is under CaseStudies/tourist-arrivals/exercises/exercise2.py.

2.3.2   *From information to knowledge*

Turning information into knowledge means adding context to the extracted information. Usually, context answers the following question: Why does this situation happen? If possible, extend your search, or look at the same dataset to answer this question. In our case, the question is as follows: Why has Portugal experienced such an incredible increase in tourist arrivals over 25 years? To answer this question, you researched and discovered that Portugal (and other countries) introduced many low-cost flights in the early 1990s (Dobruszkes, 2013). This allowed for a much greater development of tourism.

This finding is the context for your graph; thanks to the introduction of low-cost flights, Portugal has experienced an increase in tourist arrivals of over 200% in 25 years, even surpassing the increase in the other countries. Add this discovery as an annotation in the chart, as described in the script from-information-to-knowledge/chart.py.

Listing 2.14   Adding the context to the chart

```
text_comm = f"""Thanks to the introduction
of low-cost flights,
Portugal has experienced
an increase
in tourist arrivals
of over 200% in 25 years,
even surpassing the increase
in the other countries."""

df_commentary = pd.DataFrame([{'text' : text_comm}])

commentary = alt.Chart(df_commentary
).mark_text(
    lineBreak='\n',
    align='left',
    fontSize=20,
    y=0,
    x=0,
    color='#81c01e'
).encode(
    text='text:N'
).properties(
    width=200
)
```

NOTE   First, build a DataFrame with the text to add to the chart. Then, use `mark_text()` to draw the text. Also, specify the line break symbol (`line-Break='\n'`), how to align the text (`align='left'`), the font size (`fontSize=20`),

the position (y=100), and the color (color='orange'). Finally, combine the
text with the previous chart through the + operator.

To reinforce the text we have inserted into the chart, let's add an additional, very lean
chart, showing passenger traffic in Portugal over time. Use a World Bank dataset relating to air traffic (https://mng.bz/RZRP, released under the CC BY-4.0 license) in the
various countries of the world. The following listing shows how to add the line chart to
your main chart.

Listing 2.15   Adding the call to action to the chart

```
df_airports = pd.read_csv('../source/airports.csv')

df_airports = df_airports.melt(id_vars='Country Name', var_name='Year',
    value_name='Value')
df_airports.dropna(inplace=True)
df_airports = df_airports[df_airports['Country Name'] == 'Portugal']

airports = alt.Chart(df_airports).mark_line(
    color='#81c01e',
    strokeWidth=6
).encode(
    x=alt.X('Year', axis=alt.Axis(labels=False, ticks=False, grid=False),
     title='Year range: 1994-2018'),
    y=alt.Y('Value', axis=alt.Axis(labels=False, ticks=False, grid=False),
     title='Air Passengers range: 4.3M-17.3M')
).properties(
    title='Air passengers in Portugal',
    width=200,
    height=200
)


chart = ((commentary & airports) | (chart + annotation)
)

chart.save('chart.html')
```

> **NOTE**   First, load the airport dataset as a pandas DataFrame. Next, format the
> dataset to be consumable by Altair. Then, use the mark_line() method to
> build a trend line. Remove the labels, ticks, and grid from the x-axis and y-axis.
> Finally, combine the airport chart and the text using the & operator, which
> corresponds to a vertical alignment. Also, use the | operator to combine
> charts horizontally.

Figure 2.12 shows the chart resulting from the second step, turning information into
knowledge.

Thanks to context, the chart answers why this happens. Now that you have turned
information into knowledge, it's time to move to the final step of the DIKW pyramid:
turning knowledge into wisdom.

Thanks to the introduction
of low-cost flights,
Portugal has experienced
an increase
in tourist arrivals
of over 200% in 25 years,
even surpassing the increase
in the other countries.

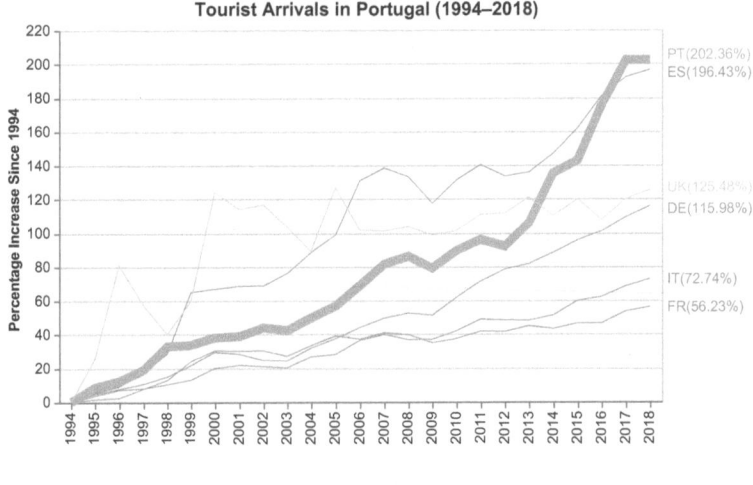

Figure 2.12   The chart produced after the second step: turning information into knowledge

2.3.3   *From knowledge to wisdom*

Turning knowledge into wisdom means adding a call to action to the chart and proposing the next steps. In this chapter, we focus only on adding a call to action. We'll see how to add the next steps in the following chapters. The call to action invites the audience to do something. In your case, the call to action should answer your boss's original question: Should I build a new swimming pool?

Given the results of the previous analysis, formulate the following call to action: since the number of tourists arriving in Portugal is constantly expanding, you can think about building a new swimming pool! Add the call to action below the main chart, as shown in the following listing and the from-knowledge-to-wisdom/chart.py script.

Listing 2.16   Adding the call to action to the chart

```
text_cta = f"""Since the number of tourists arriving in Portugal is
    constantly expanding,
you can think about building a new swimming pool!"""
df_cta = pd.DataFrame([{'text' : text_cta}])

cta = alt.Chart(df_cta
).mark_text(
    lineBreak='\n',
    align='left',
    fontSize=20,
    y=0,
    x=0,
    color='#81c01e'
```

```
).encode(
    text='text:N'
).properties(
    width=200
)


final_chart = ((commentary & airports) | ((chart + annotation) & cta)
)

final_chart.save('chart.html')
```

NOTE   Add a new textual annotation to the chart to formulate the call to action.

You can run your code locally or on a web server if it includes images (as we will see in the next chapters). If you don't have a web server, you can run a local and temporary web server from the command line, running the following command in the directory containing the produced HTML file: `python -m http.server`. The server should listen at port 8000 and should serve all the files contained in the directory from which it is started. Point your browser to http://localhost:8000/chart.html to access the chart.html file. Figure 2.13 shows the chart produced by the final step, turning knowledge into wisdom.

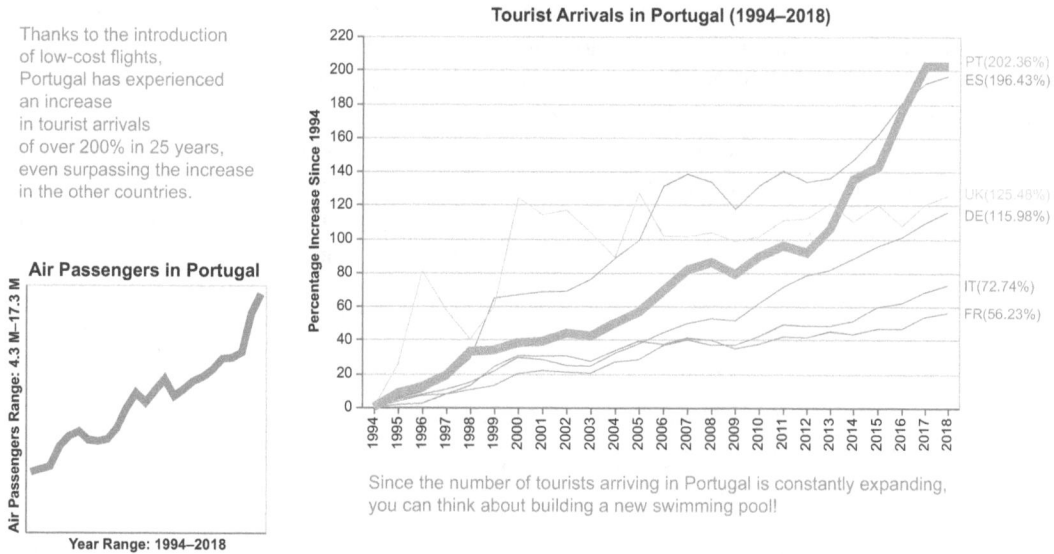

Figure 2.13   The chart produced by the final step: turning knowledge into wisdom

The chart contains the answer to your boss's question. If you compare this chart with figure 2.5, you will probably notice a big difference: the charts in figures 2.9 and 2.10

show that the audience doesn't have to do any calculations or processing. They can just read the result. If, on the other hand, they look at the chart in figure 2.5, the audience has to extract the information by themselves.

Some may argue that a lot of information was lost in the process, such as what happened to the other European countries and what happened between 1994 and 2018. Thus, the boss wouldn't want to make a big investment based only on the numbers shown in the chart. This observation is correct. However, we have simplified the example to show the workflow used to turn data into a story. In a more complex scenario, you must consider additional factors and build an appropriate story, where each factor is a single scene of your data story.

To get an idea of the differences between Altair and Matplotlib, we will build the same chart using Matplotlib in the next section. If you are unfamiliar with Matplotlib, jump directly to section 2.4, which covers the second approach: using Copilot.

2.3.4   *Comparing Altair and Matplotlib*

You might wonder why there is a section on Matplotlib in a book about Altair. The answer is simple: the methodology described in this book, including the techniques of using generative AI and the DIKW pyramid, can be applied to all data visualization libraries, not just Altair. Indeed, you could also think about using them in other languages, such as R. As an example, in this section, we see a brief comparison between Altair and Matplotlib, highlighting the differences and similarities.

Matplotlib is a popular Python library for data visualization. Quite likely, it was the first Python library you used to plot your data quickly. If you use Matplotlib as an alternative to Altair, consider that Matplotlib is an imperative library, so you must define all the steps to produce the chart. In Altair, you focus only on the desired output. Regardless, all the concepts related to data storytelling described in this book are also valid for Matplotlib, so use it if you prefer the imperative approach.

You can find the complete code in Matplotlib in the book's GitHub repository under CaseStudies/tourist-arrivals/matplotlib/chart.py and in the following listing.

Listing 2.17   Generating the chart in Matplotlib

```
import matplotlib.pyplot as plt

df = pd.read_csv('../source/tourist_arrivals_countries.csv',
    parse_dates=['Date'])
df = pd.melt(df, id_vars='Date', value_name='Tourist Arrivals',
    var_name='Country')
```
Extracts year from date   Groups by year and country
```
df.loc[:, 'Year'] = df['Date'].dt.year   ←┘

df = df.groupby(['Year', 'Country'])['Tourist Arrivals'].sum().reset_index()   ←
```
Filters out years before 1994 and after 2018
```
df = df[(df['Year'] >= 1994) & (df['Year'] <= 2018)]   ←

for country in df['Country'].unique():
    current = df[df['Country'] == country]['Tourist Arrivals']
```

```
    base = df[(df['Country'] == country) & (df['Year'] == 1994)]['Tourist
     Arrivals'].values[0]
    df.loc[df['Country'] == country, 'PI'] = (current - base)/ base*100    ⊲─┐
```

Creates ┌─▷
a figure
and axis

Adds a new column containing the difference
for each country between the number of
tourist arrivals in the current year and 1994

```
    fig, ax = plt.subplots(figsize=(10, 6))

    for country in df['Country'].unique():
        current = df[df['Country'] == country]['Tourist Arrivals']
        base = df[(df['Country'] == country) & (df['Year'] == 1994)]['Tourist
         Arrivals'].values[0]
        pi = (current - base)/ base*100
        ax.plot(df[df['Country'] == country]['Year'], pi, label=country)     ⊲─┘
```

Plots the data

```
    ax.set_title('Tourist Arrivals in Portugal (1994-2018)')
    ax.set_xlabel('')
    ax.set_ylabel('Percentage Increase since 1994')
```

Sets the title
and axis labels

```
    ax.legend()                    ⊲──── Adds a legend

    plt.savefig('chart.png', dpi=300)       ⊲──── Saves the plot
```

NOTE Build a separate line for each country.

Figure 2.14 shows the produced chart in Matplotlib. The chart could be improved by adding chart lines, data points, or labels for easy reading and good presentation.

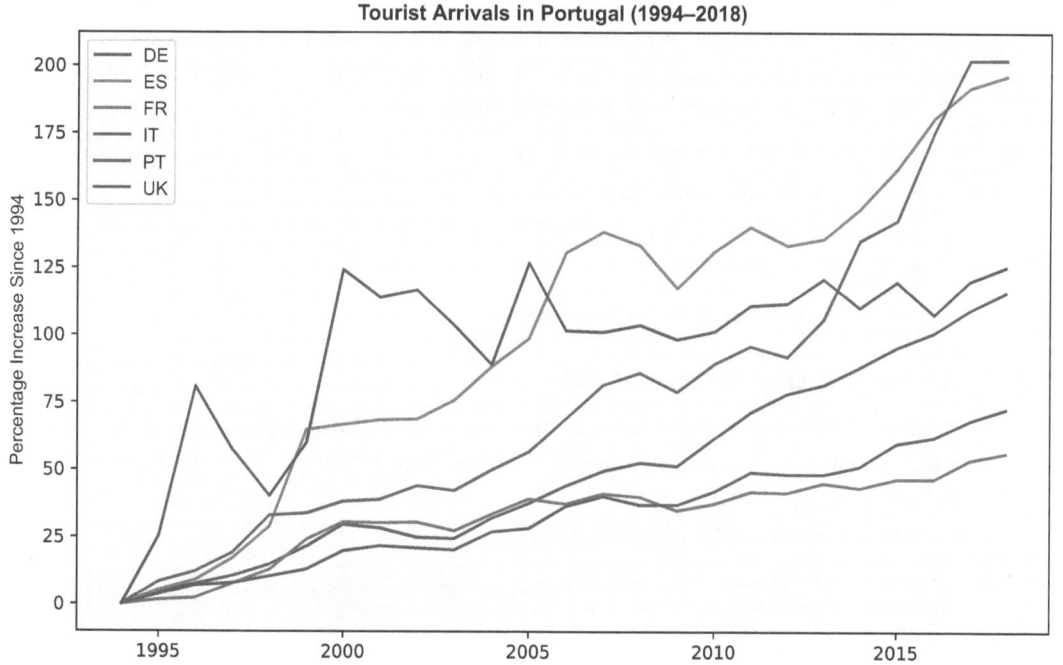

Figure 2.14   The produced chart in Matplotlib

The Altair code is more verbose than that in Matplotlib. However, in Matplotlib, you must draw each line or text separately, while in Altair, you pass the DataFrame and the library draws a line or a mark in general for you. Now that you have turned your data into wisdom by hand and seen the differences between Altair and Matplotlib, we will move to the alternative method: using GitHub Copilot to build the chart semi-automatically.

2.4   *A second approach: Copilot*

Copilot enables you to write code to implement something automatically. We'll enable Copilot as an extension of Visual Studio Code (VSC). To install VSC, refer to its official documentation (https://code.visualstudio.com/docs/introvideos/basics), and to configure Copilot, please refer to appendix A.

To use Copilot, it is sufficient to describe the sequence of operations to be implemented in an orderly manner, and Copilot will do the rest for us, even proposing different implementation options. In this section, we will use Copilot to build the framework of the graph (i.e., its main parts). Then, we will add the details by hand, using colors and font size. The main difficulty in using Copilot is describing what the code needs to implement in natural language. The clearer the text is, the more consistent Copilot will be in implementation.

In this section, we'll describe how to translate the code written in the previous sections into a sequence of texts understandable by Copilot. We will focus only on translating the framework of the graph into the natural language without considering the details, such as the font size and colors used. Find the code generated by GitHub Copilot in the book's repository under CaseStudies/tourist-arrivals/copilot/chart.py.

We will break the problem down into steps:

1 Loading and cleaning the dataset
2 Calculating the percentage increase
3 Plotting the basic chart in Altair
4 Enriching the chart

We'll cover each step in the coming sections.

2.4.1   *Loading and cleaning the dataset*

Before using the dataset, you must load the required libraries, open the dataset as a pandas DataFrame, and calculate the percentage increases. The idea is to prepare data to build the chart in figure 2.13.

Open VSC, and write the following comments.

> **Listing 2.18   Loading and filtering the dataset**

```
# Import the necessary libraries.
# Read the following dataset into a pandas DataFrame, '../source/tourist_
    arrivals_countries.csv', and parse the Date field as a date.
```

```
# Filter out rows before 1994 and after 2018.
# Extract the year from the Date field and create a new column called Year.
# Group the data by Year and calculate the sum of tourist arrivals for each year.
```

> **NOTE** Describe the sequence of operations to load and filter the dataset.

If you press Enter at the end of the last comment and start writing `import`, Copilot will propose to complete the statement. Just press Enter to accept the generated code. Alternatively, hover over the proposed text with the mouse, and open Copilot, as shown in figure 2.15.

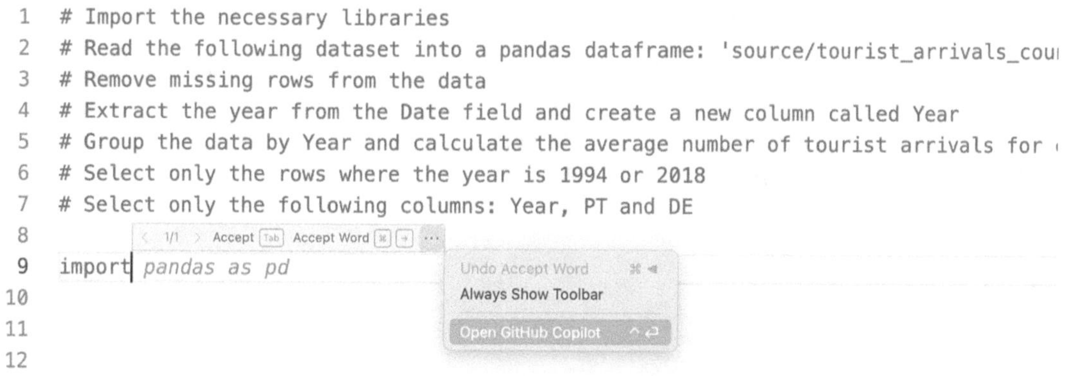

```
 1   # Import the necessary libraries
 2   # Read the following dataset into a pandas dataframe: 'source/tourist_arrivals_cou
 3   # Remove missing rows from the data
 4   # Extract the year from the Date field and create a new column called Year
 5   # Group the data by Year and calculate the average number of tourist arrivals for
 6   # Select only the rows where the year is 1994 or 2018
 7   # Select only the following columns: Year, PT and DE
 8            < 1/1 >  Accept Tab  Accept Word ⌦ →  ···
 9   import pandas as pd          Undo Accept Word      ⌘ ◂
10                                Always Show Toolbar
11                                Open GitHub Copilot    ⌃ ↵
12
```

Figure 2.15   How to open Copilot in VSC

Depending on the generated code, Copilot may propose different solutions, as shown in figure 2.16. Click Accept Solution to accept a solution. In our case, the second solution implements the described steps. Note that every time you run Copilot, it may generate different solutions, even if you rerun the same text.

2.4.2   Calculating the percentage increase

The next step involves building a new dataset, starting from the previously cleaned dataset. The new dataset will contain the percentage increase values for Portugal and Germany. Write the following operation.

Listing 2.19   Building a new DataFrame with percentage increase

```
# Add a new column to the DataFrame called PI containing the difference for
each country between the number of tourist arrivals in the year and 1994.
```

> **NOTE** Describe how to calculate the percentage increase.

As in the previous case, Copilot could propose multiple solutions. Open Copilot to check the possible solutions, and accept the one that best implements your operations.

```
 ◈ GitHub Copilot  ×

 1  Synthesizing 10/10 solutions (Duplicates hidden)
 2
 3  =======
 4
    Accept Solution
 5  import pandas as pd
 6
 7  =======
 8
    Accept Solution
 9  import pandas as pd
10
11  df = pd.read_csv('source/tourist_arrivals_countries.csv', parse_dates=
12  df = df.dropna()
13  df['Year'] = df['Date'].dt.year
14  df = df.groupby('Year').mean()
15  df = df.loc[[1994, 2018]]
16  df = df[['PT', 'DE']]
17
18  =======
19
    Accept Solution
20  import pandas as pd
21  import altair as alt
```

Figure 2.16   The different solutions proposed by Copilot in VSC

If you don't find any solution that implements your text, rewrite the text by adding more details. If, even after rewriting the text, you can't find a solution that implements your operations, accept the solution closest to your requests and modify it according to your needs.

2.4.3   Plotting the basic chart in Altair

Now, we'll instruct Copilot to build a chart similar to what we saw in figure 2.12 (except the airport dataset chart) regarding the percentage increase.

Listing 2.20   Building the chart in Altair

```
# Use the Altair library to plot the PI column for each country versus the Year.
```

NOTE   Describe how the hart should look.

Copilot could propose multiple solutions. Open Copilot to check the possible solutions, and accept the one that best implements your operations.

2.4.4 Enriching the chart

Instruct Copilot to add the context. In addition, save the resulting chart into an HTML file named output.html.

Listing 2.21   Enriching the chart

```
# Create a new chart with the following text: 'Thanks to the introduction\n
    of low-cost flights,\nPortugal has experienced\nan increase\nin tourist
    arrivals\nof over 200% in 25 years,\neven surpassing the increase\nin
    the other countries.' Use the \n as a line break to format the text. Set
    the font size to 14.
# Place the two graphs side by side. Set title to 'Tourist Arrivals in
    Portugal (1994-2018)'.
# Save the chart as an HTML. Name the file chart.html.
```

NOTE   Write the annotation and title to show in the chart.

Once the code is complete, save and run the script. Figure 2.17 shows the resulting chart.

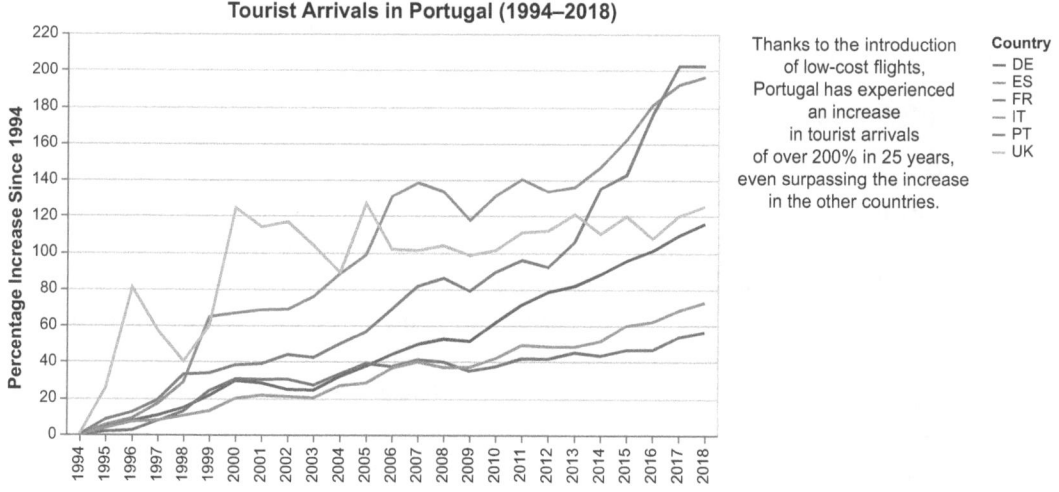

Figure 2.17   The chart produced by Copilot

The chart is a good starting point, but it still needs improvement. To improve the chart, you can run the following steps manually:

1 Increase the font size of the annotation.
2 Increase the stroke width of the PT line.

Using Copilot may seem pointless, since some elements still need to be managed manually. However, with practice, you will realize how valuable Copilot is for speeding up the writing times of the simplest parts of your code. Copilot will remember your previous code and adapt to your programming style as you use it. With practice, it will likely become your programming companion, just like it is for me!

In this chapter, you have learned how to turn a raw dataset into a data story using Altair and Copilot. In the next chapter, we will review Altair's basic concepts.

Summary

- Altair is a declarative library for data manipulation and visualization. It provides three main elements to build a chart: a chart, a mark, and encodings. Use the chart element to specify the dataset to plot. Use the mark to specify the type of chart to draw, and encodings to set the channels, such as the x- and y-axes.
- To turn raw data into a data story, use the DIKW principles progressively. Start by cleaning your dataset, and then extract only meaningful information. Draw the extracted information. Then, add a context and a call to action as annotations.
- Copilot is a powerful tool for speeding up code generation. To make Copilot generate your code, split the problem into small steps and describe each step using natural language. Although Copilot is not perfect, it can assist you in building the framework of your code.
- Use Copilot to automatically generate your code's framework, and then manually improve the code.

References

EDA

GENERAL APPROACH

- Gupta, D., Bhattacharyya, S., Khanna, A., and Sagar, K. (Eds.). (2020). *Intelligent Data Analysis: From Data Gathering to Data Comprehension.* John Wiley & Sons.
- Rasmussen, R., Gulati, H., Joseph, C., Stanier, C., and Umegbolu, O. (2019). *Data Analyst: Careers in Data Analysis.* BCS Publishing.
- Khalil, M. (2024). *Effective Data Analysis: Hard and Soft Skills.* Manning Publications.

PYTHON

- Mukhiya, S. K. and Ahmed, U. (2020). *Hands-On Exploratory Data Analysis with Python.* Packt Publishing.
- Oluleye, A. (2023). *Exploratory Data Analysis with Python Cookbook.* Packt Publishing.

Tools and libraries

- *Altair: Declarative Visualization in Python.* (n.d.). https://altair-viz.github.io/
- *GitHub.* (n.d.). *GitHub Copilot: Your AI Pair Programmer.* https://github.com/features/copilot.

Other

- Dobruszkes, F. (2013). The Geography of European Low-Cost Airline Networks: A Contemporary Analysis. *Journal of Transport Geography, 28,* 75–88.

Reviewing the basic concepts of Altair

It may, at first glance, seem counterintuitive for the topic of running a data story in Altair (chapter 2) to be covered before the basics of Altair (chapter 3). The reason is that now that you have a general understanding of how Altair works, you're ready to see all the details. If you had read this chapter right away, you probably would have gotten bored and skipped it. Instead, now, you are ready to read it calmly. In this chapter, we will review the basic concepts underlying Vega and Vega-Lite, the visualization grammars upon which Altair is built. Then, we'll focus on the Altair main components: encodings, marks, conditions, compound charts, and interactivity. In the last part of the chapter, we'll implement a practical example.

3.1 Vega and Vega-Lite

Vega and Vega-Lite are visualization grammars used by Altair. A *visualization grammar* is a set of rules and principles defining how to represent data visually, much like how grammar functions in a spoken language. A visualization grammar includes a vocabulary of visual elements, such as points, lines, and bars, as well as rules for combining and arranging these elements to create meaningful visualizations. Using a visualization grammar allows you to create clear and effective data visualizations that convey insights and tell stories.

Vega and Vega-Lite provide declarative language for creating interactive visualizations. Learning Vega and Vega-Lite before Altair is important because Altair is built on top of these two visualization libraries. By first mastering these foundational tools, you can gain a deeper understanding of how Altair works and take advantage of its full potential. Additionally, learning Vega and Vega-Lite enables you to create custom visualizations that may not be possible with Altair alone, allowing for more flexibility and creativity in data exploration and communication. In the following section, we'll explore the main components of a Vega and Vega-Lite specification.

3.1.1   Vega

Vega is a visualization grammar used to define the visual aspects and interactive features of a chart by writing code in JSON format. Vega is built at the top of D3.js (https://d3js.org/), a very popular JavaScript library for data visualization. Using Vega, you can generate web-based views that utilize HTML5 Canvas or SVG to display the resulting chart. HTML5 Canvas is a bitmap-based drawing technology that enables you to render dynamic graphics and animations on the web. SVG is a vector-based graphics format for building scalable and resolution-independent graphics.

A Vega JSON file contains the specifications for the visual appearance and interactive behavior of a chart. Listing 3.1 shows the basic structure of a JSON specification. For testing, you can use the Vega Editor (https://vega.github.io/editor/#/) to render the chart produced by Vega. We will produce a basic Vega line chart gradually, throughout this section.

> **Listing 3.1   The basic structure of a Vega JSON file**

```
{
  "$schema": "https://vega.github.io/schema/vega/v5.json",
  "description": "A basic example in Vega",
  "width": 600,
  "height": 400,

  "data": [],
  "scales": [],
  "axes": [],
  "marks": [],
  "signals": []
}
```

NOTE   First, define the version of the schema (`$schema`), a `description`, the `width`, and the `height` of the chart. Then, specify the main sections (`signals`, `data`, and so on). Listing 3.1 shows only the structure of a Vega JSON file and will not produce any chart if pasted in the Vega Editor.

The main sections of a Vega JSON file include `data`, `scales`, `axes`, `marks`, and `signals`. In the remainder of this section, we'll describe an overview of the main sections of a Vega specification. For more details, please refer to the Vega official documentation. (https://vega.github.io/vega/docs/).

DATA

This section defines the data source to use. It specifies the data format, where the data is located, and how it should be loaded and transformed. The following listing shows an example of a `data` section.

Listing 3.2 An example `data` section

```
"data": [
    {
      "name": "myData",
      "values": [
         {"x": 10, "y": 20},
         {"x": 40, "y": 60},
         {"x": 70, "y": 40},
         {"x": 90, "y": 80},
      ]
    }
  ],
```

NOTE Specify the list of data to use in the visualization. For each element of data, specify the name and content. We will use `myData` to refer to the associated data in the specification. We can specify more than one data source. The example specifies data values directly. However, you can also retrieve the data from a CSV file, by specifying the URL and the format as follows: `"url":` `"/path/to/csv/myfile.csv"`, `format": {"type": "csv"}`. Alternatively, you can retrieve data from other formats or embed them in the JSON directly.

SCALES

This section defines the scales that map the data to visual properties. Scales map a data domain (input range) to a visual range (output range). The following listing shows an example of a `scales` section.

Listing 3.3 An example `scales` section

```
"scales": [
    {
      "name": "xScale",
      "type": "linear",
      "domain": {"data": "myData", "field": "x"},
      "range": "width"
    },
    {
      "name": "yScale",
      "type": "linear",
      "domain": {"data": "myData", "field": "y"},
```

```
        "range": "height"
    }
],
```

NOTE   Specify the list of `scales` to use in the visualization. For each scale, specify at least the name, range, and domain. The example defines two scales, one for the x-axis and the other for the y-axis. Specify the data to use as the value of the domain attribute.

AXES

This section defines the axes of the chart. Use `axes` to define tick marks and labels along an axis. The following listing shows an example of an `axes` section.

Listing 3.4   An example `axes` section

```
"axes": [
    { "orient": "bottom", "scale": "xScale" },
    { "orient": "left", "scale": "yScale" }
]
```

NOTE   Specify the list of `axes` to use in the visualization. For each axis, specify at least the scale and how to orient it. The example defines two axes, one for the x-axis and the other for the y-axis.

MARKS

This section defines the visual `marks` that represent the data. Marks include points, lines, rectangles, areas, and other shapes. Use visual properties, like size, color, opacity, and shape, to style marks. The following listing shows an example of a `marks` section.

Listing 3.5   An example `marks` section

```
"marks": [
    {
        "type": "line",
        "from": {"data": "myData"},
        "encode": {
            "enter": {
                "x": {"scale": "xScale", "field": "x"},
                "y": {"scale": "yScale", "field": "y"},
                "stroke": {"value": "red"}
            }
        }
    }
],
```

NOTE   Specify the list of `marks` to use in the visualization. For each `mark`, specify at least the `type` (line, symbol, rect, and so on), the source of data (`from`), and how to encode the data (`encode`). Use the `enter` block to define the initial properties of the visualization. The example defines a line chart representing data contained in the `myData` variable defined in the `data` section. It also defines two encoding channels, `x` and `y`, and the color of the line stroke.

SIGNALS

This section defines the interactive signals you can use to modify the visualization, such as sliders, dropdowns, and check boxes. Every signal is composed of two parts: the *listener*, which is identified by the `signals` section, and the *handler*, which is defined by the keyword `signal` within the `marks` section. The signal listener responds to signal changes, and the signal handler determines how the visualization reacts. For example, a signal handler might change the color or size of data points based on a signal's value. The following listing provides an example that changes the color of a selected element in a chart when the user hovers over the bar with the mouse (figure 3.1).

Listing 3.6 An example `signals` section

```
"signals": [{
    "name": "changeColor",
    "on": [
        {"events": "mouseover", "update": "datum"},
        {"events": "mouseout", "update": "null"}
    ]
}]
```

NOTE `signals` are used to capture events or changes in the visualization. For each signal, specify its name (`changeColor`, in the example), the property to modify (`value`, in the example), and the list of events to listen (`on`, in the example).

To use the signal, you must specify an `update` block within your `mark` section. Remember that you used the `enter` block to specify the initial condition of the chart. Now, you can use the `update` block to specify a change. The following listing specifies a change in the line color.

Listing 3.7 An example `update` section

```
"marks": [{
    "encode": {
      "enter": {
          ...
      },
      "update": {
        "stroke": {
            "signal": "changeColor === datum ? 'red' : 'blue'"
            }
          }
      }
    }],
```

NOTE Under the `marks` section, specify a handler associated with the signal. This handler will be triggered when the specified event (like mouseover) associated with the signal occurs. Within the update property of the marks section, a handler is defined. This handler checks whether the signal corresponds to the mouseover event (e.g., changeColor === datum). If the condition is met (i.e., if the mouse is over the line), the handler performs an action, such as changing the color of the line accordingly. Refer to the Vega documentation (https://vega.github.io/vega/docs/signals/) for more details.

Figure 3.1 shows the resulting line chart. You can find the complete example in the book's GitHub repository under 03/vega/json/line-chart.json.

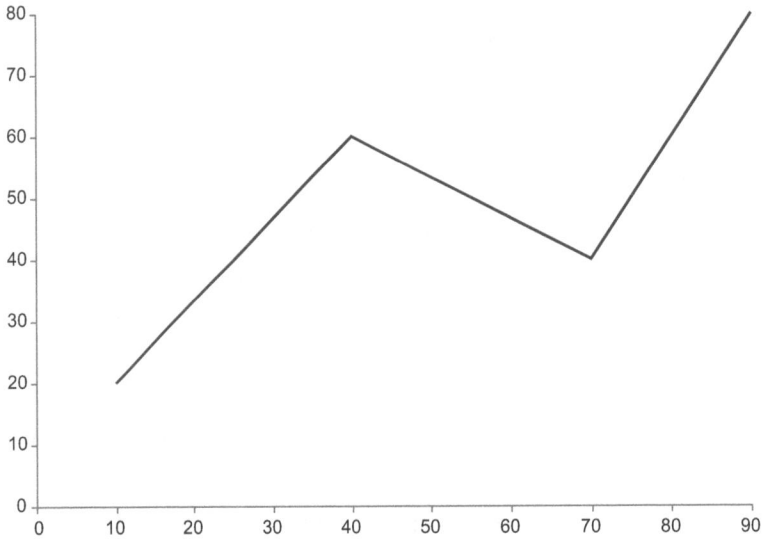

Figure 3.1   The line chart written in Vega

EXERCISE 1
Draw a line chart in Vega as follows:

 1   Load data from the following URL: https://mng.bz/ZEyZ.
 2   Add points to the line. To do this, add a new mark (in addition to the line), using symbol as a type.

You can find the solution to this exercise in the GitHub repository for the book under 03/vega/json/spec.json.

EXERCISE 2
Draw a bar chart in Vega as follows:

 1   Load data from the following URL: https://mng.bz/Ad6W.
 2   Use band as the type for the x scale.
 3   Use rect as the mark type.

4 Optionally, set a signal that changes the bar color while hovering over it with the mouse.

You can find the solution to this exercise in the book's GitHub repository under 03/vega/json/spec2.json. Now that you have learned the basic concepts behind Vega, let's move to the next step: Vega-Lite.

3.1.2 Vega-Lite

Vega-Lite is a concise JSON of a Vega visualization. Vega-Lite still maintains the specifications already defined in Vega, but it represents them through a more compact syntax. Compared to Vega, Vega-Lite offers a more streamlined and concise syntax, ideal for quickly generating simple visualizations with less code complexity, making it more accessible to users seeking rapid creation of common charts. While Vega-Lite provides a high-level abstraction for creating visualizations, it may lack the flexibility to customize every aspect of visualization. It may also perform worse than Vega. The basic sections of a Vega-Lite JSON specification are `data`, `encoding`, `mark`, and `layer`.

DATA

This section specifies the data source for the visualization. The following listing provides an example of a `data` section in Vega-Lite.

Listing 3.8 An example `data` section

```
"data": {
    "values": [
        {"x": 10, "y": 20},
        {"x": 40, "y": 60},
        {"x": 70, "y": 40},
        {"x": 90, "y": 80}
    ]
},
```

NOTE This specifies the data to use in the visualization. Refer to the Vega-Lite documentation (https://vega.github.io/vega-lite/docs/data.html) for more details.

ENCODING

This section maps the data fields to visual properties, such as position, color, and size. The following listing shows an example of an `encoding` section in Vega-Lite. Use the Vega Editor to test your code in Vega-Lite, ensuring you select Vega-Lite from the top-left drop-down menu.

Listing 3.9 An example `encoding` section

```
"encoding": {
    "x": {"field": "X", "type": "quantitative"},
    "y": {"field": "Y", "type": "quantitative"}
},
```

> **NOTE**   Specify the encodings to use in the visualization. The example defines two channels, x and y, and specifies the data type as quantitative. Vega and Vega-Lite also support other data types, such as ordinal, for sequential data; nominal, for text; and time, for temporal data.

MARK

This section defines the type of visual mark, such as bars, points, or lines. The following listing shows an example of a mark section in Vega-Lite.

Listing 3.10   An example mark section

```
"mark": {"type": "line"},
```

> **NOTE**   Specify the mark to use in the visualization. The example defines a line chart.

LAYER

This section combines multiple marks in a single chart. The next listing shows an example of a layer section in Vega-Lite.

Listing 3.11   An example layer section

```
"layer": [
    {"mark": {"type": "line", }},
    {"mark": {"type": "point", "shape": "circle", "size": 100}}
]
```

> **NOTE**   Specify the list of marks to combine. This combines two charts: a line chart and a point chart.

INTERACTIVITY

Vega-Lite simplifies the way of managing signals by using params. Instead of defining a signal, you define a param.

Listing 3.12   How to specify params in Vega-Lite

```
"params": [
    {
      "name": "changeColor",
      "select": {"type": "point", "on": "mouseover"}
    },
    {
      "name" : "default",
      "select": {"type": "point", "on": "mouseout"}

    }
  ],
```

> **NOTE**   Specify the name of each signal using the params keyword. For each param, specify the name and how to trigger it. The type can be set to point, to select a single point in the chart or interval.

After defining the `params`, you can use them in the marks, as specified in the following listing. You can find the complete example in the GitHub repository for the book under 03/vega-lite/json/line-chart.json.

Listing 3.13 How to use `params` in a `mark` in Vega-Lite

```
"stroke": {
        "condition": [
        {
            "param": "changeColor",
            "empty": false,
            "value": "red"
        },
        {

            "param": "default",
            "empty": false,
            "value": "blue"
        }
        ],
        "value" : "blue"
    }
```

NOTE The example in listing 3.13 uses the `params` defined in listing 3.12 to set the stroke color based on a condition. If the `changeColor` param is triggered, then the color line is set to `red`; if the `default` param is triggered, then the color line is set to `blue`. The default stroke color is `blue`.

EXERCISE 3
Convert the result of exercise 1 into Vega-Lite. The solution to the exercise is provided in the book's GitHub repository under 03/vega-lite/json/spec.json.

Now that you have learned the basic concepts behind Vega and Vega-Lite, let's see how to render a Vega/Vega-Lite visualization using HTML and JavaScript. If you are unfamiliar with HTML and JavaScript, you can skip the next section and move on to section 3.2.

3.1.3 *How to render a Vega or Vega-Lite visualization*

Both Vega and Vega-Lite are JSON objects that need a renderer to be shown, such as an external JavaScript library, to render the visualization from an HTML page. Vega-Lite provides a JavaScript API that automatically builds both Vega and Vega-Lite visualizations.

To render a Vega or Vega-Lite JSON, import the following JavaScript libraries into your HTML file:

- https://cdn.jsdelivr.net/npm/vega@5.22.1
- https://cdn.jsdelivr.net/npm/vega-lite@5.6.0
- https://cdn.jsdelivr.net/npm/vega-embed@6.21.0

The library versions may vary. In our case, we import version 5.22.1 for Vega, 5.6.0 for Vega-Lite, and 6.21.0 for Vega-Embed, as we usually do to import JavaScript libraries. Then, you can wrap the code described in listing 3.11 to render your Vega/Vega-Lite JSON file.

THE JAVASCRIPT CODE TO RENDER A VEGA OR VEGA-LITE JSON SPECIFICATION

For example, to render the chart described in listing 3.7, use the following HTML specification.

> **Listing 3.14   The complete HTML file**

```html
<!DOCTYPE html>
<html>                                                    Imports Vega and
  <head>                                                  Vega-Lite libraries
    <script src="https://cdn.jsdelivr.net/npm/vega@5.22.1"></script>
    <script src="https://cdn.jsdelivr.net/npm/vega-lite@5.6.0"></script>
    <script src="https://cdn.jsdelivr.net/npm/vega-embed@6.21.0"></script>
  </head>
  <body>                                         Creates a div to
    <div id="vis"></div>                         contain the chart
    <script type="text/javascript">              Builds the chart
      var request = new XMLHttpRequest();
      request.open('GET', '/path/to/json/file', false);
      request.send(null)
      var data = JSON.parse(request.responseText);
      vegaEmbed('#vis', data);
    </script>
  </body>
</html>
```

NOTE   First, import the Vega and Vega-Lite libraries in the header section. Then, create a `div` that will contain the chart. Finally, build the chart. Use an external JSON file to define the chart specifications. In the `script` body, specify the JavaScript code to import your Vega JSON file. Open a new `XMLHttpRequest()` to load the JSON file, and then use the `vegaEmbed()` function to render the JSON. Note that you must provide a valid external URL to the request object. You can't use a local JSON file.

You can find the complete example and other examples in the book's GitHub repository, section 03/vega (https://mng.bz/x2NB) and 03/vega-lite (https://mng.bz/Vxj5). Now that you have learned the basic concepts underlying Vega and Vega-Lite, let's move to the next topic: the basic components of an Altair chart.

3.2   *The basic components of an Altair chart*

In this book, we mostly use the *just-in-time teaching* methodology, which combines theory and practice to teach concepts at the exact moment you need them. However, in this chapter, we will not use this methodology. This is because we want to establish a strong foundation of fundamental concepts about Altair before diving into the more advanced topics covered later in the book. By deviating from the just-in-time teaching methodology for this chapter, we can provide a comprehensive overview and in-depth understanding of the underlying principles that will serve as building blocks for the subsequent material. Altair is built on top of Vega-Lite and offers a user-friendly API to build charts. You learned how to build a basic chart in Altair in chapter 2, and in this chapter, we'll dive deep into the basic components of an Altair chart.

The basic Altair components include the following:

- *Encodings*—These define how data is mapped to visual properties, such as color, shape, and size.
- *Marks*—These refer to the visual elements representing the data, such as bars, points, and lines.
- *Conditions*—These enable us to create more complex visualizations by specifying rules for when we should use certain encodings or marks.
- *Compound charts*—These combine multiple visual elements to create more complex and informative visualizations.
- *Interactivity*—This enables us to interact with the visualization and explore the data.
- *Configurations*—These define general properties for compound charts.

Let's analyze each component separately, starting with the first one: encodings. You can find some sample code in the 03/altair directory (https://mng.bz/rVXX) of the book's GitHub repository.

3.2.1 Encodings

Altair utilizes *encodings* to indicate where to show data. An encoding defines a mapping between a column in the dataset and an encoding channel. Each encoding channel corresponds to a specific column of the DataFrame that can be mapped to a visual feature of the plot. The encoding process determines where to display the data in the chart, according to the chosen channel type. Altair identifies several key types of channels, including the following:

- *Position*—This specifies the location of data in the chart.
- *Mark property*—This determines the chart's appearance such as color, size, and shape.
- *Text and tooltip*—These offer supplementary annotations for the graph.

Table 3.1 provides a brief description of the main encodings. For more details, please refer to the Altair documentation (https://mng.bz/d6RQ).

Refer to chapter 2 to build a chart using encodings.

Table 3.1 The main encodings in Altair

Type	Channel	Description
Position	`x, y`	Horizontal and vertical positions
	`longitude, latitude`	Geographical coordinates
	`xError, xError`	Error values for x and y
	`x2,y2, longitude2, latitude2`	Second positions for ranges
	`theta, theta2`	Start and end arc angles
Mark property	`angle, radius`	The angle and the radius of the mark
	`color, fill`	The color and the fill of the mark
	`opacity, fillOpacity`	The opacity and the fill opacity of the mark
	`shape, size`	The shape and the size of the mark
	`stroke, strokeDash, strokeOpacity, strokeWidth`	Stroke properties of the mark
Text and tooltip	`text`	The text of the mark
	`tooltip`	The tooltip of the mark

Each column in the dataset is associated with a data type, which defines the kind of values a variable can store. A data type includes primitive types, like integers and characters, and complex types, like arrays and objects. Altair supports the following data types:

- *Nominal*—Data that can be divided into distinct categories, without any specific order, such as the names of different fruits in a dataset
- *Ordinal*—Data that can be divided into distinct categories with an inherent order, such as the days of the week, where the order of the days is essential
- *Quantitative*—Numerical and continuous data, such as the number of product sales
- *Temporal*—Data with a temporal (i.e., date or time) component, such as the date on which a sale occurred
- *Geojson*—Data with a geographic component, such as latitude and longitude coordinates

When we build a chart in Altair, we must associate each column involved in the chart with a data type. Altair defines two ways to specify a data type:

- The type property of the channel, such as `x=alt.X('category', type='quantitative')`, where `category` is the column name

- A shorthand code following the : symbol, after the column name, x=alt.X
  ('category:Q')

Now that you have learned the concept of encodings, let's move on to the next concept, marks.

3.2.2   Marks

Marks enable you to represent data in visualizations. Examples of marks include bar charts, line charts, heat maps, and box plots, among others. To specify the type of mark, use the mark_<type>() function, where <type> represents the specific type of mark. For instance, to create a bar chart, use mark_bar(). Additionally, you can provide a list of attributes as input parameters for each mark property defined in table 3.1. When using a mark property within the mark_<type>() function, the value of that property is fixed to a constant value. For example, mark_circle(size=5) will draw circles with a constant size of 5. However, if you want to vary the size of the circles depending on a specific column of your DataFrame, use the size channel within the encoding function. Table 3.2 shows the main mark types provided by Altair.

Table 3.2   The main mark types provided by Altair

Method	Description
mark_arc()	A pie chart
mark_area()	A filled-area chart
mark_bar()	A bar chart
mark_circle()	A scatter plot with filled points
mark_geoshape()	A geographic shape
mark_line()	A line chart
mark_point()	A scatter plot with configurable points
mark_rect()	A filled rectangle for heatmaps
mark_rule()	A vertical or horizontal line spanning the axis
mark_text()	A text

Refer to chapter 2 to build a chart using marks.

3.2.3   Conditions

Altair provides several ways to create *conditions* in visualizations, which enable you to customize and control the appearance of your charts based on specific criteria. One common way to create conditions in Altair is to use the *if-else statements* in the encoding channels of a chart. To define a condition in Altair, use the `alt.condition()` method, which takes three arguments:

- A condition
- A value to use when the condition is true
- A value to use when the condition is false

For instance, if we want to change the color of the bars in a bar chart based on a specific condition, write the code shown in the following listing and the GitHub repository for the book under 03/altair/condition.py. The final chart is rendered in 03/altair/condition.html.

Listing 3.15 Using the `alt.condition()` method

```
import altair as alt
import pandas as pd

df = pd.DataFrame(
    {"Country": ["Japan", "USA", "Germany", "Spain", "France", "Italy"],
     "Medals": [4, 6, 10, 3, 7, 8],
     "Region":["Asia","Americas","Europe","Europe","Europe","Europe"]})

chart = alt.Chart(df).mark_bar().encode(
    x='Medals',
    y='Country',
    color=alt.condition(
        alt.datum.Region == 'Europe',
        alt.value('red'),          ⟵─┐  Color to use when
        alt.value('blue')          ⟵─┘  condition is true
    )
)
                                        Color to use when
chart.save('chart.html')                condition is false
```

NOTE The condition in the example checks if the Region is Europe. If the condition is true, the bars will be colored red, and if the condition is false, the bars will be colored blue. Use the datum variable to directly access the Data-Frame column name within a condition.

In addition to using if-else statements with the `alt.condition()` method and the type-checking functions, Altair provides two other ways to create conditions in visualizations: *transformations* and *selections*. You will see transformations in chapter 5 and selections in chapter 6. You can use transformations and selections to build interconnected charts that update synchronously.

3.2.4 Compound charts

Compound charts enable you to combine visualizations in a single chart. Altair supports the following types of compound charts: layering, horizontal concatenation, vertical concatenation, and repeated charts. To explain how each type of compound chart works, consider the two charts described in the following listing and 03/altair/layering.py.

Listing 3.16 Building two charts in Altair

```
base = alt.Chart(df
).encode(
    x='Country',
    y='Medals'
).properties(
    width=300
)

chart1 = base.mark_bar(
    color='#636466'
).properties(
    title='A bar chart'
)

chart2 = base.mark_line(
    color='#80C11E'
).properties(
    title='A line chart'
)
```

NOTE Build a base chart, with the encodings and properties shared among all the charts, and then use it to build the other charts, a bar chart (`chart1`), and a line chart (`chart2`). Use `properties()` to set the chart properties, such as the width and the title.

LAYERING

Layering lets you combine charts on top of each other, as shown in figure 3.2. In Altair, use the `alt.layer(chart1,chart2)` function to layer two charts, or the `+` operator (`chart1 + chart2`). If each chart has a title and you don't specify any general title for the compound chart, layering will set the title of the layered chart as that of the first chart.

HORIZONTAL CONCATENATION

Horizontal concatenation enables you to concatenate charts horizontally, as shown in figure 3.3. In Altair, use the `alt.hconcat(chart1,chart2)` function to layer two charts or, alternatively, the `|` operator. Refer to 03/altair/hconcat.py for the complete example.

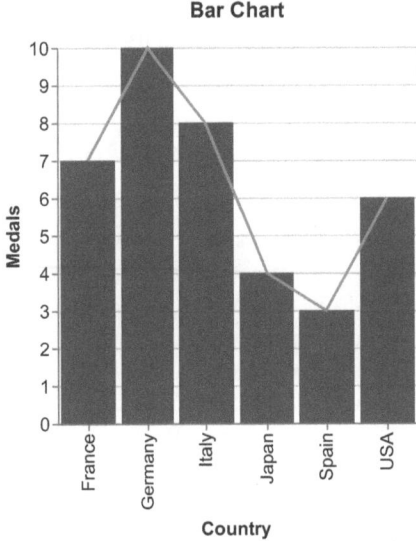

Figure 3.2   A bar chart layered with a line chart

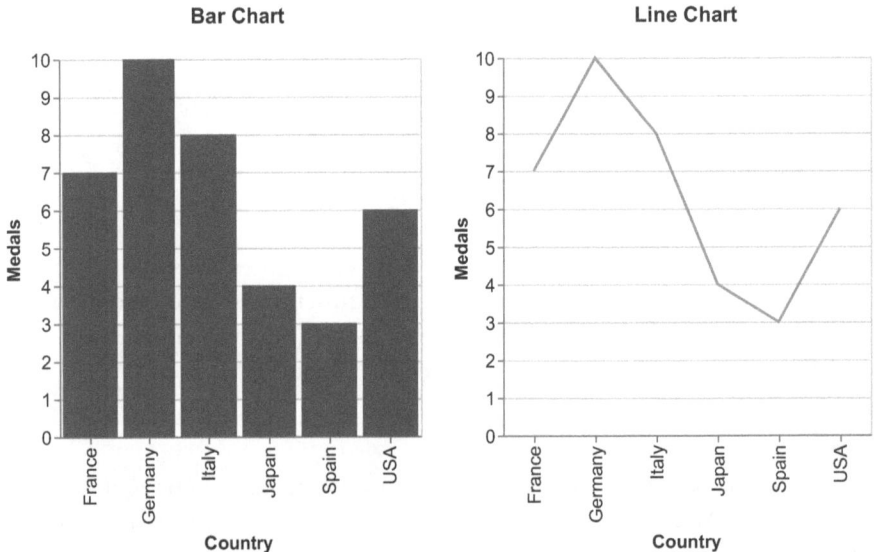

Figure 3.3   A bar chart concatenated horizontally with a line chart

VERTICAL CONCATENATION

Vertical concatenation enables you to concatenate charts vertically, as shown in figure 3.4. In Altair, use the `alt.vconcat(chart1,chart2)` function to layer two charts or the `&` operator. Refer to 03/altair/vconcat.py for the complete example.

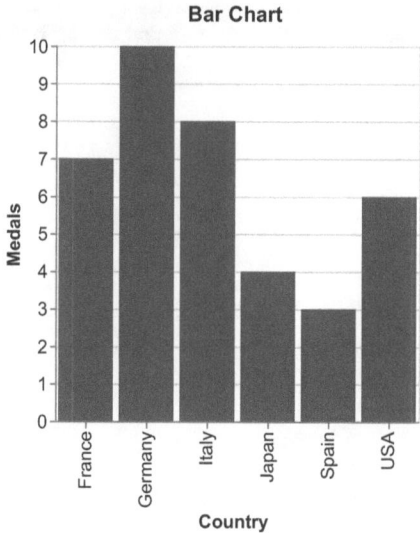

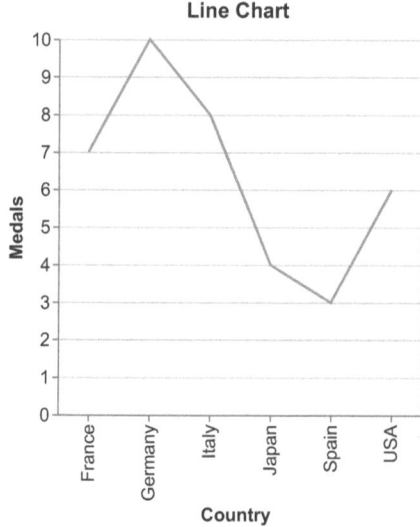

Figure 3.4   A bar chart concatenated vertically with a line chart

REPEATED CHARTS

A *repeated chart* displays similar data views in a single visualization. With a repeated chart, you can create a set of charts, each displaying a subset of the data, by repeating a base chart using different data subsets or by changing a visual encoding.

Use the `repeat()` function to create a repeated chart in Altair. This function takes the list of fields for the repeat, as shown in the following listing and under 03/altair/repeat.py.

Listing 3.17   Building a repeated chart in Altair

```python
import pandas as pd
import altair as alt

df = pd.DataFrame({
'X' : [1,2,3,4],
'Y' : [2,4,5,6],
'Z' : [3,4,5,6],
'H' : [5,6,8,9],
'M' : [3,4,5,3],
'Country' : ['USA', 'EU', 'EU', 'USA']
})


fields = df.columns.tolist()
fields.remove('Country')

chart = alt.Chart(df).mark_circle(color='#80C11E').encode(
    alt.X(alt.repeat("column"), type='quantitative'),
    alt.Y(alt.repeat("row"), type='quantitative')
).properties(
    width=100,
    height=100
).repeat(
    row=fields,
    column=fields
)

chart.save('repeat.html')
```

> **NOTE**   First, create the pandas DataFrame. Then, extract the list of fields to repeat and store them in the `fields` variable. Next, define the `repeat()` method with the rows and columns to repeat. Finally, use the `row` and `column` variables just defined as a variable of the chart encodings.

Figure 3.5 shows the output of listing 3.17. Altair has built 25 charts, one for each combination of the columns of the DataFrame specified in the `fields` variable.

3.2.5   *Interactivity*

Interactivity refers to the ability to manipulate visualizations through user input, such as hovering over data points, clicking legends, or dragging sliders. In Altair, use the `interactive()` method to enable features like zooming, panning, and hovering over data points to display tooltips. Listing 3.18 shows an example of how to make a chart interactive, and figure 3.6 shows the resulting chart. You can also find the complete example under 03/altair/interactive.py.

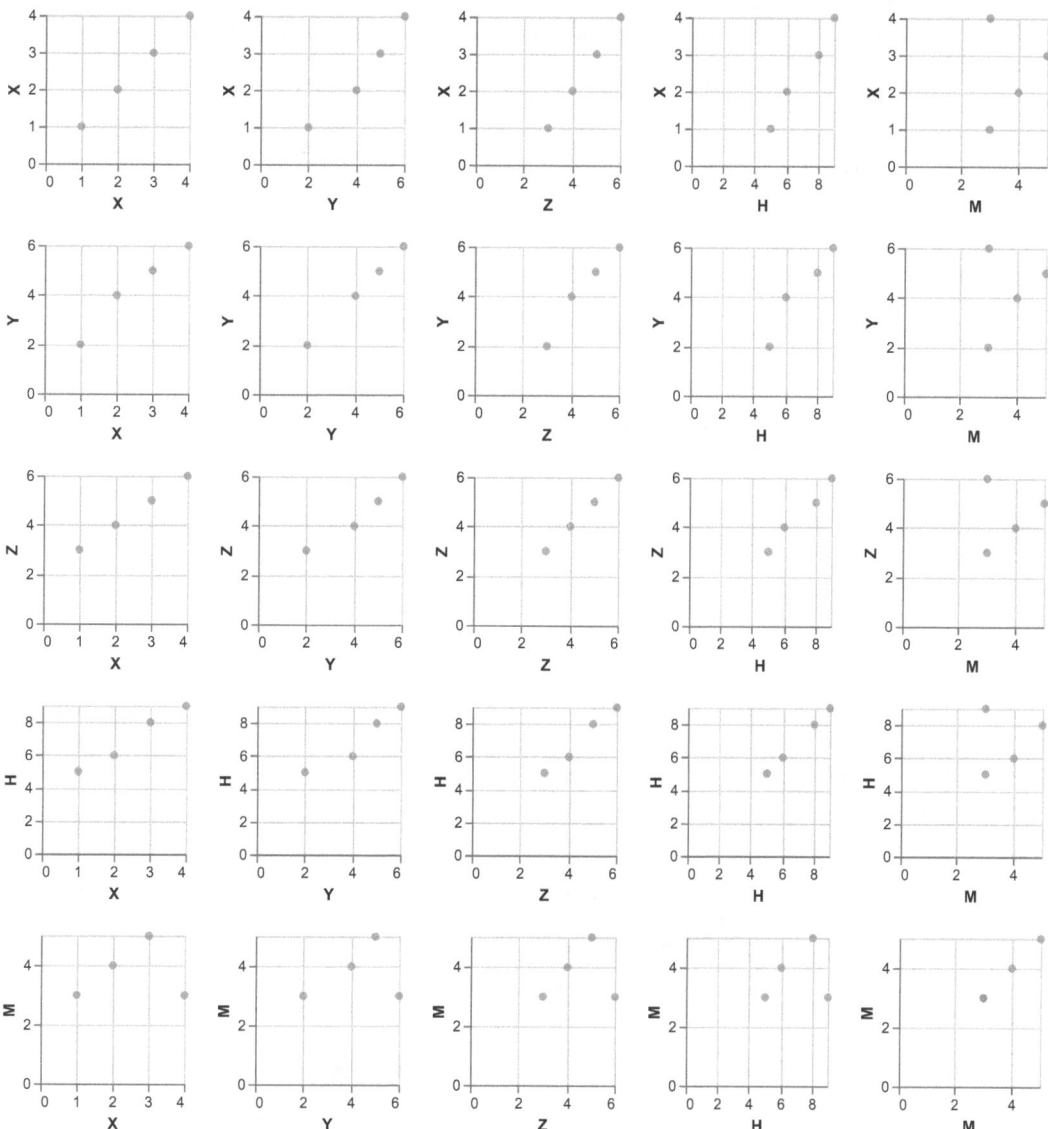

Figure 3.5   A repeated chart varying the encodings

```
import altair as alt
import pandas as pd

df = pd.DataFrame(
    {"Country": ["Japan", "USA", "Germany", "Spain", "France", "Italy"],
```

```
        "Medals": [4, 6, 10, 3, 7, 8],
        "Region":["Asia","Americas","Europe","Europe","Europe","Europe"]})

chart = alt.Chart(df).mark_bar(color='#636466').encode(
    x='Country',
    y='Medals',
    tooltip=['Country', 'Medals', 'Region']
).properties(
    width=300,
    title='A bar chart'
).interactive()

chart.save('interactive.html')
```

> **NOTE** First, create the pandas DataFrame. Then, build the Altair chart, append-
> ing the `interactive()` method at the end. Also, add a tooltip, which receives
> the list of DataFrame columns to show.

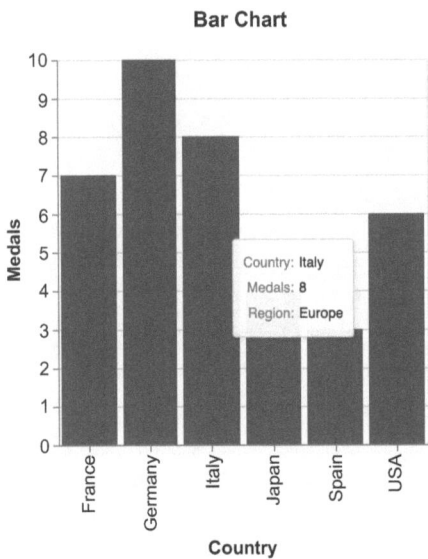

Figure 3.6   A bar chart with
interactivity enabled

3.2.6 *Configurations*

To configure the basic properties of a single chart, use the `properties()` method. For
more complex properties and to configure a compound chart, instead, you must use
global configurations. Altair supports many global configurations. In this section, we will
describe axes, title, and view. We will wait to describe the other configurations until
later in the book, when we will require them.

To show how configurations work, we will start with the chart described in figure 3.7
and generated through the code in the following listing.

Listing 3.19 Defining the basic chart

```python
import pandas as pd
import altair as alt

df = pd.DataFrame(
    {"Country": ["Japan", "USA", "Germany", "Spain", "France", "Italy"],
     "Medals": [4, 6, 10, 3, 7, 8],
     "Region":["Asia","Americas","Europe","Europe","Europe","Europe"]})


chart = alt.Chart(df).mark_bar(color='#636466').encode(
    x='Country',
    y='Medals'
).properties(width=300, title='A bar chart')

chart.save('chart.html')
```

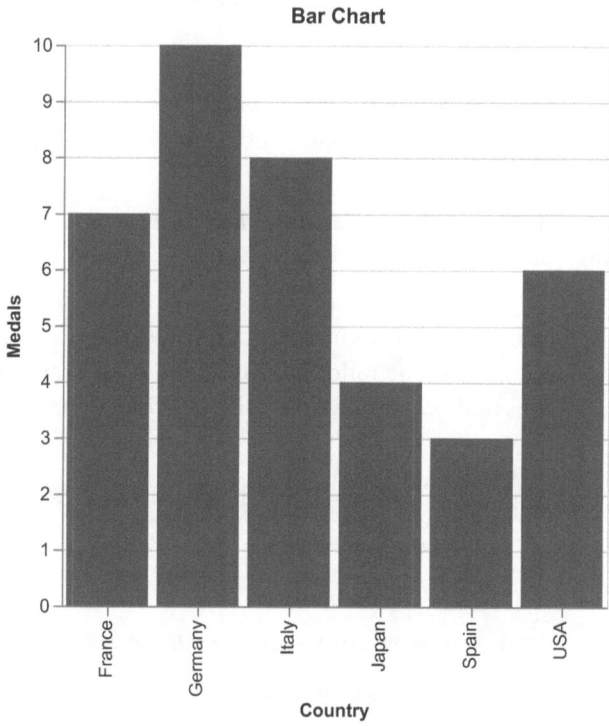

Figure 3.7 The basic bar chart built by the default configuration

Now, let's start modifying the basic chart by configuring axes.

CONFIGURING AXES

To configure axes at a global level, Altair supports three methods:

- `configure_axis()`—Configures both axes.
- `configure_axisX()`—Configures the x-axis.
- `configure_axisY()`—Configures the y-axis.

Configuring axes enables you to customize several properties, such as labels, tick marks, and scales. The following listing shows an example, which you can also find under 03/altair/configure-axis.py.

Listing 3.20   Configuring axes

```
chart = chart.configure_axis(
    labelAngle=0,              Sets the label angle
    titleFontSize=14,          Sets the font size of the axis title
    labelFontSize=12
)                              Sets the font size of the axis labels

chart.save('configure-axis.html')
```

> **NOTE** Use `configure_axis()` to configure labels and titles in axes. As an alternative, use `configure_axisX()` to configure the x-axis and `configure_axisY()` configure the y-axis.

Figure 3.8 shows the difference between the layouts when using `configure_axis()` (on the left) and not using `configure_axis()` (on the right).

CONFIGURING TITLE

Use the `configure_title()` method to customize the title properties of your chart, such as color, font size, and font family. The following listing shows an example of how to use the `configure_title()` method, which you can also find under 03/altair/configure-title.py.

Listing 3.21   Configuring `title`

```
chart = chart.configure_title(     Sets the title font size
    fontSize=20,
    color='#80C11E',               Sets the title color
    offset=30,                     Sets the distance between the title and the chart
    anchor='start'
)                                  Sets where to anchor the title

chart.save('configure-title.html')
```

> **NOTE** Use `configure_title()` to configure the title properties. The anchor must be either `None`, `start`, `middle`, or `end`.

Figure 3.9 shows the difference between the layouts when using `configure_title()` (on the left) and not using `configure_title()` (on the right).

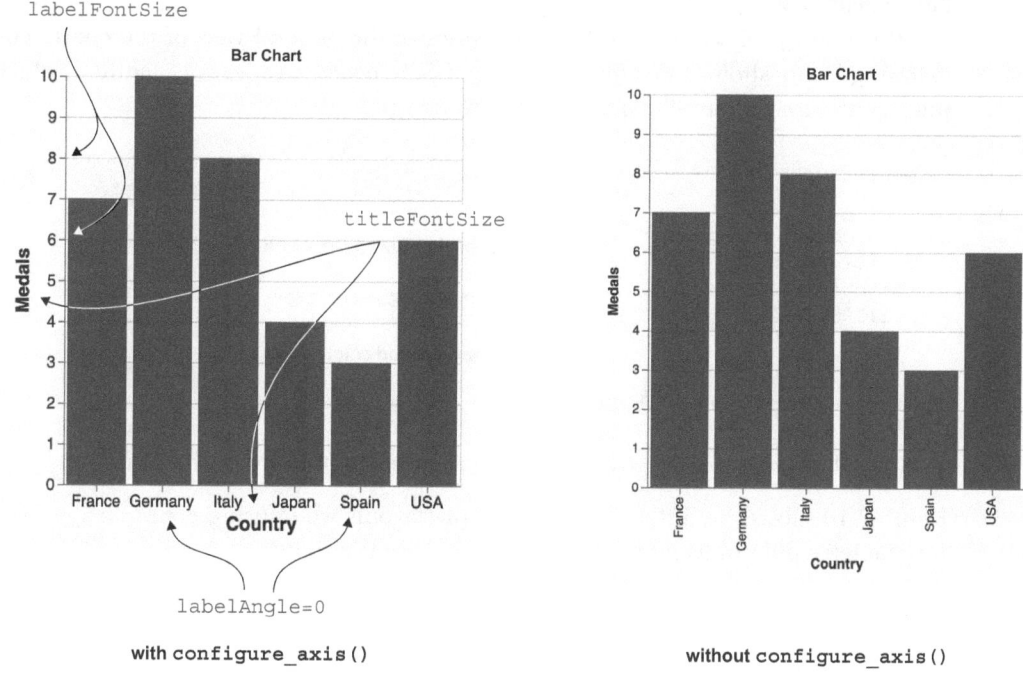

Figure 3.8 The difference between the layouts with `configure_axis()` **and without** `configure_axis()`

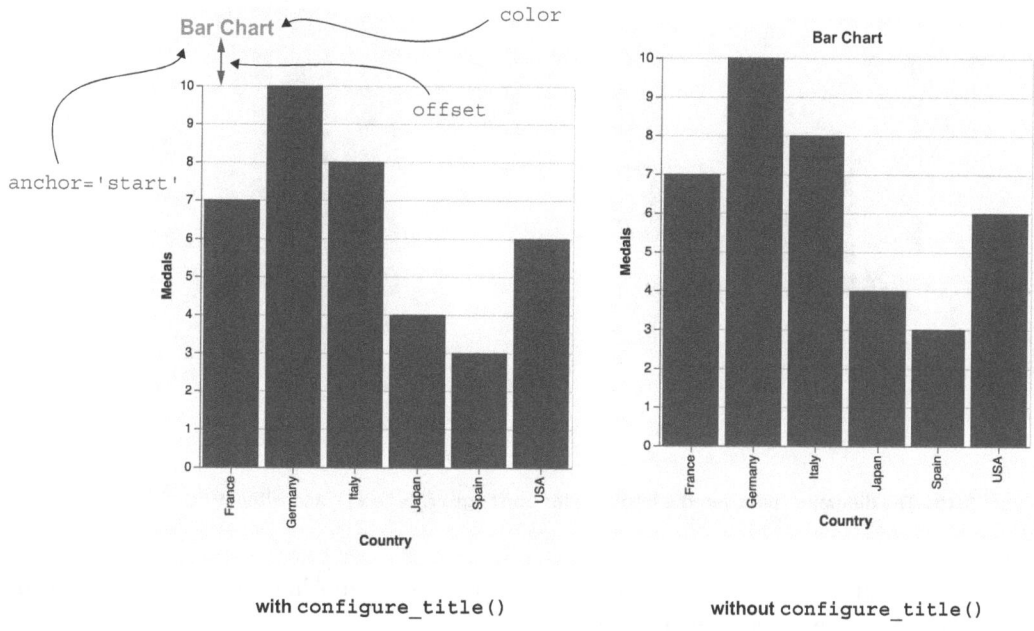

Figure 3.9 The difference between the layouts with `configure_title()` **and without** `configure_title()`

CONFIGURING VIEW

Use the `configure_view()` method to customize the general view of the chart. The following listing shows an example of usage of the `configure_view()` method, which you can also find under 03/altair/configure-view.py.

Listing 3.22   Configuring `view`

```
chart = chart.configure_view(
    strokeWidth=0,          ⟵──┐  Sets the border stroke width
    stroke='#80C11E',       ⟵──┐  Sets the border stroke color
    fill='#E0E0E0'          ⟵──┐
)                                  Sets the background color

chart.save('configure-view.html')
```

NOTE   Use `configure_view()` to configure the general view of the chart.

Figure 3.10 shows the difference between the layouts when using `configure_view()` (on the left) and not using `configure_view()` (on the right).

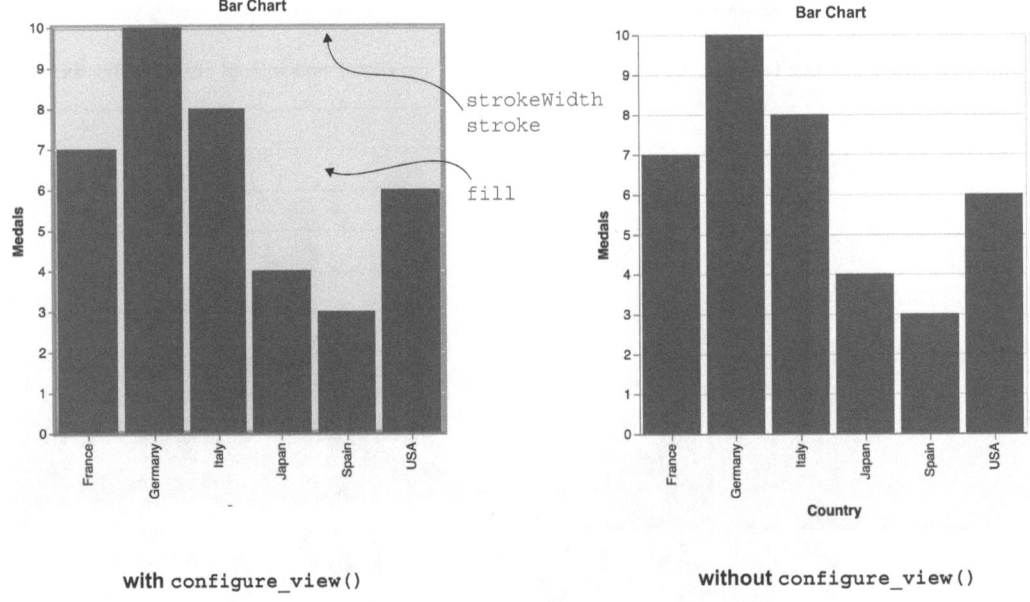

with `configure_view()`   without `configure_view()`

Figure 3.10   The difference between the layouts with `configure_view()` and without `configure_view()`

Now that you have learned the basic components of an Altair chart, let's move to the next step: implementing a case study.

3.3 Case study

Let us imagine you work as a researcher at the Global Demographic Analysis Center (GDAC), who wants to study population growth in North America. For this study, you will be using the World Bank's population dataset (released under the CC BY-4.0 license, https://data.worldbank.org/indicator/SP.POP.TOTL); you can find a cleaned version of the dataset in the GitHub repository for the book under CaseStudies/population/source/population.csv. The dataset contains the population from 1960 to 2021 for all the countries of the world, as shown in table 3.3. The Country Name column also contains continents. You can find this case study's complete code provided as a Jupyter Notebook in the book's GitHub repository under CaseStudies/population/population.ipynb. In this case, simply execute a cell with the chart variable to visualize an Altair chart.

Table 3.3 The population dataset

Country Name	1960	...	2021
Aruba	54,608		106,537
...			
Zimbabwe	3,806,310		15,993,524

You start by drawing a raw chart in the following listing, showing all the trend lines from 1960 to 2021 for all the countries and continents. The Country Name column is dirty, so the resulting chart is very confusing, as shown in figure 3.11.

Listing 3.23 Drawing the raw chart

```
import pandas as pd
import altair as alt
alt.data_transformers.disable_max_rows()        ◁──── Uses this Altair function
                                                       for datasets with more
                                                       than 5,000 rows
df = pd.read_csv('data/population.csv')
df = df.melt(id_vars='Country Name',            ◁──── Unpivots the dataset
             var_name='Year',
             value_name='Population')
df['Year'] = df['Year'].astype('int')

chart = alt.Chart(df).mark_line().encode(
    x = 'Year:Q',
    y = 'Population:Q',
    color = 'Country Name:N'
)                                  ◁──── Draws the chart

chart
```

> **NOTE** Use `mark_line()` to draw the chart. To plot the different countries, use the `color` channel.

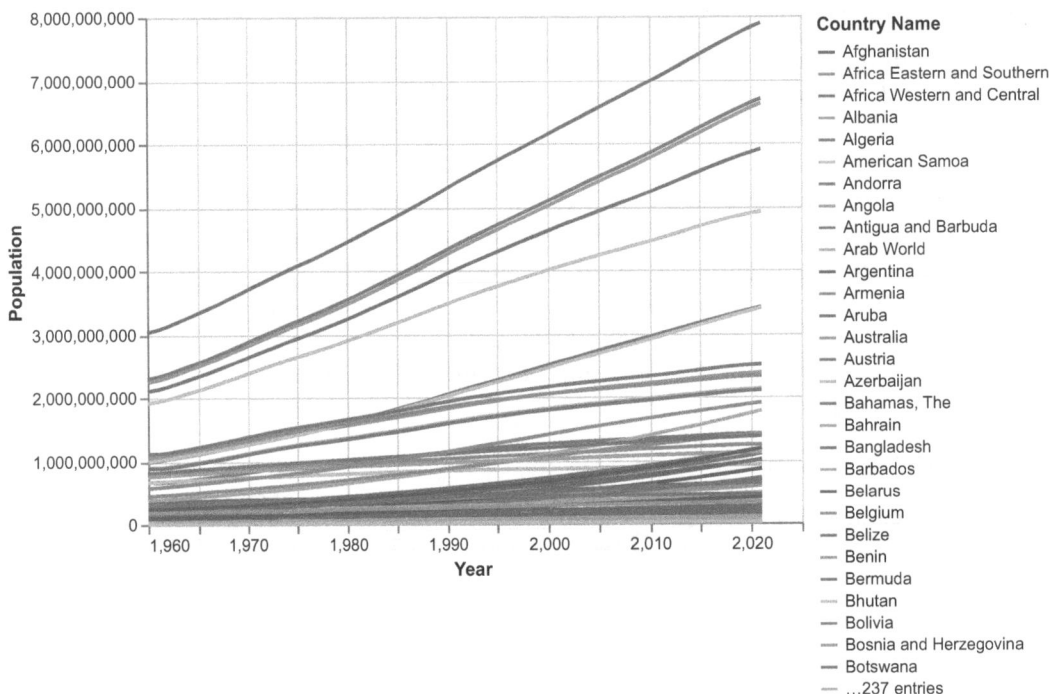

Figure 3.11   The chart with raw data

The chart is very confusing because it presents the following problems: there are too many countries, there are too many colors, and it does not focus on North America. To solve these issues, let us use the DIKW pyramid, starting from the first step: turning data into information.

3.3.1   *From data to information*

Turning data into information involves extracting insights from data. From the point of view of a chart, it means making the chart readable and understandable. Start by grouping countries by continent. The dataset already contains values for continents. List the countries using the `unique()` method provided by the pandas DataFrame applied to the Country Name column. For more details on the `unique()` method, please refer to appendix B. Note that the list of countries also contains the continents, so we can build a list of the continents and filter the DataFrame based on that list. Then, plot the resulting chart.

Listing 3.24   Decluttering the chart

```
continents = ['Africa Eastern and Southern',
              'Africa Western and Central',
              'Middle East & North Africa',
```

```
                        'Sub-Saharan Africa',
                        'Europe & Central Asia',
                        'Latin America & Caribbean',
                        'North America',
                        'Pacific island small states',     ──┐  Defines the list
                        'East Asia & Pacific']             ◁──┘  of continents

    df = df[df['Country Name'].isin(continents)]       ◁──┐  Filters the
                                                           │  DataFrame on the
                                                           │  basis of the list
    chart = alt.Chart(df).mark_line().encode(
        x = 'Year:Q',
        y = 'Population:Q',
        color = 'Country Name:N'
    )                                     ◁───── Draws the chart

    chart
```

NOTE   Group data by continents to reduce the number of lines.

Figure 3.12 shows the resulting chart, which is more readable. However, it still presents some problems: there too many colors and it does not focus on North America. In practice, it is difficult to compare North America with the other continents.

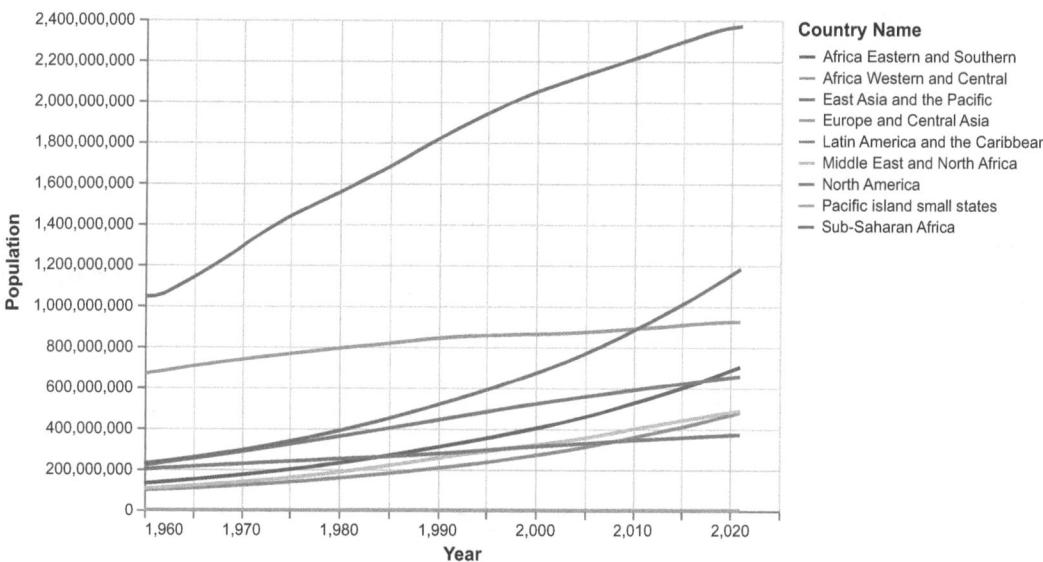

Figure 3.12   The chart with a focus on continents

Change the chart by focusing on North America and grouping the other continents by calculating their average value, as shown in the following listing.

Listing 3.25   Focusing on North America

```
mask = df['Country Name'].isin(['North America'])
df_mean = df[~mask].groupby(by='Year').mean().reset_index()

df_grouped = pd.DataFrame({
    'Year' : df[mask]['Year'].values,
    'North America' : df[mask]['Population'].values,
    'World': df_mean['Population'].values
})

df_melt = df_grouped.melt(id_vars='Year', var_name='Continent',
    value_name='Population')
```

Defines a mask that selects only North America

Calculates the average value for the continents not in the mask

Builds a new DataFrame with North America and the average value of the other continents

NOTE Focus on North America, and calculate the average value of the other continents.

Then, draw the chart.

Listing 3.26   Drawing the chart

```
colors=['#80C11E', 'grey']

chart = alt.Chart(df_melt).mark_line().encode(
    x = alt.X('Year:Q',
              title=None,
              axis=alt.Axis(format='.0f',tickMinStep=10)),
    y = alt.Y('Population:Q',
              title='Difference of population from 1960',
              axis=alt.Axis(format='.2s')),
    color = alt.Color('Continent:N',
                      scale=alt.Scale(range=colors),
                      legend=None),
    opacity = alt.condition(alt.datum['Continent'] == 'North America',
      alt.value(1), alt.value(0.3))
).properties(
    title='Population in the North America over the last 50 years',
    width=400,
    height=250
).configure_axis(
    grid=False,
    titleFontSize=14,
    labelFontSize=12
).configure_title(
    fontSize=16,
    color='#80C11E'
).configure_view(
    strokeWidth=0
)

chart
```

Formats x labels

Formats y labels

Does not show the legend

Sets the range of colors

Disables grid

NOTE To format the x-axis labels, use the `format` parameter. Use `.0f` to show only the integer part. Also, use the `tickMinStep` parameter to set the distance between two labels. The example sets the value to `10`, thus labeling the decades. To format the y-axis labels, set the format parameter to `.2s`, using scientific notation with two significant figures.

Figure 3.13 shows the resulting chart. We have omitted the legend for now and will add it later. The chart is clearer; however, the audience needs to do some calculations to understand the difference between the two lines.

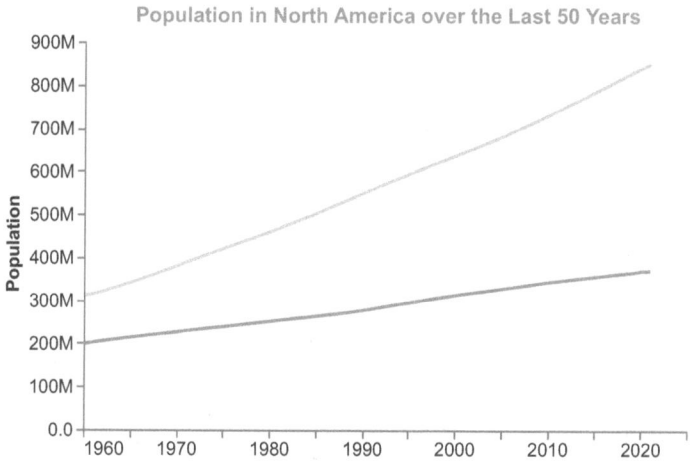

Figure 3.13   The chart focuses on North America.

When drawing a chart, do the calculations for the audience to serve them the ready-made dish. Let us calculate the difference from the starting year (1960) for both lines, thus making them comparable. The following listing shows how to calculate the difference.

Listing 3.27   Calculating the difference

```
baseline = df_melt[df_melt['Year'] == 1960]

continents = ['North America', 'World']
for continent in continents:
    baseline_value = baseline[baseline['Continent'] ==
      continent]['Population'].values[0]
    m = df_melt['Continent'] == continent
    df_melt.loc[m, 'Diff'] = df_melt.loc[m,'Population'] - baseline_value
```

NOTE Calculate the baseline value for 1960, and then for each continent, calculate the difference between each year and the baseline.

Having calculated the difference, draw the chart. The code is similar to that in listing 3.23. Change only the y-axis column to Diff. Figure 3.14 shows the resulting chart.

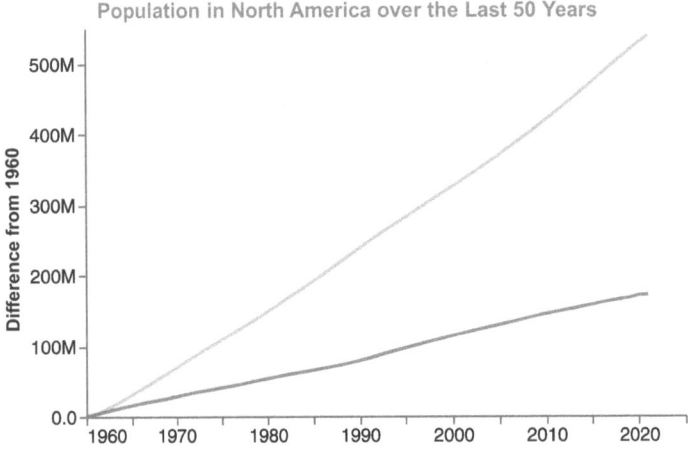

**Figure 3.14   The chart with a focus on the difference in population from
1960**

The chart clearly shows the difference between the two lines, although it is not
explicit. To complete the chart, add a label for each line as an additional textual mark.

Listing 3.28   Adding labels

```
mask = df_melt['Year'] == 2021
na = df_melt[mask]['Diff'].values[0]
oth = df_melt[mask]['Diff'].values[1]

df_text = pd.DataFrame({'text' : ['Rest of the world','North America'],
        'x' : [2023,2023],                          Sets the x position
        'y' : [oth,na]})
                                          Sets the y
                                          position
text = alt.Chart(df_text).mark_text(fontSize=14, align='left').encode(
    x = 'x',
    y = 'y',
    text = 'text',
    color = alt.condition(alt.datum.text == 'North America',
     alt.value('#80C11E'), alt.value('grey'))
)
```

NOTE   Use mark_text() to set the line labels. We label the lines by placing a
text annotation next to each line. We could have used the legend to label the
lines, but as Cole Nussbaumer Knaflic says in her book, *Storytelling with Data: A
Data Visualization Guide for Business Professionals* (John Wiley & Sons, 2015), it's
clearer to use text annotations for better clarity. To set the y text position, use

the coordinates of the last point in each line. For the x, instead, slightly increase the coordinates of the last point in each line (e.g., 2023 in the example, while the last point is 2021).

Now, combine the textual mark and the original chart.

Listing 3.29   Combining the chart and the text labels

```
total = (chart + text).configure_axis(
    grid=False,
    titleFontSize=14,
    labelFontSize=12
).configure_title(
    fontSize=16,
    color='#80C11E'
).configure_view(
    strokeWidth=0
)

total
```

> **NOTE**   Use + to combine the two layers. Then, use `configure_axis()`, `configure_title()`, and `configure_view()` to set the global chart configurations.

Figure 3.15 illustrates the resulting chart at the end of the information step.

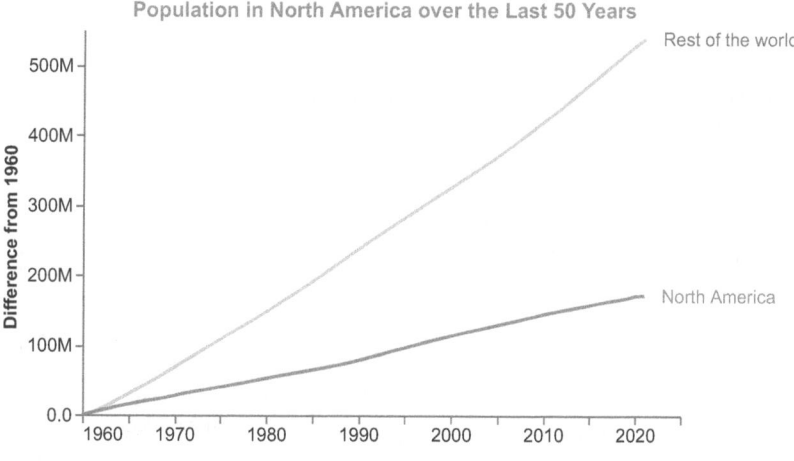

Figure 3.15   The chart at the end of the information step

The next step of the DIKW pyramid is turning information into knowledge. So let us proceed.

3.3.2   From information to knowledge

Turning information into knowledge involves adding context to the chart. We add three types of context: an annotation showing the gap between the two lines, an annotation containing the data source, and a text explaining the causes of this gap.

CHART ANNOTATION

Chart annotation aims to ease the audience's life by making explicit the gap between the two lines. Figure 3.16 shows the annotation we are going to add to the chart. The annotation comprises two elements: a vertical line showing the gap between the two lines and a "big number" (BAN), a large number containing the gap value.

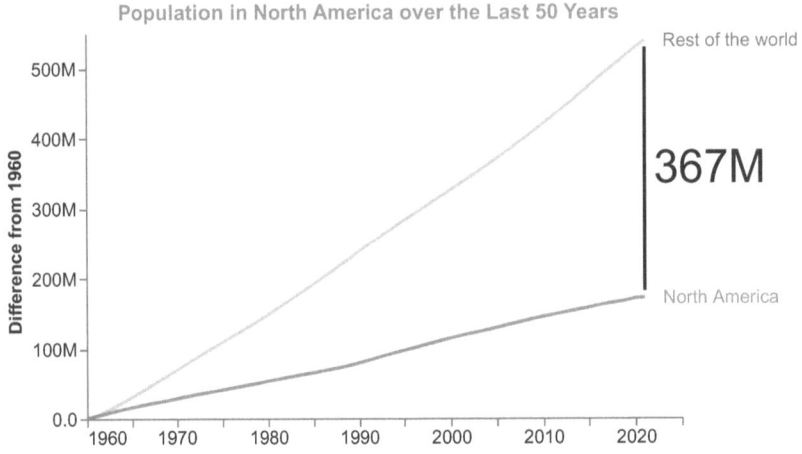

Figure 3.16   The chart with an explicit annotation on the gap between North America and the rest of the world

To draw the chart annotation, use `mark_line()` to draw the vertical line and `mark_text()` to draw the text. The following listing describes how to render the previous annotations.

Listing 3.30   Adding the annotation

```
offset = 10000000
mask = df_melt['Year'] == 2021
na = df_melt[mask]['Diff'].values[0]
oth = df_melt[mask]['Diff'].values[1]

df_vline = pd.DataFrame({'y' : [oth - offset,na + offset],
                         'x' : [2021,2021]})
```

Calculates the last value
for North America line

Calculates the last value
for the other line

Defines the distance between the vertical
line and the last point in the country line

```
line = alt.Chart(df_vline).mark_line(color='black').encode(
    y = 'y',
    x = 'x'
)                      ⟵── Draws the vertical line

df_ann = pd.DataFrame({'text' : ['367M'],
        'x' : [2022],                    │  Places the BAN at the
        'y' : [na + (oth-na)/2]})     ⟵─┘  middle of the vertical line

ann = alt.Chart(df_ann).mark_text(fontSize=30, align='left').encode(
    x = 'x',
    y = 'y',
    text = 'text'
)                      ⟵── Draws the BAN

total = chart + text + line + ann        ⟵── Combines the charts
total = total.configure_axis(
    grid=False,
    titleFontSize=14,
    labelFontSize=12
).configure_title(
    fontSize=16,
    color='#80C11E'
).configure_view(
    strokeWidth=0
)

total
```

> **NOTE** Use the last data points of the lines of the continents as y-coordinates for the vertical line and the BAN. We have hardcoded the BAN value for simplicity. However, you can dynamically calculate it in a complex scenario.

DATA SOURCE

Place the data source as a textual annotation immediately under the title.

Listing 3.31 Adding the data source

```
df_subtitle= pd.DataFrame({'text' : ['source: World Bank'],
                'href' : '
    https://data.worldbank.org/indicator/SP.POP.TOTL' })

subtitle = alt.Chart(df_subtitle
                ).mark_text(
                    y=0
                ).encode(
                    text='text',
                    href='href'
                )
```

> **NOTE** Use the mark_text() mark to add a text to the chart. Also, set the href channel to the link to the data source.

TEXT ANNOTATION

Place the text annotation to the left of the graph to prepare your audience for the chart. The text explains the possible causes of the gap between North America and the other continents. The following listing describes how to add a text annotation to the chart.

Listing 3.32   Adding a text annotation

```
df_context = pd.DataFrame({'text' : ['Why this gap?',
                            '1. Lower Fertility Rate',
                            '2. Lower Immigration Rate',
                            '3. Higher Average Age'],
                           'y': [0,1,2,3]})

context = alt.Chart(df_context).mark_text(fontSize=14, align='left',
    dy=50).encode(
    y = alt.Y('y:O', axis=None),
    text = 'text',
    stroke = alt.condition(alt.datum.y == 0, alt.value('#80C11E'),
     alt.value('black')),
    strokeWidth = alt.condition(alt.datum.y == 0, alt.value(1), alt.value(0))
)

total = (context | (chart + text + line + ann + subtitle)).configure_axis(
    grid=False,
    titleFontSize=14,
    labelFontSize=12
).configure_title(
    fontSize=16,
    color='#80C11E'
).configure_view(
    strokeWidth=0
)

total.show()
```

> **NOTE**   Use `mark_text()` to draw the text. Then, combine the text annotation with the chart.

Figure 3.17 shows the resulting chart at the end of the knowledge step.

The chart is almost ready. Let us proceed to the last step: turning knowledge into wisdom.

3.3.3   *From knowledge to wisdom*

Turning knowledge into wisdom involves adding a call to action. In the previous chapters, we added a simple call to action in the title, inviting the audience to do something. However, in a real scenario, the call to action must include a proposal of possible next steps, justified with appropriate resources. The next step invites the audience to do something concrete; thus, the situation depicted in the chart changes. For example, you can add the call to action as an additional chart, which describes the percentage

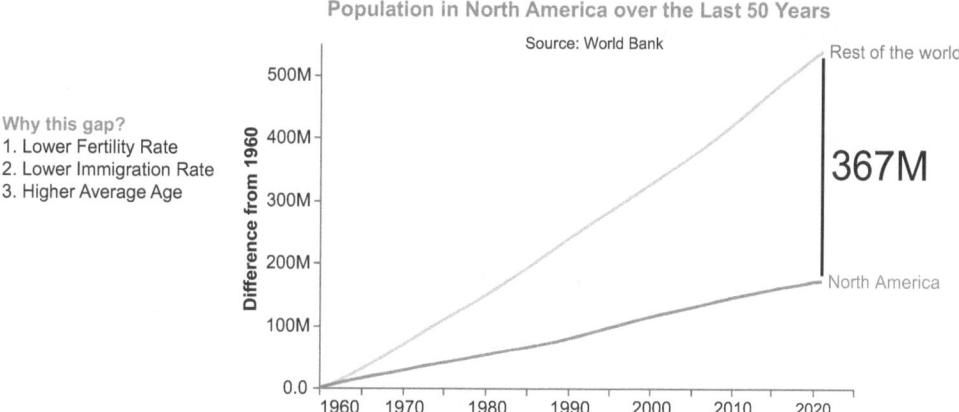

Figure 3.17   The chart at the end of the knowledge step

of success of each possible alternative. In our example, we can add a bar chart with possible strategies to improve population growth in North America. The percentages in the example are imaginary and do not reflect reality—we've added them just to show how a call to action works. The following listing shows how to build the call to action as a bar chart and add it to the main chart.

Listing 3.33   Adding a call to action

```
df_cta = pd.DataFrame({
    'Strategy': ['Immigration Development', 'Enhance Family-Friendly
    Policies', 'Revitalize Rural Areas'],
    'Population Increase': [20, 30, 15]   # Sample population increase
    percentages
})

cta = alt.Chart(df_cta).mark_bar(color='#80C11E').encode(
    x='Population Increase:Q',
    y=alt.Y('Strategy:N', sort='-x', title=None),
    tooltip=['Strategy', 'Population Increase']
).properties(
    title='Strategies for population growth in North America',
)

total = alt.vconcat((context | (chart + text + line + ann)),
    cta,center=True).configure_axis(
    grid=False,
    titleFontSize=14,
    labelFontSize=12
).configure_title(
    fontSize=20,
    color='#80C11E',
    offset=10
).configure_view(
```

```
    strokeWidth=0
).configure_concat(
    spacing=50
)

total.show()
```

NOTE   Draw a chart specifying the possible alternatives and their percentage of success.

Figure 3.18 shows the final chart. Following the steps specified by the DIKW, we have transformed the raw chart of figure 3.11 into a data story.

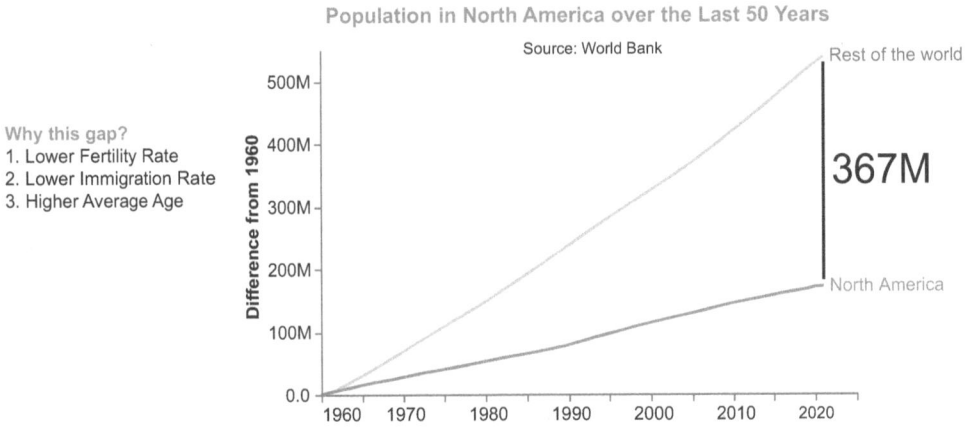

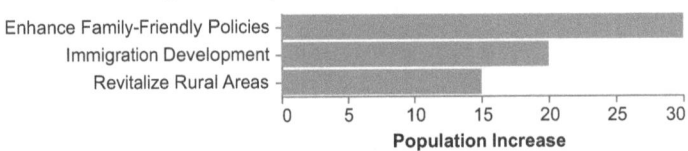

Figure 3.18   The final chart at the end of the wisdom step

In this chapter, you have learned the basic concepts of Altair. In the next chapter, you'll review the basic concepts of generative AI.

Summary

- Vega and Vega-Lite are visualization grammars to define the visual aspects and interactive features of a chart by writing code in JSON format.
- Altair is built at the top of Vega-Lite and offers a user-friendly API to build charts. The basic components of an Altair chart are encodings, marks, conditions, compound charts, interactivity, and configurations.

- Encodings define how data attributes (like x-coordinates, y-coordinates, color, and size) are visually represented in the chart.
- Marks are the visual elements in the chart, such as bars, points, lines, or areas.
- Conditions specify how data attributes are mapped to visual properties based on different criteria.
- Compound charts involve combining multiple charts into a single view.
- Interactivity refers to the ability of the chart to respond or change based on user input or interactions.
- Configurations encompass various settings to alter the appearance of the chart, such as adjusting axes, scales, or layout, to better convey the intended information or aesthetics.

References

Data visualization

- Belorkar, A., Guntuku, S. C., Hora, S., and Kumar, A. (2020). *Interactive Data Visualization with Python.* Packt Publishing.
- Abbott, D. (2024). *Everyday Data Visualization.* Manning Publications.

Vega, Vega-Lite, and D3.js

- Kats, P. and Katz, D. (2019). *Learn Python by Building Data Science Applications.* Packt Publications.
- Meeks, E. and Dufour, A.M. (2024). *D3.js in Action.* Manning Publications.
- *Vega: A Visualization Grammar.* (n.d.). https://vega.github.io/vega/.
- *Vega-Lite: A Grammar of Interactive Graphics.* (n.d.). https://vega.github.io/vega-lite/.

Other

- Knaflic, C. N. (2015). Storytelling with Data: A Data Visualization Guide for Business Professionals. John Wiley & Sons.

Generative AI tools for data storytelling

This chapter covers

- Basic concepts, including artificial intelligence, machine learning, deep learning, and generative AI
- The basic structure of a ChatGPT prompt
- The basic structure of a DALL-E prompt
- GitHub Copilot

We've finally gotten to the point: how to use ChatGPT and DALL-E to do data storytelling. You might wonder why you had to wait until now to see these tools in action. I will give you an answer during this chapter, in which we will review the basic concepts behind generative AI tools and how to use them in the context of data storytelling. The chapter does not describe the fundamentals of generative AI. Instead, the first part gives a quick overview of AI, generative AI, and generative AI tools. Next, the chapter describes how to incorporate three main tools into data storytelling: ChatGPT, DALL-E, and GitHub Copilot. Finally, it shows a practical use case, which demonstrates how you can combine the three tools to build data stories quickly.

4.1 Generative AI tools: On the giants' shoulders

Some time ago, I participated in a webinar entitled *Considerations on GPT Technology* (*Considerazioni sulla tecnologia GPT*, 2023, https://www.youtube.com/watch?v=So48Y MYyl58), where panelists discussed the (possible) benefits introduced by generative AI. During the discussion, one particular intervention by Dr. Michele Monti caught my attention. He drew an interesting analogy between using ChatGPT to assist you while writing and using a satellite navigator on a smartphone to assist you while going to a destination.

In the past, when satellite navigators were not readily available, travelers relied on paper maps to reach a destination, with the notable difference that a paper map doesn't need any infrastructure or energy to be operational. However, with the advent of satellite navigators, the convenience and efficiency they offered made paper maps almost obsolete. Nowadays, very few people would consider using a paper map when they have the option of using a satellite navigator.

A similar analysis could be made of generative AI tools, especially ChatGPT. Before the emergence of generative AI tools, one needed to write texts manually. This could be time consuming and labor intensive. As generative AI advances, it may increasingly become the primary mode of assistance when writing and generating images, although one should never become totally dependent on it.

In my original idea, this book was only supposed to describe how to do data storytelling with Altair and the DIKW pyramid. It was my editor (whom I sincerely thank), in light of the rapid emergence of generative AI at the time of writing, who suggested also including generative AI. At first, I only thought about Copilot, but then, as I wrote the examples, I made more and more use of other newly available generative AI tools to complete the data stories. And that's why I decided to also describe in the book how to do data storytelling using generative AI tools.

You might wonder why you had to wait until chapter 4 to read about ChatGPT and DALL-E. The real reason is that to apply the generative AI tools correctly, it is necessary first to understand the context in which you work. If I had immediately introduced ChatGPT and DALL-E, I would have risked readers delegating everything to these tools without building a solid foundation. Knowledge of how to do data storytelling in Python would inevitably be superficial. Imagine the concepts described in this book as a house where the foundations are data storytelling concepts, the floors are Altair, and the roof and walls are generative AI. Just as you cannot build a house from the roof and the walls, you cannot start immediately with generative AI. However, to not leave you hungry for generative AI in the first chapters, I still wanted to mention GitHub Copilot because, as they say, appetite comes with eating.

This book does not describe how to automate data storytelling but how to use generative AI tools as assistants in developing our ideas. Before seeing how to use these tools to do data storytelling, let's try to understand what lies behind the concept of generative AI.

Generative artificial intelligence (AI) is a subfield of AI; more precisely, it is a subfield of deep learning. Describing the foundations of generative AI is out of the scope of this book, but a quick overview of it is mandatory to understand the general context before we delve into its use. At the time of writing this book, there was a lot of debate surrounding the use of generative AI tools. On the one hand, some people argue that these tools are dangerous, as they could potentially replace human jobs, leading to widespread unemployment. Additionally, they often produce hazardous, biased, and hallucinated results, among other issues. There are worries about the ethical implications of AI decision making, as well as the potential misuse or manipulation of these tools for malicious purposes. On the other hand, other people highlight their benefits, recognizing the level of automation introduced by these tools under the supervision of a responsible user. In this book, I do not take a position, leaving for you the responsibility to use these tools consciously.

In this book, we'll try to have a conservative and responsible attitude toward generative AI tools. We will use them as assistants that help us better define our content but which will never replace our supervision and ideas. By "conservative and responsible attitude," we mean that we will not use the output produced by generative AI tools as the absolute truth. Instead, we will adopt the human-in-the-loop strategy, by always checking the produced output.

This section gives a quick overview of what AI, machine learning, deep learning, and generative AI are. If you are feeling impatient and want to get straight to the main topic—generative AI in data storytelling—skip this section and jump to the next one directly.

4.1.1   What is artificial intelligence?

Artificial intelligence is a discipline aiming to construct intelligent agents capable of independent reasoning, learning, and decision making. AI focuses on developing theories and methodologies to create agents that can emulate human-like thinking and behavior. Figure 4.1 shows the main subfields of AI.

The figure also highlights the subfield in which generative AI falls. Generative AI is a subfield of deep learning, which is a subfield of machine learning, which is a subfield of artificial intelligence. Now that you have learned the main subfields of AI, let's focus on machine learning.

4.1.2   What is machine learning?

Machine learning (ML) is a subfield of AI that enables computers to acquire knowledge and improve performance by learning from examples. Consider, for example, the case of assessing whether an outfit is fashionable or not. Without ML, you would have to code every line description. Instead, using ML will require collecting and labeling a substantial dataset of outfit descriptions and fashion elements. Then, you will use it to train an ML model to recognize whether the outfit is fashionable or not. Different from traditional programming, where you must write the code line by line to provide

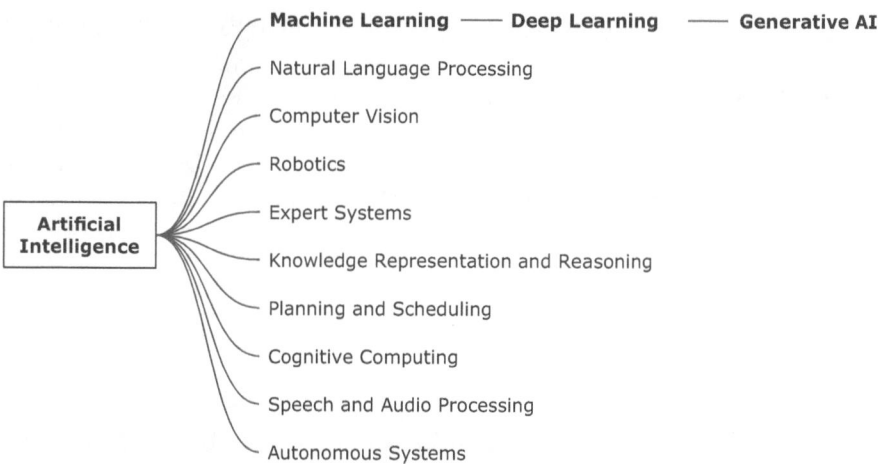

Figure 4.1   The main subfields of AI, with a focus on generative AI

instructions, ML allows machines to learn from data and make predictions or take actions based on that learned information. This approach empowers machines to discover patterns, relationships, and insights that may not be immediately apparent to human programmers. ML relies on different types of learning. The most popular are supervised learning and unsupervised learning.

SUPERVISED LEARNING

Supervised learning starts with a problem to solve and a dataset of labeled data, which already contains the solved problem for a subset of samples. First, within the labeled data, input features must be extracted. Input features represent the distinct attributes of the data that the model will analyze to learn patterns and relationships. In supervised learning, there is a preliminary phase, called *training*, during which algorithms learn from labeled data. During training, each sample in the training dataset is paired with its corresponding label or output. The algorithm learns how to map input features to their respective outputs by generalizing from the provided labeled data. During the training phase, the algorithm optimizes its parameters to minimize discrepancies between the predicted outputs and the true labels. Once the training phase is completed, the algorithm can make predictions or classify new, unseen instances based on the patterns it has learned from the labeled data. Use this technique for tasks such as image classification, sentiment analysis, spam detection, and speech recognition.

In supervised learning, there are two main types of tasks:

- *Classification*—This involves predicting a categorical or discrete class label for an input. For instance, given an input (like an image or a set of features), the algorithm predicts which category or class it belongs to. The classes are distinct and predefined. Examples include spam detection (classifying emails as spam or not spam) and image recognition (identifying objects like cats or dogs in images).

- *Regression*—Here, the goal is to predict a continuous numerical value. Instead of predicting a class label, regression algorithms predict a quantity. For example, predicting house prices based on features like size, location, and number of rooms is a regression problem. The output is a range of values rather than distinct classes.

Both classification and regression are part of supervised learning because they rely on labeled data, where the algorithm is trained on input–output pairs to learn the relationship between inputs and their corresponding outputs.

> ### Challenge: Identifying classification vs. regression in healthcare
> Consider the following scenario. A healthcare system aims to predict whether a patient has a particular disease based on various medical test results. The system takes blood pressure, cholesterol levels, and other health indicators as input and outputs whether the patient is diagnosed with the disease. Can you identify if this scenario is an example of classification or regression?

Figure 4.2 shows an example of a classification model. First, the model must be trained with sample data and class labels. As a result of the training process, the model has learned how to discriminate between two classes: circles and squares. When the model receives new data as input, it predicts the class associated with each sample as the output.

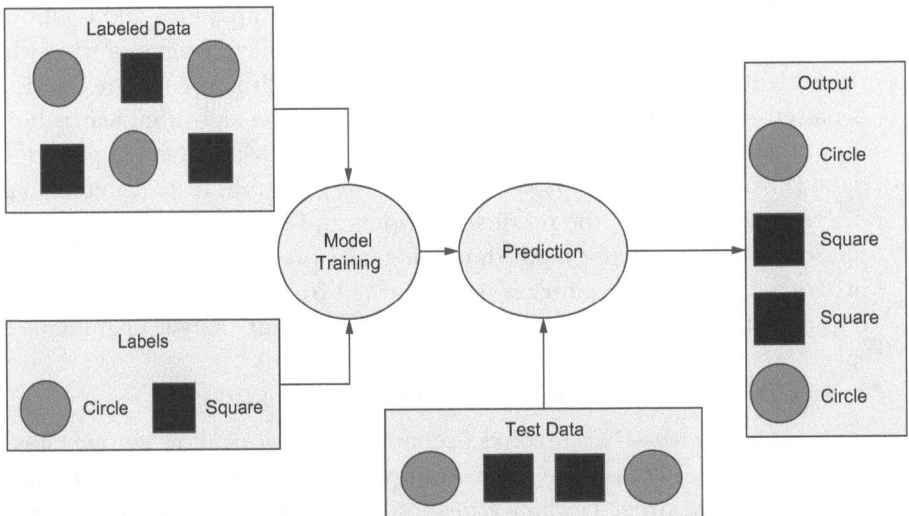

Figure 4.2 An example of a classification model

UNSUPERVISED LEARNING

In *unsupervised learning*, algorithms identify the hidden structure and patterns in unlabeled data, where no specific outputs or labels are provided during training. Common unsupervised learning techniques include clustering, dimensionality reduction, and anomaly detection.

Figure 4.3 shows an example of a clustering model. The model has not been previously trained with data. Instead, it applies an internal formula to group data. The model receives a dataset of squares and circles as an input and groups them into two clusters, one for each shape.

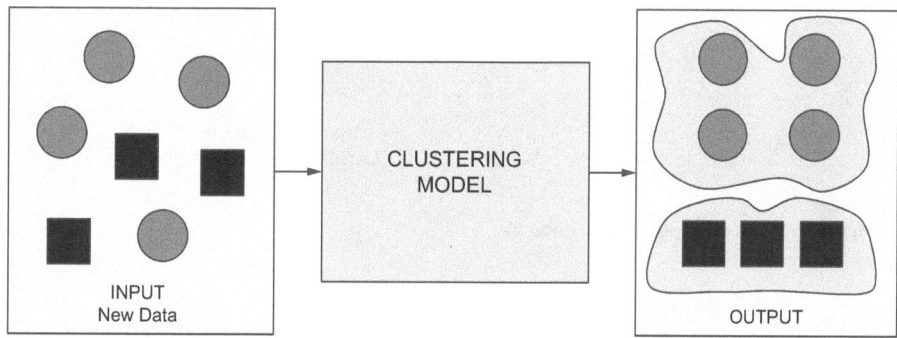

Figure 4.3   An example of a clustering model

You've now learned the main types of ML. Next, let's move a step further and focus on deep learning.

4.1.3   *What is deep learning?*

Deep learning (DL) is a subfield of ML that focuses on training artificial neural networks to process and understand complex data. Inspired by the human brain, DL models consist of multiple hidden layers of interconnected nodes, known as *artificial neurons* or *units*. The greater the number of layers, the more complex the data structure the network can recognize. DL is quite resource intensive because it requires high-performance GPUs and large amounts of training data. You can use DL models to solve tasks such as image and speech recognition, natural language processing, and even game playing.

Figure 4.4 shows an example of a DL network with one input layer with three nodes, three hidden layers, and two outputs.

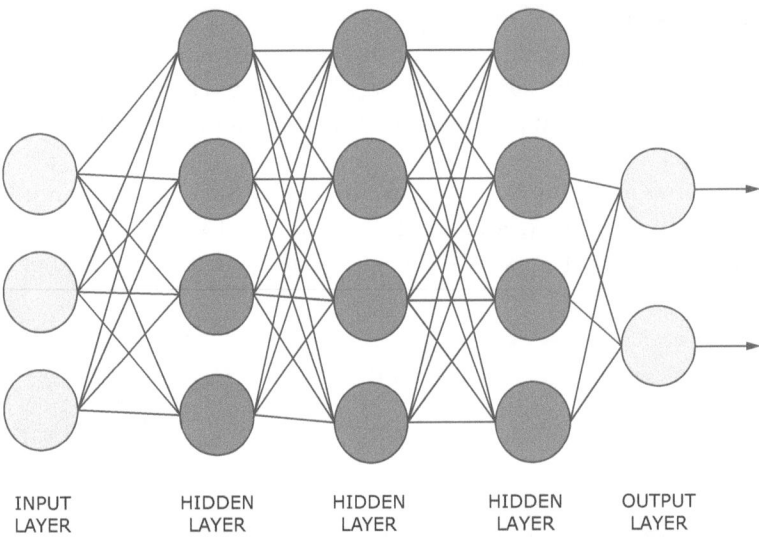

INPUT
LAYER   HIDDEN
LAYER   HIDDEN
LAYER   HIDDEN
LAYER   OUTPUT
LAYER

Figure 4.4   **An example of a DL network**

Let's break it down:

- *Input layer*—This is the first layer of the DL network where data is fed into the
  network. In this case, there is one input layer with three nodes. Each node represents a feature or input variable.
- *Hidden layers*—These are the intermediate layers between the input and output layers where the network learns patterns and representations from the
  input data. In this network, there are three hidden layers, each with four
  nodes. The number of nodes in each hidden layer can significantly affect the
  network's learning capacity and performance.
- *Output layer*—This is the final layer of the network that produces the output.
  In this case, there are two output nodes, which means the network is designed
  to produce two outputs.

Now that you have seen an overview of DL, let's move on to the next step: defining generative AI.

4.1.4   *What is generative AI?*

Generative AI is a subfield of DL aimed at creating new content based on what it has
learned from existing content. The result of the learning process from existing content is a statistical model (generative model), which is used to generate new content.

Generative models try to understand data distribution. Suppose we have a dataset
of pictures of circles and squares. Generative models learn the overall patterns and
characteristics of both circles and squares. Then, using this understanding, they generate new pictures that look similar to the ones they have seen.

Figure 4.5 shows an example of a generative model. The model takes some text asking it to generate a new circle as an input and produces a new circle as the output.

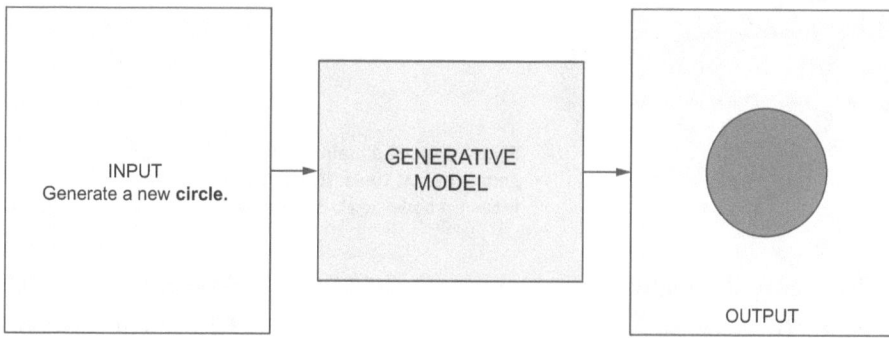

Figure 4.5   An example of a generative model

The model depicted in figure 4.5, from text to images/code, is an example of a generative model specific to this book. In general, *generative* means that we can do some interpolation in the feature space and get a somewhat meaningful output in the label space.

A generative model takes a *prompt* as an input. A prompt is a piece of text used to direct output quality. It serves as a guide for the generative model to generate responses or outputs that align with a specific desired outcome. The basic prompt is a text asking the tool to perform a specific action. In chapter 2, you have written some basic prompts to make Copilot generate the code to build an Altair chart.

Prompt engineering is a discipline aimed at carefully crafting and refining the prompts to achieve the desired results. The goal of prompt engineering is to elicit responses that are coherent, relevant, and aligned with the desired outcome, while minimizing any unintended biases or undesired behavior. Now that you have learned what generative AI is, let's move on to describing the generative AI tools landscape.

4.1.5   Generative AI tools landscape

You can use generative AI in several contexts, including for generating text, images, code, speech, and much more. For each specific task, you can use a different tool. In all cases, you start with a text prompt, which the tool converts into the desired output (text, image, code, and so on) based on its specific nature. For most services, you must pay a fee to use generative AI tools. However, some generative AI tools have free-of-charge versions with downgraded capabilities.

Figure 4.6 shows the main application fields of generative AI tools (Madhvani, 2023). If not properly used, these tools may produce wrong or biased information. Thus, you must carefully monitor and adjust the generated output to ensure accuracy and fairness.

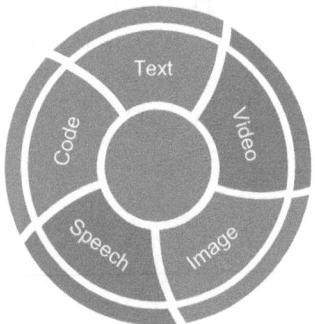

Figure 4.6   The main application fields of generative AI tools. In this book we will focus on code, text, and image.

The figure also highlights the application fields used in this book:

- *Text-to-text tools*—Use them to generate text, such as for content creation, chatbots, virtual assistants, creative writing, and automating repetitive tasks like writing emails or generating reports. Examples of text-to-text tools include Open AI ChatGPT, Google Bard, and Claude AI, developed by Anthropic. In this book, we'll focus on ChatGPT.
- *Text-to-image tools*—Use them to generate visual representations based on textual input. In this book, we'll use Open AI DALL-E. If you signed up to DALL-E before April 6, 2023, you will be granted free credits. New users must buy credits for a small fee to use the service.
- *Text-to-code tools*—Use them to generate code. In this book, we'll focus on GitHub Copilot, which requires an active subscription.

You may ask why you should pay to use a tool. The main answer is that quality generative AI tools are only available for a fee; free tools are either not up to par or are very limited until you pay or subscribe. Fortunately, the fees are fairly small. Fully open source tools are free, but the computing resources they require are not free. On the other hand, paid generative AI tools are expensive, but computing is factored into the price.

We could have used ChatGPT to generate code. However, we prefer to use Copilot, primarily for the following three reasons:

- *Code-specific expertise*—Copilot is specifically designed for coding tasks and is trained on a vast amount of code from GitHub repositories.
- *VSC integration*—Copilot is integrated directly into Visual Studio Code (VSC), allowing developers to access its suggestions and code completions while working on their projects within their programming editor.
- *Focus on code generation*—Copilot primarily focuses on generating code snippets and assisting developers during the coding process. It's optimized to assist with writing functions, classes, and entire blocks of code, providing quick suggestions based on the context of the code being written.

While ChatGPT can offer assistance with programming tasks, its primary function isn't specifically tailored to coding. It's intended for general-purpose use and can assist with a wide range of tasks, which are not solely limited to programming. Additionally, in my personal experience, ChatGPT has an occasional tendency to invent an imaginary Python method or function (this is a case of hallucination), as shown in figure 4.7.

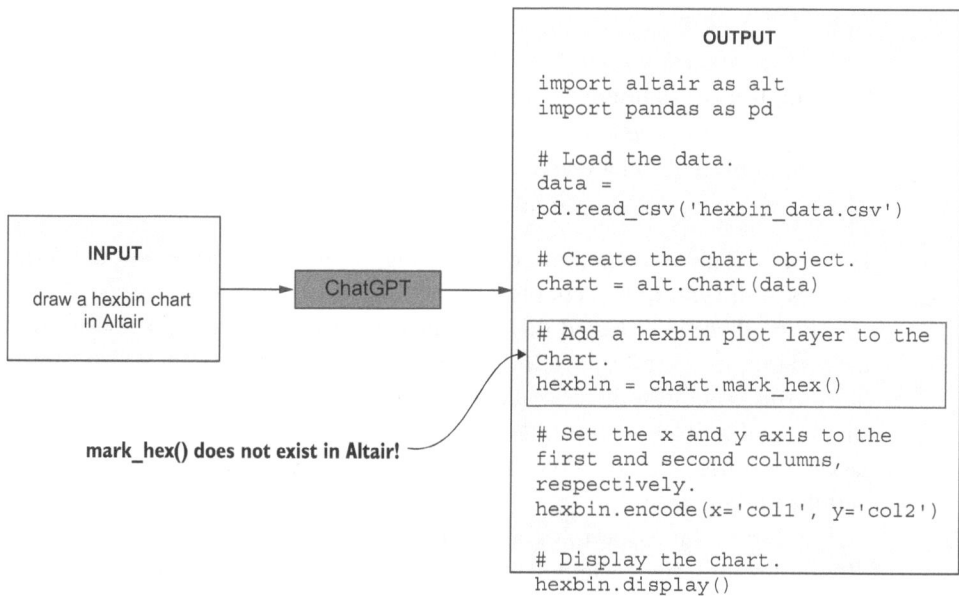

Figure 4.7 An example of a bad output produced by ChatGPT to generate code

We asked ChatGPT to generate a hexbin chart, and ChatGPT used the `mark_hex()` method, which is not currently implemented in Altair. This is an example of *hallucination*, a topic we'll discuss more deeply in chapter 10. For now, it is sufficient to know that a generative AI model can hallucinate, meaning it could produce outputs that convincingly resemble real data, even though these outputs may not be explicitly grounded in reality.

We have chosen specific tools to generate text, images, and code for data storytelling. However, you can easily adapt the described concepts to other tools with minimal effort. In all cases, it's crucial to provide accurate and concise instructions to ensure the tool generates the correct output. Write specific and unambiguous instructions, using simple language and avoiding complex terminology. Also, write consistent instructions regarding syntax, language, and format. Inconsistent instructions can confuse the tool, leading to incorrect outputs.

Now that we have quickly defined the context of generative AI tools, we can move on to the next step: using generative AI tools for data storytelling. We will start with ChatGPT; then describe DALL-E; and, finally, cover GitHub Copilot.

4.2   *The basic structure of a ChatGPT prompt*

To interact with ChatGPT, you must write an input text (prompt) that defines the instructions to be performed. You can use the ChatGPT web interface or the Open AI API, as appendix A describes.

There are several different ways to structure a prompt for ChatGPT. In this book, we consider a prompt to be composed of three main consecutive texts:

- *Tell ChatGPT to act in a specific role.* For example, "You are an examiner looking at high school students' English papers."
- *Tell ChatGPT to tailor its outputs to an intended audience.* For example, "Explain your gradings in a way that can be understood by high schoolers."
- *Define the task.* For example, "Grade this text, and explain your reasoning."

For more details about prompt engineering, refer to the resources outlined in the references at the end of this chapter.

Although we have tested this prompt structure, sometimes, it doesn't produce the desired output, so you can refine the prompt using different iterations. In the remainder of this section, we'll describe all the parts of a prompt separately. Although the structure of a prompt should be *role, audience, task*, we'll start by describing first the task to perform, to show how the produced prompt changes when varying the roles and the audience.

We'll show some examples of produced texts. If you try your own tool, you'll obtain different results, since ChatGPT generates new content every time you use it.

4.2.1   *Defining the task*

A task is a specific objective ChatGPT must fulfill. A task includes clear instructions to guide ChatGPT in generating responses relevant to the given task. Examples of tasks include answering questions, generating creative content, translating language, and more. In the context of data storytelling, we'll focus on the following tasks:

- Generating a title
- Generating a subtitle
- Summarizing a text
- Translating a text
- Proposing ideas

We'll analyze each task in the next chapters in more detail.

Imagine a scenario in which you must engage an audience striving for success in their personal and professional lives. You have prepared the following text, which shows the contrast between appearing strong and perfect and showing one's fragility:

> *In a world that glorifies strength and perfection, remember that fragility is the essence of our humanity. Our struggles and imperfections make us human, yet we often conceal them out of fear and shame. However, those who embrace their weaknesses with courage*

and authenticity cultivate remarkable resilience. Those who fear their fragility risk becoming rigid, trapped in a facade of unattainable perfection. For it is in our vulnerabilities that we grow and truly live. (Adapted from Oliani, 2023, "Ho Paura di Essere Debole [I'm Afraid of Being Weak]," Messaggero di Sant'Antonio, issue 6, p. 53).

Suppose your task is to generate a title for the previous text. Formulate the following prompt for ChatGPT: *Generate the title for the text in angular brackets <text>*. Replace the word *text* with the previous text about fragility and vulnerabilities. Examples of titles ChatGPT may generate include the following:

- *Embracing Fragility: The Path to Remarkable Resilience and Authentic Living*
- *Fragility as the Gateway to Authenticity and Resilience in a Perfect-Obsessed World*
- *Fragility: Embracing Imperfections for Authentic Resilience and Growth*

All of these titles are general purpose. The next step is adding a role to the prompt to make ChatGPT generate a more specific output.

4.2.2   Acting as a role

You will likely have one of the following goals when conveying a message:

- *Entertainment*—The focus is on providing enjoyment, amusement, or relaxation. You may want to make people laugh or evoke emotions.
- *Education*—The focus is on providing knowledge, insights, or skills to the audience to broaden the audience's knowledge base and encourage learning.
- *Information*—The focus is on keeping the audience informed about a specific topic, such as news updates or current events.
- *Inspiration*—The focus is to motivate the audience, such as sparking creativity or encouraging personal growth. Inspirational messages often share success stories or showcase role models.
- *Invitation to action*—The focus is to motivate the audience to engage in a specific behavior or event.
- *Relationship building*—The focus is to foster connections with the audience through personal anecdotes and shared experiences. The message aims to build trust, strengthen existing relationships, or cultivate a sense of community.

Depending on the goal, you'll cover a different role, which uses specific storytelling techniques. In some cases, in the same situation, you may cover different roles, but generally, the role depends on the target audience.

Consider again the scenario where you must engage an audience striving for success in their personal and professional lives. Use ChatGPT to adapt the text to your role. Use the following syntax to formulate the prompt for ChatGPT: *Act as a [role]. Generate the title for the text in angular brackets <text>*. Table 4.1 shows the outputs produced for all the described roles.

Table 4.1   The results of the prompt for all roles

Role	Output
Entertainer	Unmasking Our Humanity: Embracing Fragility and Thriving in Imperfection
Educator	Nurturing Resilience: Embracing Fragility and Embracing Imperfection for Personal Growth and Well-Being
Informer	The Power of Embracing Fragility: Celebrating Imperfections and Cultivating Resilience
Inspirer	Embrace Your Fragile Strength: Unleashing Resilience and Embracing Imperfections for a Fulfilling Life
Inviter to action	Embrace Your Fragile Humanity: Rise Above Fear and Embrace Imperfection to Unleash Your Resilient Potential
Relationship builder	Embracing Fragility Together: Fostering Authentic Connections and Building Resilient Relationships

Each output has the same structure ([title] [colon] [subtitle]), but their goal changes based on the role. Now that you have learned how to influence the output using a different role, let's move on to how to influence the output based on the different target audiences.

4.2.3   *Tailoring to an audience*

Depending on the context in which you will tell your data story, your audience changes. For example, you could have an audience of stakeholders, businesspeople, children, or even philosophers. More generally, there are the following types of audiences:

- *General audience*—A broad group of people with varied interests, backgrounds, and demographics
- *Target audience*—A specific group with well-defined interests, backgrounds, and other factors
- *Niche audience*—A relatively small group of people with specific interests, preferences, or needs, such as enthusiasts of specific hobbies or professionals in specific fields
- *Mass audience*—A large and diverse group of people who consume appealing content. Compared to a general audience, a mass audience is the target of media and advertising campaigns.

Consider again the scenario where you must engage an audience striving for success in their personal and professional lives. Use ChatGPT to adapt the text to your audience. Use the following syntax to formulate the prompt for ChatGPT: *Act as a [role] tailored to [the audience]. Set the title for the text in angular brackets <text>*. Table 4.2 shows the outputs for all the described audience types for the relationship-builder role.

Table 4.2 The result of the prompt for all the audience types in the relationship-builder role

Audience	Output
General audience	Embracing Fragility: The Path to Resilience and Authentic Living
Target audience: businessmen	Embracing Vulnerability: The Power of Authenticity and Resilience in Business Relationships
Niche audience: football enthusiasts	The Beautiful Game: Celebrating Fragility as the Heart of Football Passion
Mass audience: young people	Embracing Your Imperfections: Unleashing Your Inner Strength as a Young Person

All the outputs have the same structure ([title] [colon] [subtitle]), but their meaning is adapted to the specific audience. Consider again the title generated specifying only the role: *Embracing Fragility Together: Fostering Authentic Connections and Building Resilient Relationships*. This title is similar to that generated for a general audience. When you add a specific audience, the title is tailored to it. Now that you have learned the basic concepts for defining a prompt for ChatGPT, let's move on to the next step: defining a prompt for DALL-E.

4.3 The basic structure of a DALL-E prompt

To interact with DALL-E, you must write a prompt specifying the instructions. You can use either the DALL-E web interface or the Open AI API, as described in appendix A.

There are several different ways to structure a prompt for DALL-E. In this book, we consider a prompt composed of the following:

- *A subject*
- *A style*

In the remainder of this section, we'll describe all the parts of a prompt separately, starting with the subject. Remember that if the output is not satisfying, you can refine it or use the Edit Image tool, as described later in this section.

4.3.1 Subject

The subject of an image is the main focus that captures the viewer's attention. Examples of subjects include persons, objects, or scenes. When defining the subject of your image, be as specific as possible. A generic subject will generate a generic image. For each input, DALL-E generates four output images. In the remainder of the section, we'll show some examples of produced images. If you try out the tool for yourself, you'll obtain different examples.

Figure 4.8 shows the output produced by DALL-E for the prompt *a person*. Since the input is very generic, the output is also very generic. The four generated images are different from each other.

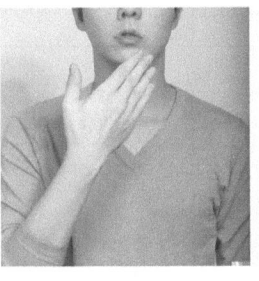   

Figure 4.8   The output for the prompt *a person*

To improve the focus of the image, you can add the following elements:

- An adjective
- An emotion
- Context
- Color

Let's start by adding an adjective that qualifies the subject. We will give a specific example of generating images related to *poor people*. I chose this theme to show the potential of inserting images into a data story to arouse emotions in the audience. In fact, data storytelling also gives voice to the people behind data.

Figure 4.9 shows the outputs for the prompt *a poor person*, where *poor* is the qualifying adjective. All the produced images feature a person with a hand on their head but not necessarily person who is poor. However, with respect to the generic output, we have obtained a better result.

Figure 4.9   The output for the prompt *a poor person*

To be more specific, add an emotion describing the person's mood. Figure 4.10 shows the output produced by the prompt *a sad, poor person*.

The figure still needs to be improved. Let's add a background, creating context for the subject. Figure 4.11 shows the output produced by the prompt *a sad, poor person, with rubbish in the background*.

Figure 4.10   The output for the prompt *a sad, poor person*

Figure 4.11   The output for the prompt *a sad, poor person, with rubbish in the background*

Finally, let's add color. Figure 4.12 shows the output for the prompt *a sad, poor person, wearing a torn brown shirt with rubbish in the background* (to see this in color, visit the e-book version).

Figure 4.12   The output for the prompt *a sad, poor person wearing a torn brown shirt with rubbish in the background*

Now, the output looks satisfying. We can proceed to the next step: adding style.

4.3.2   Style

Style defines the type of art that will be output, including photographs, digital art, pencil drawings, and so on. In addition, you can specify a particular artist's style, such as Van Gogh or Caravaggio. You just have to indulge your imagination!

Figure 4.13 shows the output for the prompt *a sad, poor person, with rubbish in the background, digital art,* and figure 4.14 shows the output for the prompt *a sad, poor person, with rubbish in the background, black and white photograph.*

Figure 4.13   The output for the prompt *a sad, poor person wearing a torn brown shirt with rubbish in the background, digital art*

Figure 4.14   The output for the prompt *a sad, poor person wearing a torn brown shirt with rubbish in the background, black and white photograph*

The prompt structure described in this chapter is only an entry point to drawing images using DALL-E. You can also define the environment (indoor or outdoor), the hour of day, the angle, and more. You can find more information in *The DALL·E 2 Prompt Book* (https://mng.bz/0GNx).

4.3.3   The Edit Image tool

If you are not completely satisfied with a generated image, you can modify it using the Edit Image tool provided by the DALL-E interface. To access this tool, select the image, and then click Edit. You can erase a part of the image (as shown in figure 4.15), add a new generation frame, or upload an image.

Figure 4.15   Using the eraser

Then, you can write a new prompt for the selected part. Let's add a dog to the image of figure 4.15 using the following prompt: *a small black dog eating a piece of bread, black and white photograph.* Figure 4.16 shows the produced output.

Figure 4.16   The output of the modified image using the eraser

We finally have all the pieces: ChatGPT, DALL-E, and GitHub Copilot (from the previous chapter). Now, it's time to implement a practical example, using all the tools. But before that, let's dust off GitHub Copilot, defining a more systematic approach, to make the best use of it.

4.4 *Using Copilot to build the components of an Altair chart*

When writing instructions for Copilot, split them into single steps. Also, provide the context surrounding what you want Copilot to do, and provide specific details on the requirements of the code you are writing. Finally, remember that Copilot is a tool designed to assist developers, not replace them. Copilot is powerful, but it can't replace you as a developer. Always use your expertise to evaluate the code suggestions provided by Copilot, and ensure they are appropriate for your specific needs.

In this section, you'll see how to write the instructions to make Copilot build an Altair chart. We'll focus on prerequisites, marks, encodings, conditions, compound charts, and interactivity.

4.4.1   *Prerequisites*

Before drawing a chart, ask Copilot to import the required libraries and load the dataset as a pandas DataFrame.

> **Listing 4.1   Importing the required libraries and loading the dataset**

```
# Import the required libraries.
# Load the file '/path/to/data.csv' into a pandas DataFrame.
```

> **NOTE**   The first instruction specifies importing the required libraries. Copilot will calculate which libraries to import automatically, based on the subsequent instructions. The second instruction specifies the exact path of the dataset to load.

EXERCISE 1
Write the instructions for Copilot to load the data.csv, available in the GitHub repository of the book under 04/data/data.csv. You can find the solution for this exercise and the next ones in this section under 04/copilot/example.py.

4.4.2   *Marks*

To define a mark, simply instruct Copilot to draw it in Altair.

> **Listing 4.2   Defining marks**

```
# Draw a <type of mark> named chart in Altair
```

> **NOTE**   Use the keywords Draw and Altair to specify that you want to build a chart in Altair. Replace <type of mark> with the type of mark you want to build. For example, write bar chart if you want to build a bar chart.

4.4.3   *Encodings*

To set an encoding, add the keyword with: to the instruction defined in listing 4.2. In addition, add a bullet point for each encoding channel you want to define.

> **Listing 4.3   Setting encodings**

```
# Draw a bar chart named bar in Altair with:
# * The category column as the x-axis
# * The value column as the y-axis
# * The country column as the color.
```

NOTE   After defining the mark, use the keyword `with:` followed by a list of encoding channels. For each encoding channel, specify the column of the dataset to use and the channel.

EXERCISE 2

Write instructions for Copilot to draw a line chart in Altair using the dataset loaded in the previous exercise. Specify to use the `x` column for the x channel and the `y` column for the y channel.

4.4.4   Conditions

To set a condition related to a specific encoding channel, use the instructions specified in the following listing.

Listing 4.4   Defining conditions

```
# * The country column as the color. Set the color to:
#   - 'red' for 'IT'
#   - 'green' for 'FR'
```

NOTE   Append the condition to the selected encoding channel. Use a list to specify the content of the conditions. The example sets the color to `red` if the country is `IT` and `green` if the county is `FR`.

EXERCISE 3

Write instructions for Copilot to do the following:

1 Load the dataset data3.csv, available under 04/data/data3.csv.
2 Draw a bar chart with the category column in the x channel and the value column in the y channel.
3 Set the color of the bar to `red` if the country column is equal to `IT`; otherwise, set it to `green`.

4.4.5   Compound charts

Build each chart separately. Then, to build a compound chart, write a specific instruction, as shown in the following listing. Be sure to give each chart a different name.

Listing 4.5   Building compound charts

```
# Build a compound chart named chart with the line and bar charts aligned
    vertically
```

NOTE   Specify the type of combination you want to draw. The example aligns the charts vertically.

EXERCISE 4

Write instructions for Copilot to vertically align the line and bar charts drawn in the previous exercises.

4.4.6   *Interactivity*

To make the chart interactive, write the simple instruction in the following listing.

Listing 4.6   Making the chart interactive

```
# Make the chart interactive.
```

NOTE   Write simple and clear instructions to make the chart interactive.

EXERCISE 5

Write instructions for Copilot to make the bar chart interactive. Note that you can insert new instructions for Copilot above or below the generated code.

Now that you have learned how to write the basic instructions to make Copilot build a chart in Altair, let's move to a practical case study that combines Altair and Copilot.

4.5   *Case study: Your training team*

Let's imagine you work in a sports company. You are training a team of young athletes in various disciplines. For each discipline, you have noted the world record and recorded the best time achieved by your team for the sake of comparison. Unfortunately, your company has limited investment funds available. Your boss asks you to distinguish which disciplines are worth continuing to train in, to hope to achieve good results in the upcoming competitions. You can find the code for this example in the GitHub repository for the book under CaseStudies/competitions. The directory contains different Python scripts, one for each step described in the remainder of this section.

Table 4.3 shows, for each discipline, the record and the best time achieved by your team. In your case, the objective is to display the difference between your team's time and the record for each training type and then choose in which training types you should continue investing.

Table 4.3   The dataset of the case study shows the record to beat, the record holder, the record time, and your team's best time for each training type.

Training Type	Record to Beat	Record Holder	Record Time (Seconds)	Team's Best Time
Sprinting	100 m	Usain Bolt	9.58	15.07
Long-Distance Running	Marathon	Eliud Kipchoge	7,299	13,261.76
Swimming	100 m freestyle	César Cielo	46.91	88.01
Cycling	Individual time trial (ITT)	Rohan Dennis	1,963.49	2,795.22
Rowing	2,000 m	Josh Dunkley-Smith	335.8	462.33

We'll use the DIKW pyramid to transform this dataset into wisdom. Let's start with the first step of the pyramid: turning data into information.

4.5.1 *Turning data into information*

The simplest graph for this purpose is a stacked bar chart, which shows the two columns as stacked series. Let's use GitHub Copilot to generate this simple chart. Start by writing the list of comments to generate the chart.

Listing 4.7   Generating the stacked bar chart

```
# Import the required libraries.
# Load '../data/competitions.csv' as pandas dataframe.
# Select only the columns Training Type, Record Time (Seconds), Our Best Time.
# Use melt to transform the dataframe into a long format, with the parameter
      id_vars set to 'Training Type'.
# Draw a bar chart named chart in Altair with:
# * The Training Type column as the x-axis
# * The value column as the y-axis
# * The variable column as the color
# Save the chart as 'competitions.html'.
```

NOTE   Describe the sequence of operations to load, manipulate, and plot the dataset.

Copilot will generate several code options. Select the code shown in the following listing.

Listing 4.8   The stacked bar chart generated by Copilot

```
import pandas as pd
import altair as alt

df = pd.read_csv('data/competitions.csv')
df = df[['Training Type', 'Record Time (Seconds)', 'Our Best Time']]
df = pd.melt(df, id_vars='Training Type')

chart = alt.Chart(df).mark_bar().encode(
    x='Training Type',
    y='value',
    color='variable'
)

chart.save('competitions.html')
```

NOTE   First, load and manipulate the dataset. Then, use `mark_bar()` to draw the stacked bar chart.

Figure 4.17 shows the resulting bar chart. In addition to being very narrow, the generated chart presents a fundamental problem: we are unable to determine from the chart which disciplines we should focus on. This can be attributed to a lack of clarity in the chart, which makes it difficult for us to interpret the data effectively. In practice, the problem is the difference in scale between the variables. Sprinting is in the range of 10 seconds, while long-distance running is in the range of thousands of seconds. As a result, we need to revise the chart design to convey information easily.

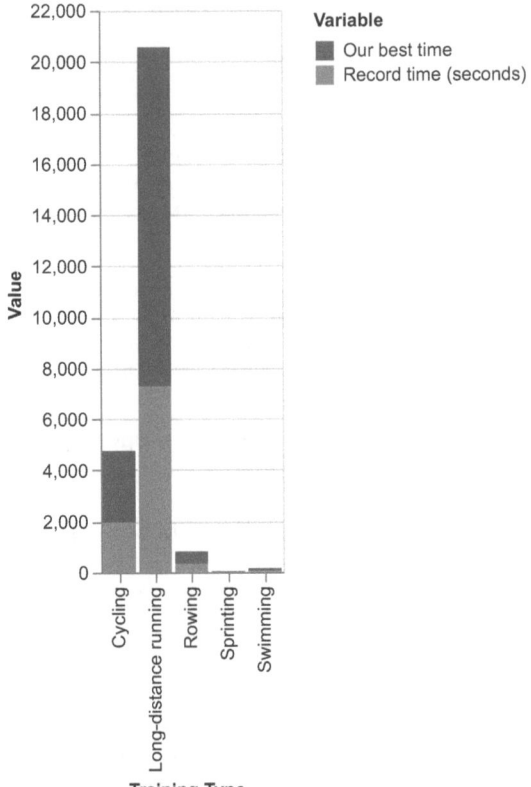

Figure 4.17   The stacked bar chart generated by Copilot

In general, when we want to compare two variables across multiple categories, it can be useful to calculate the percentage difference or the percentage increase of one variable relative to the other for each category and then show it. This approach allows us to compare the two variables and identify any patterns or trends within the data. To generate this type of chart, we can once again make use of Copilot, which can assist us in writing the necessary code. Write the list of instructions specified in the following listing. The instructions in bold show what changes from listing 4.7.

Listing 4.9   Generating the bar chart with percentage difference

```
# Import the required libraries.
# Load '../data/competitions.csv' as pandas dataframe.
```

```
# Select only the columns Training Type, Record Time (Seconds), Our Best Time.
# Calculate the percentage difference between columns Our Best Time and Record
      Time (Seconds), and store it in a new column called Percentage Difference.
# Draw a bar chart with the following encodings:
# * The Percentage Difference on the y-axis with the following properties:
#    - The domain of the Y scale to [0,100]
# * Training Type on the x-axis with the following properties:
#    - The values sorted in descending order (-y)
# Set the following properties of the bar chart:
# * width to 300 pixels
# Save the chart as 'competitions.html'.
```

NOTE   Describe the sequence of operations to load, manipulate, and plot the
dataset. Also, specify to sort data in descending order, and set the domain of
the Y scale to [0,100] to make the chart clearer. Since the chart is more complex than the previous one, add the instruction with the following encodings when you ask Copilot to draw the chart. Do the same thing for mark
properties. Also, note the nested lists to add specific property values to encoding channels.

Copilot will generate several code options. Select the code shown in the following listing.

Listing 4.10   The bar chart generated by Copilot

```
import pandas as pd
import altair as alt

df = pd.read_csv('data/competitions.csv')
df = df[['Training Type', 'Record Time (Seconds)', 'Our Best Time']]
df['Percentage Difference'] = (df['Our Best Time'] - df['Record Time
      (Seconds)']) / df['Record Time (Seconds)'] * 100

chart = alt.Chart(df).mark_bar().encode(
    x=alt.X('Training Type', sort='-y'),
    y=alt.Y('Percentage Difference', scale=alt.Scale(domain=[0,100]))
).properties(
    width=300
)

chart.save('competitions.html')
```

NOTE   First, load and manipulate the dataset. Then, add a column to the
dataset, named Percentage Difference, containing the percentage difference. Finally, use mark_bar() to draw the stacked bar chart. Copilot may not
generate the full code, so we need to adjust it based on the desired output.

Figure 4.18 shows the resulting chart.

   The chart highlights that the sports with the greatest percentage difference
between the Our Best Time and the Record Time are Swimming and Long-Distance

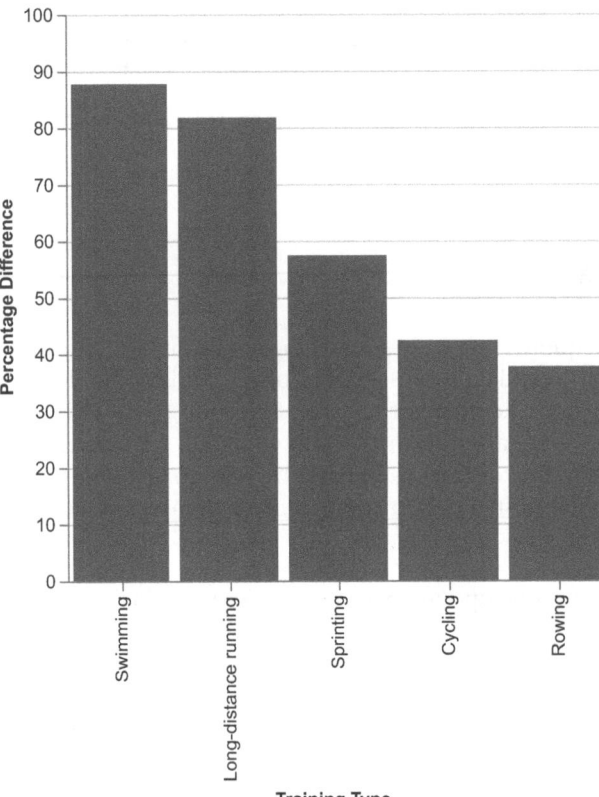

Figure 4.18   The bar chart with percentage difference generated by Copilot

Running. However, the chart may be misleading because it is not intuitive, since the largest bars are, in fact, the worst performers. To improve the readability and intuitiveness of the chart and better convey the actual performance of athletes in each sport, we can calculate the percentage improvement.

Listing 4.11   Calculating the percentage improvement

```
df['Percentage Improvement'] = 100 -
(df['Our Best Time'] - df['Record Time (Seconds)']) / df['Record Time
    (Seconds)'] * 100
```

NOTE   Calculate the percentage similarity: 100 minus the percentage difference. Replace lines 5 and 6 of listing 4.10 with the code shown in listing 4.11.

The percentage improvement represents the proportion of improvement in the performance of Our Best Time compared to the Record Time. A higher percentage improvement indicates that the athlete's performance is closer to or better than the record holder's, while a smaller percentage improvement indicates that the athlete's performance is further from the record holder's performance. This is more intuitive, as the larger bars now indicate the best performance. Figure 4.19 shows the percentage improvement. Rowing and cycling are the sports with better performances; thus, you can choose those sports for further investment.

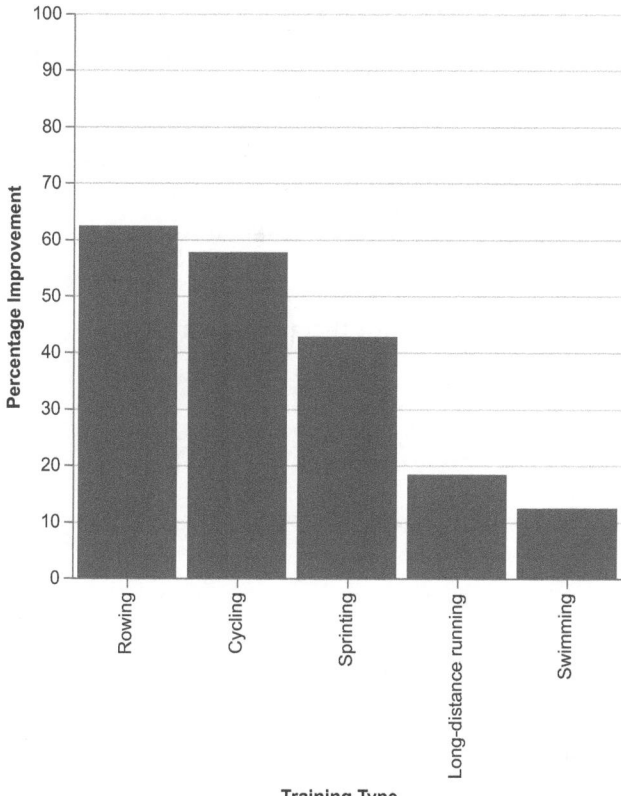

Figure 4.19   The bar chart with percentage improvement

Now that we have extracted information from the data, we can move to the next step of the DIKW pyramid. In the following section, we will dive into turning information into knowledge.

4.5.2   *Turning information into knowledge*

This step involves adding context to the information. In this case, the context may involve three aspects:

- Highlighting the two sports where you want to invest
- Adding an image that enforces the top two sports
- Adding a baseline, for example, at 50% of percentage improvement, showing why you chose those sports

Let's start from the first point, highlighting the two sports in which you want to invest.

HIGHLIGHTING THE TOP TWO SPORTS

Within the previously generated code, add the comment shown in the following listing immediately after the y channel to generate the code using Copilot.

> Listing 4.12   Highlighting the best-performing sports

```
# Add the color encoding. Set the color to:
    # - #80C11E if the Percentage Improvement is greater than 50,
    # - lightgray otherwise
```

NOTE   Describe how to highlight the sports that perform better.

The following listing shows the generated code.

> Listing 4.13   The generated code to highlight the best-performing sports

```
color=alt.condition(
        alt.datum['Percentage Improvement'] > 50,
        alt.value('#80C11E'),
        alt.value('lightgray')
    )
```

NOTE   The code generated by Copilot uses a conditional statement to select the color to apply to each bar.

Figure 4.20 shows the resulting chart. Now that we have highlighted the top two sports, we can move on to add images to the chart.

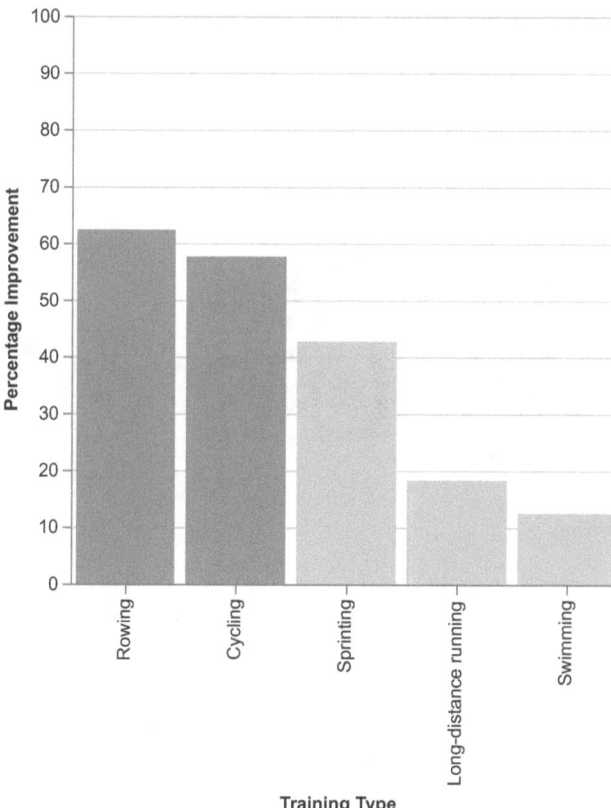

Figure 4.20   The bar chart with the top two sports highlighted

ADDING ENFORCEMENT IMAGES

To generate an image for each of the top two sports, provide DALL-E with the following prompt: *an enthusiastic athlete practicing [sport], in a white background, cartoon.* Figures 4.21 and 4.22 show the outputs for *rowing* and *cycling*, respectively.

Figure 4.21   The DALL-E output for rowing

We choose the second image in figure 4.21 and the first image in figure 4.22. If you find some imperfections in an image, you can improve it using the editing tool. To include an image in an Altair chart, you must save it into a remote repository, such as GitHub, and provide Altair with the remote URL.

Figure 4.22   The DALL-E output for cycling

Now, let's ask Copilot to write the code to add the images to the chart. The following listing describes the instructions for Copilot.

Listing 4.14   Adding images to the chart

```
# Add a new column to df called 'url' with the following value:
# * 'https://[..]/cycling.png' for Training Type = 'Cycling'
# * 'https://[..]/rowing.png' for Training Type = 'Rowing'
# * '' for all other Training Types
# Add the following image to the chart:
# * The image is a 35x35 pixel image.
# * The image is located at x='Training Type', y='Percentage Improvement'.
```

NOTE   First, add a column to the DataFrame with the URL to the image. Then, specify how to add images to the chart.

The following listing shows the output.

Listing 4.15   The code for adding images

```
df['url'] = ''
df.loc[df['Training Type'] == 'Cycling', 'url'] = 'https://[..]/cycling.png'
df.loc[df['Training Type'] == 'Rowing', 'url'] = 'https://[..]/rowing.png'
chart = chart + alt.Chart(df).mark_image(width=35, height=35).encode(
    x=alt.X('Training Type', sort='-y'),
    y=alt.Y('Percentage Improvement'),
    url='url'
)
```

NOTE   First, Copilot suggests to manipulate the DataFrame by adding a new column. Next, Copilot suggests using the mark_image() mark.

Figure 4.23 shows the images added to the columns of the top two sports. Now, we can move on to the next step: generating a baseline.

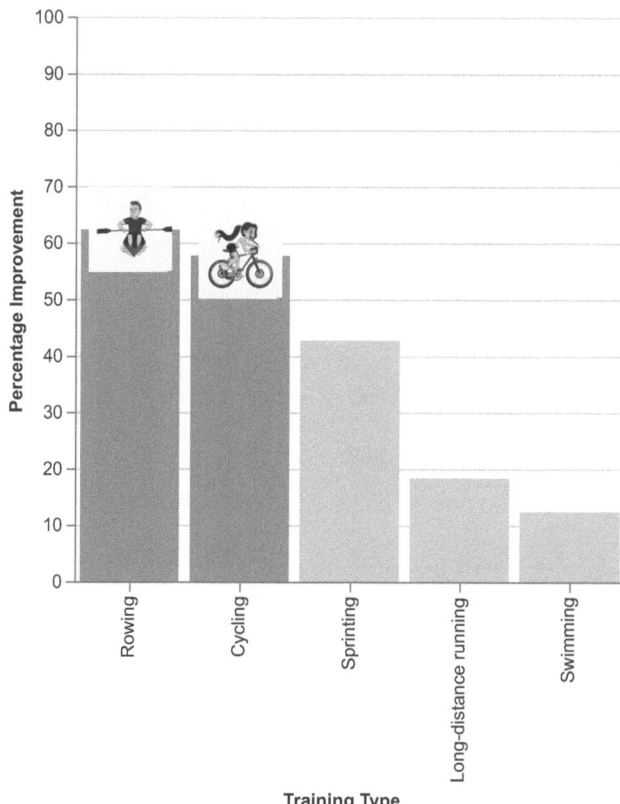

Figure 4.23   The bar chart after adding images

ADDING A BASELINE

Adding a baseline means adding a horizontal red line to the chart for a comparison. For example, we can set the value of the horizontal line to 50, which corresponds to 50%. The following listing shows the instructions for Copilot.

Listing 4.16   Generating a baseline

```
# Add a horizontal red line to the chart at y=50.
# Add the line to the chart.
```

NOTE   Describe how to generate the baseline at 50% improvement.

The following listing shows the generated code.

Listing 4.17 The generated code to create a baseline

```
line = alt.Chart(pd.DataFrame({'y':
    [50]})).mark_rule(color='red').encode(y='y')
chart = chart + line
```

> **NOTE**   The code generated by Copilot uses `mark_rule()` to generate a horizontal line and the + operator to layer the two charts.

Figure 4.24 shows the final chart. With respect to figure 4.19, you have added context that helps the reader immediately understand which sports you should invest in.

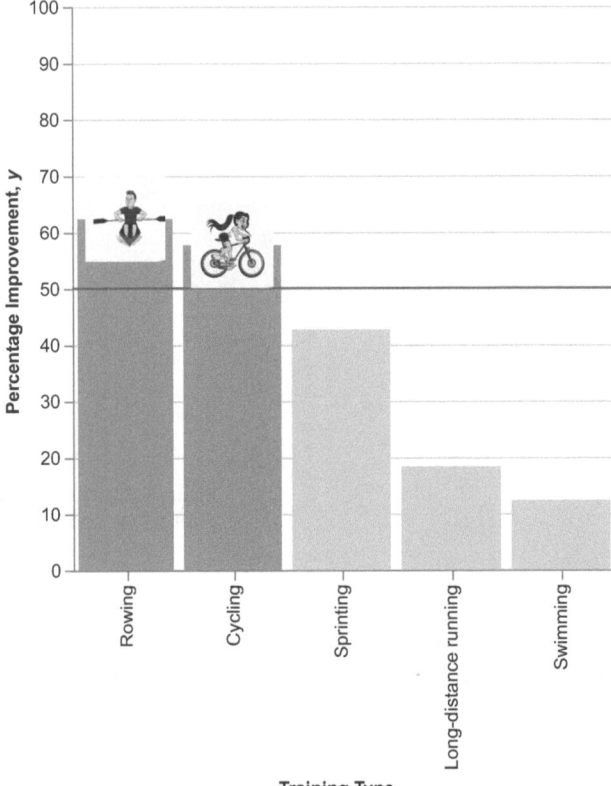

Figure 4.24 The bar chart showing the two sports that perform better than the baseline

We could further improve the chart in figure 4.24 by adjusting axes titles and applying decluttering. We'll see how to do this in the next chapters. Now that we have added context to the chart, we can move to the next step of the DIKW pyramid: turning knowledge into wisdom.

4.5.3 *Turning knowledge into wisdom*

This step involves adding an action to the chart. Let's add a title that invites the audience to invest in rowing and cycling. We can use ChatGPT to generate the title. Write the following prompt:

- *Act as an inviter to action.*
- *Tailored to the boss of a sports company.*
- *Generate a compelling title for the following text in angle brackets <You should invest funds for rowing and cycling, which are the most practiced sports>.*

An example of a generated title is Unlock the Potential: Invest in Rowing and Cycling for Maximum Returns! If you are unsatisfied with the generated title, you can ask ChatGPT to generate another one or a list of *n* titles.

Once we have defined the title, we can write the instructions for Copilot to add it to the chart, as shown in the following listing. Add the comment within the `properties()` method of the chart.

Listing 4.18   Adding a title

```
# Add the following properties to the chart:
    # * title to 'Unlock the Potential: Invest in Rowing and Cycling for
    Maximum Returns!'
```

NOTE   Add this instruction within the properties function of the chart.

The following listing shows the generated code.

Listing 4.19   The generated code to add a title

```
title='Unlock the Potential: Invest in Rowing and Cycling for Maximum Returns!'
```

NOTE   The code generated by Copilot uses the `title` property.

Figure 4.25 shows the final chart. To render this chart, you should run the generated HTML from a web server. If you don't have a web server, run a temporary web server using the following command from your directory: `python -m http.server`.

Now, you can finalize the chart. Start by rotating x-axis labels and removing the title. Modify the x channel by adding the `axis` property and the `label` angle (`axis=alt .Axis(title=None, labelAngle=0)`). Next, set the title to `Percentage Improvement` in all the charts (the main chart, images, and annotation). Finally, add the annotation text above the red line as shown in listing 4.20.

Unlock the Potential: Invest in Rowing and Cycling for Maximum Returns!

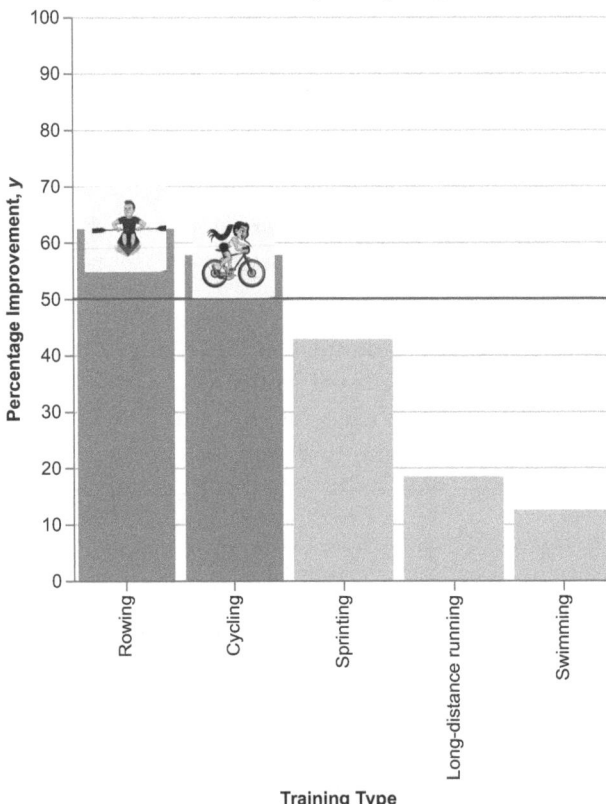

Figure 4.25   The bar chart with the call to action in the title

---

Listing 4.20   Adding a textual annotation

```
annotation_df = pd.DataFrame({'text': ['Use 50% as the benchmark for sports
    selection']})
annotation = alt.Chart(annotation_df
        ).mark_text(
            size=10,
            align='left',
            color='red',
            x=170,
            y=150,
            dy=-10
        ).encode(
            text='text'
        )

chart = (chart + annotation)
```

NOTE   First, define the annotation DataFrame, and then use it as input to the chart.

You can find the complete code for this example in the GitHub repository for the book under 04/case-study/7-bar-chart.py. Figure 4.26 shows the resulting chart.

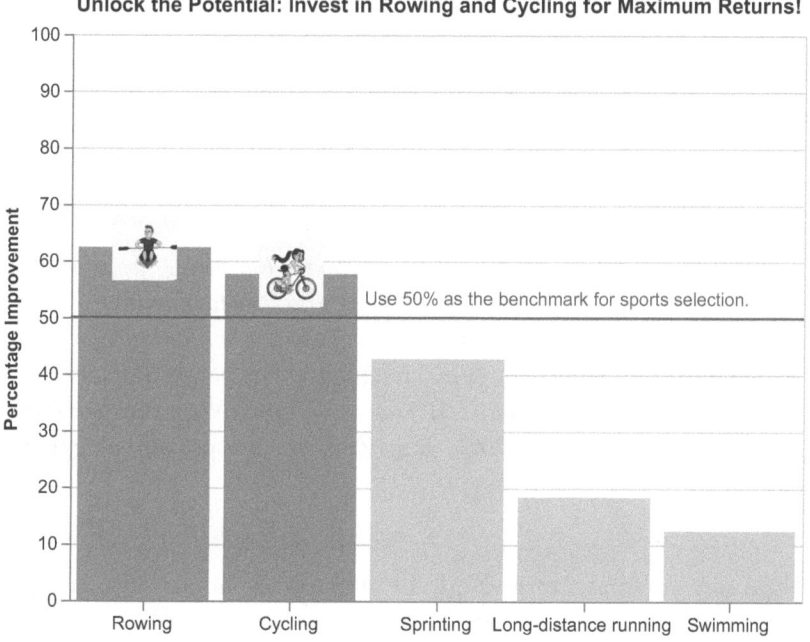

Figure 4.26   **The final bar chart with the rotated labels, adjusted title, and annotation**

In this chapter, you have learned the basic concepts behind generative AI tools, specifically focusing on ChatGPT, DALL-E, and Copilot. In the next chapter, you'll see how to combine Altair and generative AI tools to build compelling data stories, using the DIKW pyramid.

Summary

- Generative AI is a subfield of artificial intelligence, aiming at generating new content based on the trained data. You can use generative AI for different purposes, including text, image, code, video, and speech generation.
- A prompt is an input text containing the instructions for a generative AI tool.
- ChatGPT is a text-to-text generative AI tool. To write effective prompts for ChatGPT, structure the prompt in three parts: role, audience, and subject.
- DALL-E is a text-to-image generative AI tool. To write effective prompts for DALL-E, define the subject and the style.
- Copilot is a text-to-code generative AI tool. To make Copilot write code suggestions efficiently, write specific and unambiguous instructions, using simple language and avoiding complex terminology.

- Don't use Copilot as a substitute for you as a developer. Rather, use it to assist you while programming.
- Always keep in mind that generative AI cannot be a substitute for your creativity and reasoning capabilities.

References

Prompt engineering

- Barber, T. (2023). *Dall-E Prompting Tips & Tricks.* https://torybarber.com/dall-e-prompting-tips-tricks/.
- *DALL·E 2 Prompt Book.* (2022). https://mng.bz/0GNx.
- Fulford, I. and Ng, A. (n.d.). *ChatGPT Prompt Engineering for Developers.* https://www.deeplearning.ai/short-courses/chatgpt-prompt-engineering-for-developers/.
- Kemper, J. (2023). *ChatGPT Guide: Use These Prompt Strategies to Maximize Your Results.* https://the-decoder.com/chatgpt-guide-prompt-strategies/.
- Madhvani, N. (2023). *Generative AI Tools in the Creative Domains: The Power and Pressure Game Is On!* March 2023. https://www.rapidops.com/blog/generative-ai-tools/.

Other

- *Supervised Machine Learning.* (n.d.). Javatpoint. https://www.javatpoint.com/supervised-machine-learning.

Part 2

Using the DIKW pyramid for data storytelling

When you build a data-driven story, you should consider many aspects: the insight extracted from data; the contextual background behind data; and the next steps connected to data, inviting the audience to do something. The data, information, knowledge, wisdom (DIKW) pyramid helps you consider all these aspects. In this part, you'll deepen your understanding of all the steps of the DIKW pyramid.

In chapter 5, you'll consolidate the concepts related to the DIKW pyramid. The first part of the chapter centers on a case study of homelessness and how to transform its data from a raw chart into a data story using the DIKW pyramid. The chapter also introduces some general concepts related to data storytelling, such as the structure of a narrative and the data storytelling arc. In the second part of the chapter, you will implement another case study, based on fake news.

Chapters 6–9 deepen each step of the DIKW pyramid. They implement a single case study: the problem of safety in salmon aquaculture in the United States. At the end of each chapter, you'll improve this case study with the concepts described in the chapter.

Chapter 6 starts with the bottom of the DIKW pyramid and focuses on turning data into information. You'll learn some basic techniques to extract and represent insights. The chapter also illustrates how to build the most common chart families: bar charts, line charts, pie charts, and geographical maps.

In chapters 7 and 8, you'll climb the pyramid by transforming information into knowledge. You'll learn how to add context to your data story. In chapter 7, you'll focus on the textual context, and in chapter 8, on the visual context. You'll also learn advanced concepts, such as retrieval augmented generation (RAG) and large language model (LLM) fine-tuning. Since these concepts are evolving continuously, it's possible some may be outdated by the time you read this part. However, the overall theory is still valid, although the code syntax may differ.

In the last chapter of part 2, chapter 9, you'll reach the top of the DIKW pyramid by transforming knowledge into wisdom. You'll learn the three main elements of wisdom: knowledge synthesis, experience, and good judgment. You will also use ChatGPT as an additional source of experience.

Crafting a data story
using the DIKW pyramid

This chapter covers

- A homelessness tale, our scenario for this chapter
- What a data story is and how it relates to the DIKW pyramid
- How to incorporate generative AI into the DIKW pyramid

Creating stories is the most exciting part of data storytelling. It's really about following a plot with a beginning, a main point, and a conclusion—just like in the stories at the cinema or theater. Here, however, there are constraints imposed by the data. In this chapter, we will focus on using the DIKW pyramid by describing two examples. The first example will analyze the problem of homelessness in Italy, and the second will examine the problem of fake news on a hypothetical website. We will also describe the concept of a data story and how it relates to the DIKW pyramid already discussed in the previous chapters. Finally, we will describe some strategies to incorporate generative AI in the DIKW pyramid, based on how to write specific prompts for ChatGPT to generate the context of a chart and for DALL-E to generate contextual images to incorporate into the chart.

5.1 *Breaking the Ice: The homelessness tale*

Imagine that Angelica works for a humanitarian organization that wants to apply for funding from a foundation to help reduce the population of homeless people in Italy. Humanitarian interventions can be applied to up to four Italian regions. The call for funds involves preparing a data visualization chart motivating the selected regions and detailing reasons to fund the proposal.

Angelica's boss asks her to complete a study on which regions to invest in and to motivate her choice. Angelica starts a web search for possible datasets on homelessness in Italy. After several searches, she comes across the ISTAT dataset on homelessness in 2021 (https://mng.bz/Bd6J). She downloads the dataset and starts analyzing it. Table 5.1 shows a simplified version of an extract of the homelessness dataset. The original dataset column names are in Italian. They are translated into English to improve readability.

Table 5.1   An extract of the ISTAT dataset on homelessness in 2021

ITTER107	Territory	Sex	Age	Citizenship	Value
ITC1	Piemonte	M	TOTAL	ITL	4,218
ITC1	Piemonte	F	TOTAL	ITL	1,496
ITC2	Valle d'Aosta	M	TOTAL	ITL	41
ITC2	Valle d'Aosta	F	TOTAL	ITL	17

The ISTAT dataset on homelessness

The dataset has the following columns:

- *ITTER107*—The region ID
- *Territory*—The region name
- *Sex*—One of Male (M), Female (F), or Total (T)
- *Age*—One of Sum of All Ages (TOTAL), Under 17 (Y_UN17), Between 18 and 34 (Y18-34), Between 35 and 54 (Y35-54), and Greater Than 55 ('Y_GE55)
- *Citizenship*—One of Italian (ITL), Foreign (FRGAPO), or Total (TOTAL)
- *Value*—The actual number of homeless people

At the end of her analysis, Angelica produces the chart shown in figure 5.1 by writing the code in the following listing, using Altair. You can find the full code in the GitHub repository for the book under CaseStudies/homeless-people/raw-chart.py.

Listing 5.1   The map in Altair

```
import pandas as pd
import altair as alt
```

```
df = pd.read_csv('source/homeless.csv')

df['Territory'] = df['Territory'].str.replace('Trentino Alto Adige',
    'Trentino-Alto Adige/Südtirol')

df = df[(df['Age'] == 'TOTAL') & (df['Sex'] == 'T') & (df['Citizenship'] ==
    'TOTAL')]

url = "https://raw.githubusercontent.com/openpolis/geojson-
    italy/master/topojson/limits_IT_regions.topo.json"
map = alt.topo_feature(url, "regions")

chart = alt.Chart(map).mark_geoshape().encode(
    tooltip='properties.reg_name:N',
    color=alt.Color('Value:Q')
).project('mercator').properties(
    width=500,
    height=500
).transform_lookup(
    lookup='properties.reg_name',
    from_=alt.LookupData(df, 'Territory', ['Territory', 'Value'])
).properties(title='Homeless in Italy in 2021')

chart.save('raw-chart.html')
```

←┐ **Data cleaning**

← **Focus on total age, total sex, and total citizenship.**

← **Load the TopoJSON file by URL.**

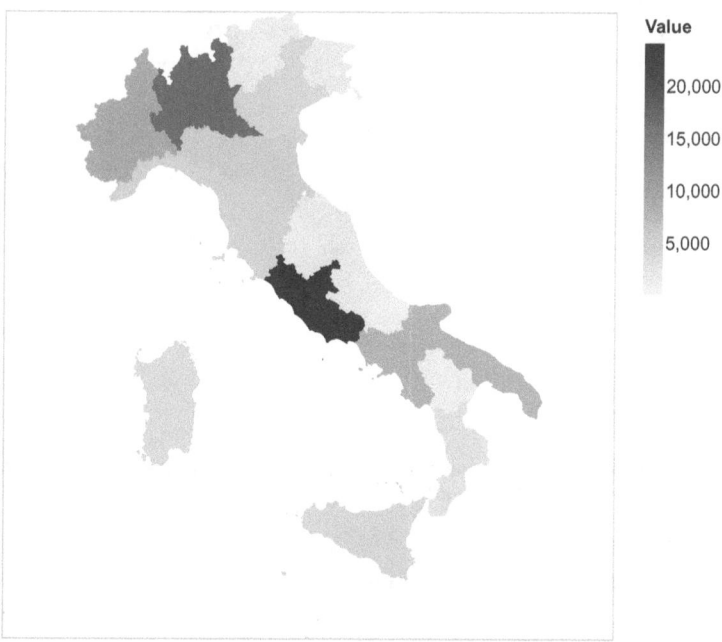

Homeless Population in Italy in 2021

Figure 5.1 A map showing the homeless people in Italy in 2021

> **NOTE** The chart builds a map using an underlying `topoJSON` object containing the map of Italy. To build the map, use the `mark_geoshape()` marker, and specify the projection through the `project()` method. In addition, combine the homeless dataset (`df`) and the map through the `transform_lookup()` method. Map the `properties.reg_name` variable of the `topoJSON` file with the `Territory` variable of the `df` DataFrame. We'll discuss how to build a geographical map more deeply in the next chapter.

Angelica shows the chart to her boss, who asks her to answer their question: *Which are the four regions for which we must apply for funding?* Angelica looks at the chart and gives this answer: *those with the darker color.* Her answer is very ambiguous because from the chart she produced, it is not immediately clear which exactly are the four regions with the darkest color. In addition, it is not clear why to select the four regions. Angelica has to admit that her chart has failed in its mission.

5.1.1   What was wrong with the chart?

The main problem with the chart is there is a large discrepancy between what Angelica's boss (audience) expected to see in the chart and what the chart actually shows. In practice, Angelica's chart does not answer the question for which it was requested (figure 5.2).

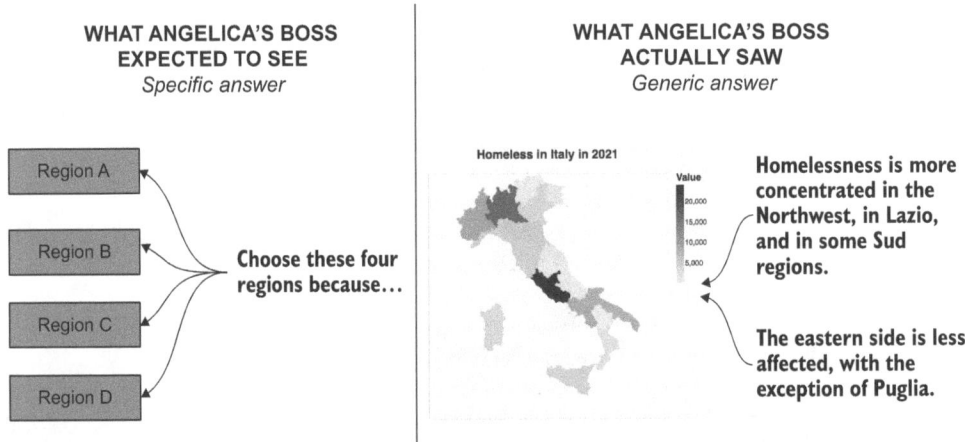

Figure 5.2   Angelica's boss (audience) requested an answer to a specific question (on the left). The chart actually answered generic questions (on the right).

Here comes the first rule when building a chart: a chart must answer exactly the question the intended audience asks. Try not to leave it up to your audience to decipher the answer. Do the work for them!

Angelica's boss asks her to redo the chart. She returns to her office with her head down and analyzes her dataset again. Angelica understands that the problem is not so much data analysis but rather how to represent data effectively. She tries this reasoning: *at a glance, the audience must understand the four regions to apply for funding. What could these four regions be? Probably the ones that have the highest value of homeless people.* What is missing in her previous chart is the *probably* part. She considers making a bar chart that shows the regions in descending order based on the number of homeless people present. After a small effort, she produces the chart in figure 5.3, writing the code in listing 5.2. You can also find the code in the GitHub repository for the book under CaseStudies/homeless-people/bar-chart.py.

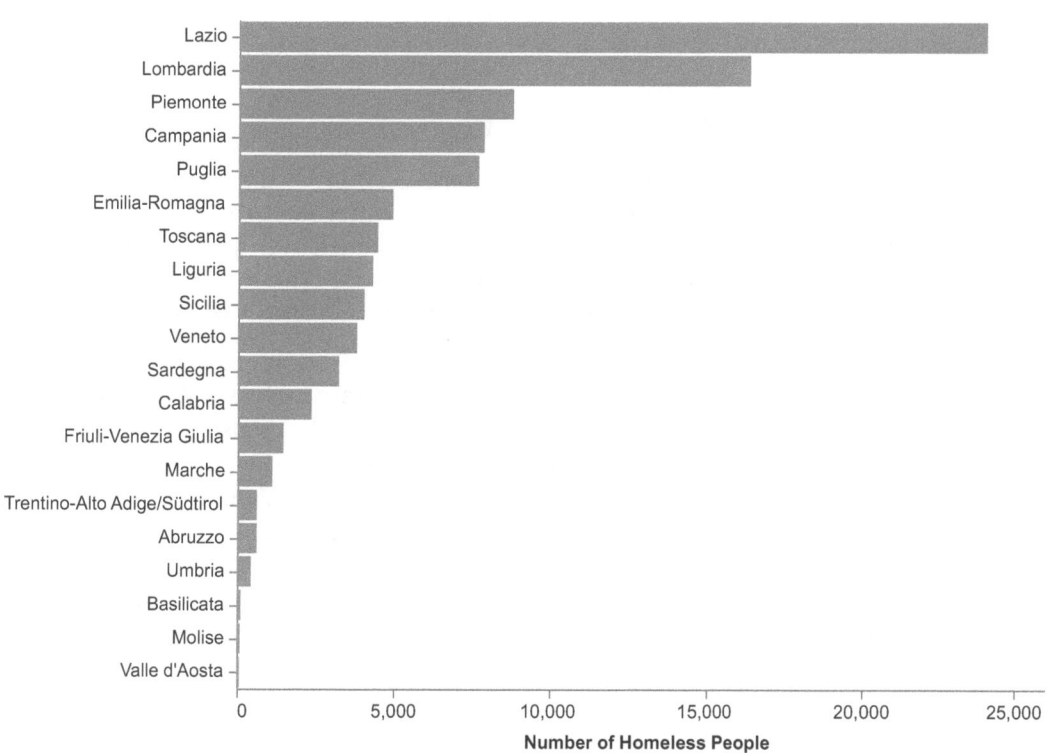

Figure 5.3 A bar chart showing the number of homeless people in Italy by region in 2021

Listing 5.2   The bar chart in Altair

```
chart = alt.Chart(df).mark_bar(
    color-'#80C11E'
).encode(
    y = alt.Y('Territory',
              sort='-x',
              axis=alt.Axis(title='')),
```

```
    x = alt.X('Value',
                axis=alt.Axis(tickCount=4,title='Number of homeless people'))
).properties(
    width=500,
    title='Homelessness in Italy in 2021'
)

).properties(title='Homeless in Italy in 2021')

chart = chart.configure_title(
    fontSize=20,
    offset=25
).configure_axis(
    grid=False
).configure_view(
    strokeWidth=0
)

chart.save('bar-chart.html')
```

> **NOTE** The code builds a bar chart using the `mark_bar()` method. It also sets the bars' color using the `color` parameter. To order the bars in descending order, use the `sort` parameter within the y channel, and specify to sort in descending order of the x encoding channel. Use `configure_title()` to configure the title parameters, such as the font size (`fontSize`) and the `offset` between the title and the chart.

The chart answers the boss's question: the four regions are Lazio, Lombardia, Piemonte, and Campania. Angelica notes that three of the four regions also correspond to the most populous regions of Italy. As Angelica reflects on this, she notices that her chart contains an underlying error. The most populous regions also have the highest number of homeless people precisely because their population is greater.

Therefore, the chart does not show the situation in a relevant way. She needs to normalize the data based on the population—or scale the numbers based on population size. Without any normalization, data could be biased. Consider, for example, the case of a region with a population of 100 people, of which 30 are homeless. Also, consider region with a population of 10 people, of which 8 are homeless. If you represent absolute values, you conclude that in the first case, there are more homeless people than in the second one. But if you pay more attention to the scenario, in the second case, 80% of the population are homeless, while in the first case, only 30% are homeless.

Figure 5.4 shows the resulting chart after the normalization process, focusing on the regions with the four largest homeless populations. Listing 5.3 describes the code written to draw the chart (script CaseStudies/homeless-people/story-chart.py).

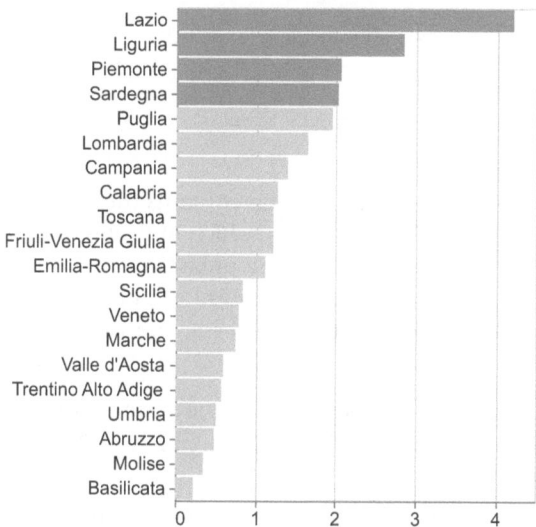

Number of Homeless People in a Population of 1,000 in 2021

Figure 5.4   A bar chart showing the number of homeless people in a region with a population of 1,000 in 2021

Listing 5.3   Listing 5.3 The improved bar chart in Altair

```python
import pandas as pd
import altair as alt


df = pd.read_csv('source/homeless.csv')

df['Territory'] = df['Territory'].str.replace('Trentino Alto Adige',
    'Trentino-Alto Adige/Südtirol')                                    Data cleaning

df = df[(df['Age'] == 'TOTAL') & (df['Sex'] == 'T') &  (df['Citizenship']
    == 'TOTAL')]
df = df[['Value', 'ITTER107']]                          Focus on total age, total
                                                        sex, and total citizenship.

df_pop = pd.read_csv('source/population.csv')
df_pop = df_pop[(df_pop['Age'] == 'TOTAL') & (df_pop['Sex'] == 'T')]

                                                        Normalize the values
df_pop = df_pop[['Value', 'ITTER107','Territory']]        by popultation.

df_tot = df_pop.set_index('ITTER107').join(df.set_index('ITTER107'),
    lsuffix='_pop', rsuffix='_hom').reset_index()
df_tot['Ratio'] = df_tot['Value_hom']/df_tot['Value_pop']*1000


chart = alt.Chart(df_tot).mark_bar().encode(
    y = alt.Y('Territory',
            sort='-x',
            axis=alt.Axis(title='')),
    x = alt.X('Ratio',
            axis=alt.Axis(tickCount=4,title='')),
```

```
color=alt.condition(alt.datum.Ratio > 2,
                    alt.value('#80C11E'),
                    alt.value('lightgray'))
).properties(
    width=500,
    title='Number of homeless people in a population of 1,000'
)
```

> **NOTE**   The code loads the population dataset, which contains the same fields as the homeless dataset, except for the Value column, which indicates the actual population. The code merges the two datasets, population (`df_pop`) and homeless (`df`), and calculates the ratio between the number of homeless and the population. Then, the chart builds a bar chart using the ratio as the x encoding channel.

Angelica's chart eventually answers her boss's question: the four regions are Lazio, Liguria, Piemonte, and Sardegna. Finally, Angelica shows her chart to her boss, who is very satisfied with her result.

To apply for funds, Angelica must send her chart to a commission, which will decide whether or not to finance her organization's proposal. However, after a careful evaluation, the foundation commission decides not to finance her project because the chart lacks motivation.

5.1.2   *What was wrong with the presentation?*

To understand why the foundation commission refused to fund Angelica's proposal, let's trace back the path she followed. She started with raw data, analyzed it, extracted an insight, and represented it through a chart before she sent it to the foundation commission (figure 5.5).

What is lacking about her chart is that it contains no context and no motivation to push the foundation commission to accept her proposal. Angelica should have added

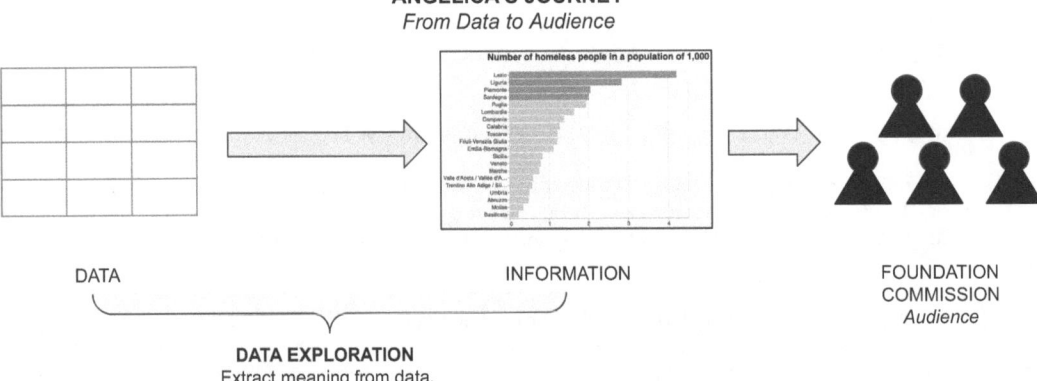

ANGELICA'S JOURNEY
From Data to Audience

DATA   INFORMATION   FOUNDATION
                                                       COMMISSION
                                                       Audience

DATA EXPLORATION
Extract meaning from data.

Figure 5.5   Angelica's journey to chart generation. Angelica presented the results of data exploration directly to her audience.

an explanatory phase to the exploratory phase of the data, where she could have engaged the audience with her data and motivated them to accept her proposal. Angelica could have used the DIKW model to move from the exploratory phase of her data to the explanatory phase (figure 5.6).

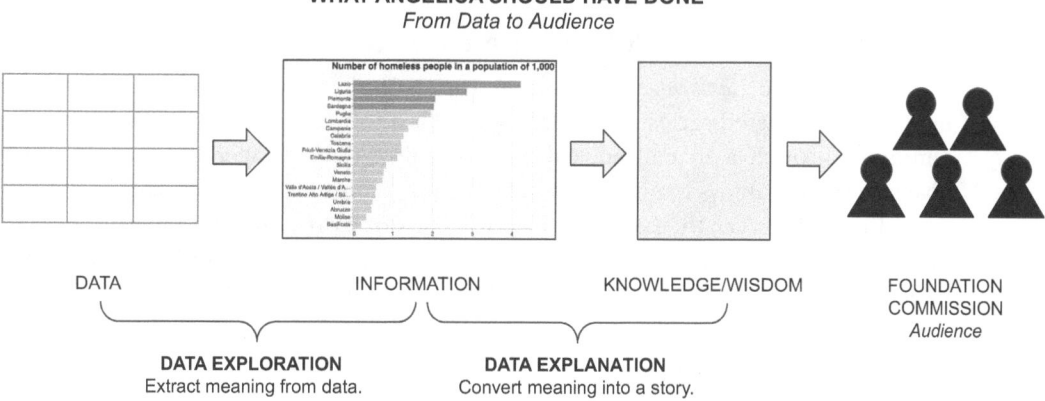

Figure 5.6   Angelica should have transformed her data exploration into a data explanation phase.

In other words, Angelica hasn't turned data into a story. In his book, *Effective Data Storytelling*, Brent Dykes says, "The formation of a data story begins with using exploratory data visualizations to discover insights. Once a meaningful insight is uncovered, explanatory data visualizations are used to tell a story." (Dykes, 2019) The chart built by Angelica constitutes only the main point of the data story. However, to have a complete story, Angelica should have also included a background and next steps (figure 5.7). The

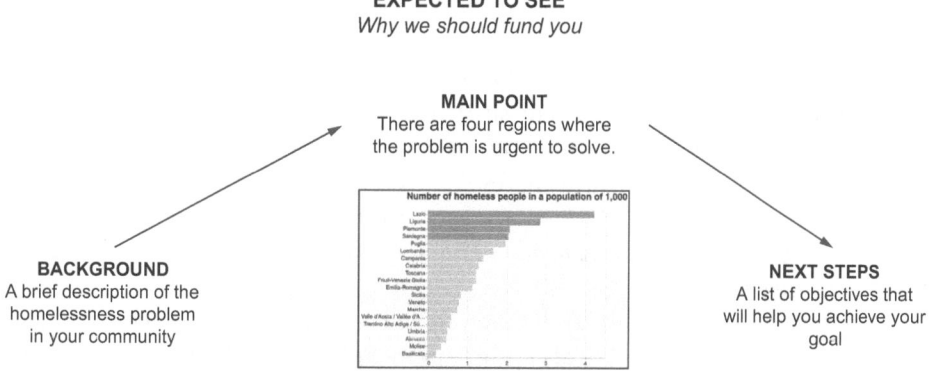

Figure 5.7   The foundation commission expected to see a story with a background, a main point, and next steps, inviting them to fund Angelica's proposal.

foundation commission expected Angelica's chart to answer this question: *Why should we fund your project?* Organizing the chart as a story would have increased her chances of receiving funding.

Now, you understand the urgency behind turning data into a story to communicate a message to an audience effectively. Next, let's move to the next step: what a data story is and its main components.

5.2   *Uncovering the narrative: What a data story is*

According to the *Cambridge Dictionary*, a story is "a description, either true or imagined, of a connected series of events." A story is a way to share information or entertainment through a structured, cohesive narrative that engages the audience. The goal of a story is always to engage its audience to communicate a message.

In his book *Poetics*, the Greek philosopher Aristotle proposed a framework to structure a tragedy, also known as the *three-act structure*. The structure consists of three main parts: the setup, the confrontation, and the resolution (figure 5.8).

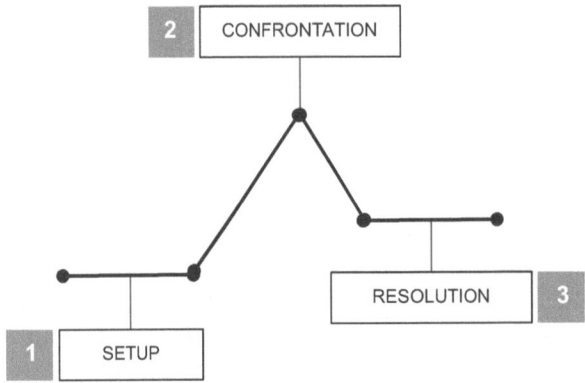

Figure 5.8   The three-act structure proposed by Aristotle to structure a tragedy

The setup introduces the audience to the characters, setting, and basic conflict of the story. In the confrontation, the main character faces obstacles and struggles to overcome them. Finally, in the resolution, the story reaches the point at which conflicts are resolved. Aristotle's tragedy structure is still widely used today in literature, film, and other forms of storytelling. Its effectiveness lies in its ability to build tension and suspense, leading to a satisfying resolution for the audience.

In *Effective Data Storytelling*, Dykes proposes an updated version of the three-act tragedy, adapted to the data storytelling scenario (figure 5.9). Dykes calls it the *data storytelling arc*.

A data story starts by (a) defining the background behind data and (b) raising the audience's interest. Dykes also proposes including a hook during this initial phase. A hook is "a notable observation that acts as a turning point in the story and begins to

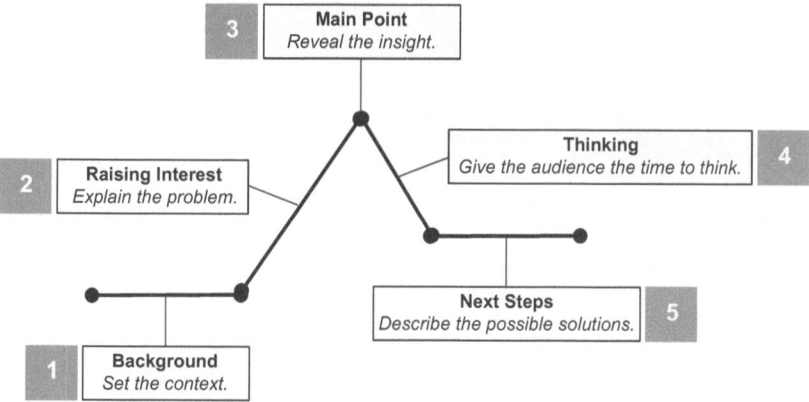

Figure 5.9   The data storytelling arc proposed by Dykes to structure a data story

reveal a problem or opportunity" (Dykes, 2019). Then, the data story culminates with the main point behind the data by (c) revealing the insight, (d) leaving the audience some time to think, and (e) terminating by proposing the next steps.

The structure proposed in figure 4.9 might seem abstract and difficult to apply when constructing a chart. Using the DIKW pyramid covered in chapter 1, you can turn a simple chart into a story.

5.2.1   *Using the DIKW pyramid to streamline a data story*

In chapter 1, we introduced the data, information, knowledge, wisdom (DIKW) pyramid and used it to transform data into wisdom in the practical example of an event dedicated to pets. We can generalize the example described in chapter 2 and use the DIKW to build any data story. Figure 5.10 shows how to map the data storytelling arc to the DIKW pyramid:

- Extracting insights from data corresponds to the main point of a data story.
- Adding context to the extracted information corresponds to defining the background and raising the audience interest.
- Adding a call to action corresponds to inviting the audience to think and follow the next steps.

It might seem strange to have knowledge before the data. However, here, by *knowledge*, we mean of all the elements needed to understand the story and place it in a specific context. The described mapping suggests the natural flow when you build a data story (figure 5.11). First, identify the main point of your story; then set the background; and, finally, set the next steps.

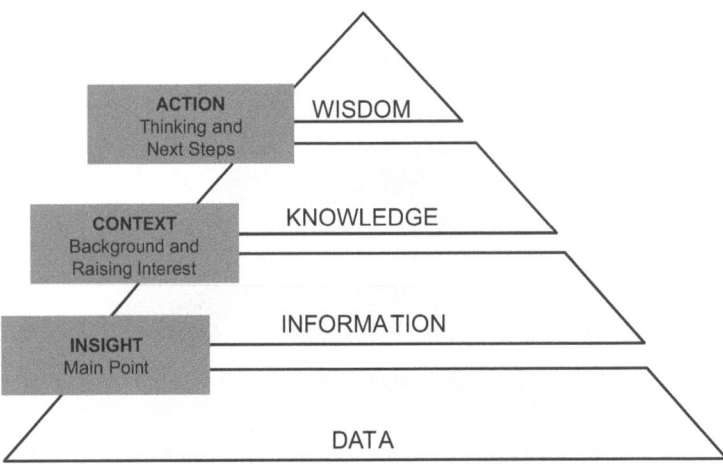

Figure 5.10   A mapping of the data storytelling arc and the DIKW pyramid

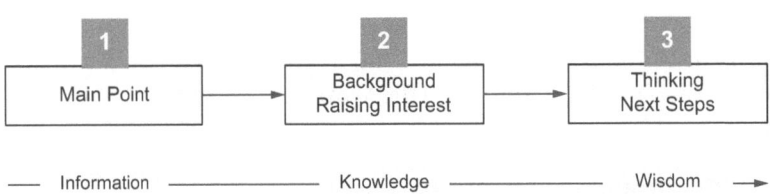

Figure 5.11   The flow to build a data story

5.2.2   *DIKW in action: Completing the homelessness tale*

The homelessness example described at the beginning of the chapter stopped at the information level of the DIKW pyramid that, in terms of a data story, consisted only of the main point. Adding background to the chart involves grabbing your audience's attention and making them interested in the problem you're describing. In chapter 2, you learned that in Altair, you can add textual annotations to a chart to transform information into knowledge. In addition, there are other ways to add context to a chart, including labels, titles, sources, images, descriptions, and comparisons. We'll describe these in more detail in chapter 6.

In the case of the homeless problem, you can add text describing the situation of these people and possibly a photo that gives a face to the people involved. We will see how to add a hero to the story in chapter 6. For now, it's enough to give a face to the people behind the data.

In listing 5.4, we add text describing the context as the chart subtitle. Use the `TitleParams()` function to specify the title properties, including the subtitle. Set the subtitle to

Homelessness is a heartbreaking reality that leaves individuals and families without a stable home, leading to devastating consequences such as poor health and social isolation.

Also, add a short description, helping the reader to focus on the data:

The chart describes the number of homeless people in Italy per 1,000 inhabitants, organized by region. The data is from 2021, the most recent year available (Source: ISTAT).

Listing 5.4 Adding the context as the subtitle

```
# Add context
chart = chart.properties(width=500,title=alt.TitleParams(
    text=["Together, Let's Make a Difference:","Support Our Project to Help
    the Homeless!"],
    subtitle=['Homelessness is a heartbreaking reality that leaves
    individuals and families without a stable home,','leading to devastating
    consequences such as poor health and social isolation.', 'The chart
    describes the number of homeless people in Italy per 1,000 inhabitants,
    organized by region. ', 'The data is from 2021, the most recent year
    available (Source: ISTAT).']],
    subtitleFontSize=18,
))
```

NOTE Within the properties() method, use the title parameter to set the chart title. Define the title using the TitleParams() function, which can receive many parameters as an input. The example passes the following parameters: title (text), subtitle (subtitle), subtitle font size (subTitleFontSize), title font size (titleFontSize).

Listing 5.5 shows how to add two photos to the chart and combine them with the original chart to build a dashboard. You must run the code on a web server to make it work. If you don't have a web server, you can run a local and temporary web server from the command line, running the following command in the directory containing the produced HTML file: python -m http.server. The server should listen at port 8000 and serve all the files in the directory from which it is started. Point your browser to http://localhost:8000/chart.html to access the file chart.html.

Listing 5.5 Adding the context as images

```
image1 = alt.Chart(pd.DataFrame({'image_url':
    ['source/homeless1.png']})).mark_image(
    width=200,
    height=200,
).encode(
    url='image_url',
    x=alt.value(0),   # pixels from left
    y=alt.value(50)   # pixels from the top
)                                    ← Build the first image.
```

```
image2 = alt.Chart(pd.DataFrame({'image_url':
    ['source/homeless2.png']})).mark_image(
    width=200,
    height=200,
).encode(
    url='image_url',
    x=alt.value(0),   # pixels from left
    y=alt.value(300)
)                              ◁──┐  **Build the second image**
image1 + image2 | chart
```

> **NOTE** Use `mark_image()` to add an image to a chart. Set the path to the image file in the DataFrame passed to the chart. Use the x and y channels to set the image position in the chart. Use the same chart to draw `image1` and `image2` (+ operator), and draw the main chart on the right with respect to `image1 + image2`.

To complete the homelessness story, add a call to action, which includes the next steps. In this example, we first modify the title: *Together, Let's Make a Difference: Support Our Project to Help the Homeless!* We'll see different strategies to add a call to action to a chart in chapter 9. Figure 5.12 shows the resulting chart.

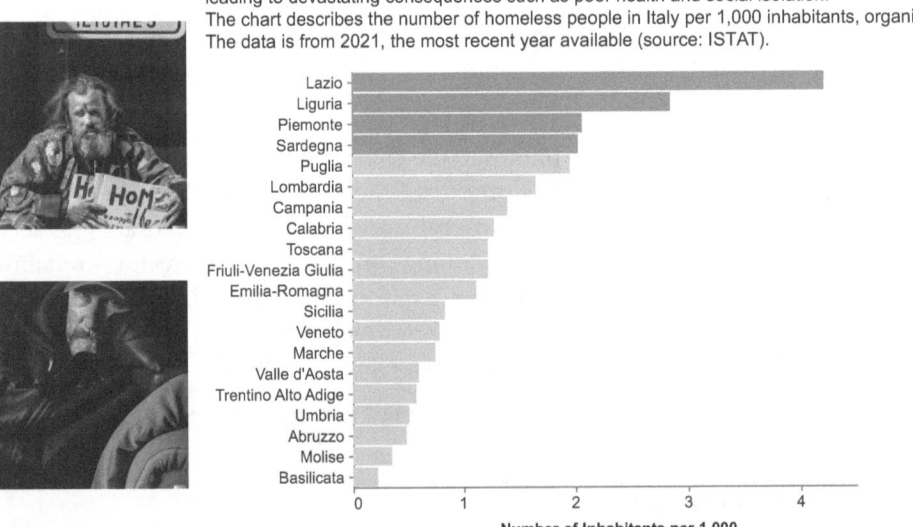

Together, Let's Make a Difference: Support Our Project to Help the Homeless!
Homelessness is a heartbreaking reality that leaves individuals and families without a stable home, leading to devastating consequences such as poor health and social isolation.
The chart describes the number of homeless people in Italy per 1,000 inhabitants, organized by region. The data is from 2021, the most recent year available (source: ISTAT).

Figure 5.12   The homelessness chart is enriched with a context (images and subtitle).

Then, we add a new part to the chart, describing how we will use funds (figure 5.13). For example, we will use 35% of the funds for shelter and housing, 25% for job training, and so on.

Our Visionary Plan to Harness Funds

Figure 5.13   A possible next step explaining how we will use funds

The following listing shows how we implemented the charts in figure 5.13. We assume
we have a DataFrame storing the percentage of funds (allocation) for each category.

Listing 5.6   Adding a next step

```
import pandas as pd
import altair as alt

donuts = None

for index, row in ns.iterrows():
    curr_ns = pd.DataFrame(
        {'Category': ['A', 'B'],
         'Value': [row['Allocation'], 100-row['Allocation']]
        }
    )

    donut = alt.Chart(curr_ns).mark_arc(outerRadius=30,
      innerRadius=20).encode(
        theta=alt.Theta("Value:Q", stack=True),
        color=alt.Color("Category:N", scale=alt.Scale(range=['green',
      'lightgray']), legend=None)
    )

    title = alt.Chart(curr_ns).mark_text(text=row['Category'], y=0, size=16)

    text = alt.Chart(curr_ns).mark_text(text=f"{row['Allocation']}%",
      color=iColor, size=16)

    donut = donut.properties(
        height=100,
        width=100
    )

    if index == 0:

        donuts = title + donut + text
    else:

        donuts = alt.hconcat(donuts, title + donut + text)

donuts = donuts.properties(title='Our visionary plan to harness the funds')
```

Creates a DataFrame for the current row to use as data for the chart

Generates the donut chart using Altair

Creates a title for the donut chart

Creates a text annotation for the donut chart

Sets the size of the donut chart

Combines the title, donut chart, and text annotation

If it's not the first iteration, horizontally concatenates the combined chart with the existing donuts

If it's the first iteration, assigns the combined chart to donuts

Sets the title for the final donuts chart

> **NOTE** Implement each donut chart as a separate chart using `mark_arc()`. For each donut, build an ad-hoc DataFrame (`current_ns`) that contains the actual allocation value and its complementary value (100 – Actual Allocation Value). Then, use different colors to plot the current allocation value and its complementary value.

Figure 5.14 shows the resulting chart. If Angelica had submitted the chart in figure 5.14 to the commission foundation, she probably would have had more opportunities to get funded. Now that you have completed the homelessness tale, let's describe how you can incorporate the power of generative AI in the DIKW pyramid.

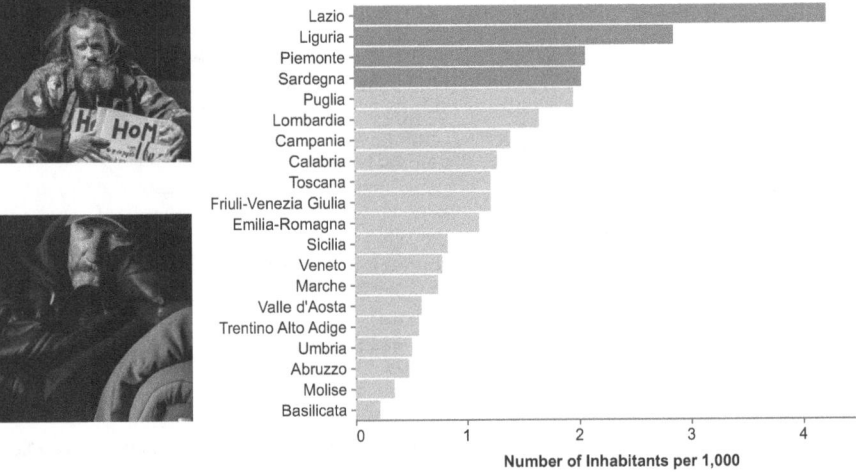

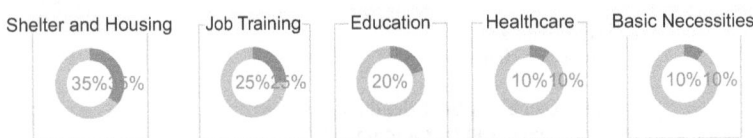

Figure 5.14   The final chart for the homeless tale

5.3   *Incorporating generative AI into the DIKW pyramid*

Generative AI, as we covered in chapter 4, can help us take things one step further by enabling the creation of sophisticated and realistic computer-generated content that can revolutionize various industries and domains. Now, armed with this knowledge, we can leverage generative AI's capabilities to augment human creativity and automate content generation processes.

Figure 5.15 shows where we can incorporate generative AI tools into each step of the DIKW pyramid:

- *Insight extraction*—Use generative algorithms to automate the creation code that builds the chart.
- *Context adding*—Use text generation to transform basic texts into more engaging ones for the audience. Add images, voice, and videos generated by AI to give the audience additional context.
- *Call to action*—Use text generation to suggest new ideas and engaging text.

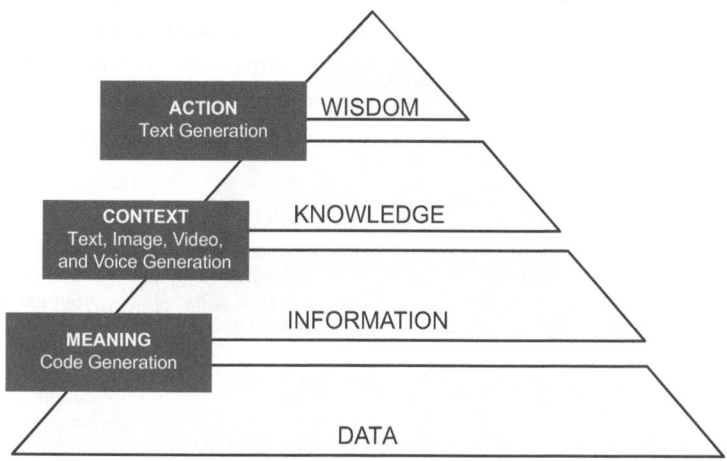

Figure 5.15   Where we can incorporate generative AI tools into the DIKW pyramid

It's important to note that the examples listed are just a few of the ways that you can incorporate generative AI into the DIKW pyramid. There are countless other ways in which you can take advantage of these tools, such as synthesizing large amounts of data, developing personalized, predictive models, and constructing personalized recommendations based on data. However, in this book, we will focus specifically on the techniques for integrating generative AI into the DIKW pyramid, as it can assist in building a data story. Anyway, I encourage you to think outside the box and explore new and innovative ways to apply generative AI in your work. With so much potential waiting to be unlocked, the possibilities are truly endless! Now that you have learned how to incorporate generative AI tools into the DIKW pyramid, let's apply this strategy to our case study: the homelessness tale.

5.4   *Behind the scenes: The homelessness tale*

In the previous chapters, you saw how to use GitHub Copilot to generate the code to build the chart. Now, it's time to go a step further.

Consider the homelessness tale again, specifically figure 5.12. The figure contains a title, defining the call to action; a subtitle, setting the context; and two photos, adding to the context. While you may have thought that we came up with the title and created or downloaded the images, in reality, it was ChatGPT and DALL-E that worked together to generate the content! Let's see how we used ChatGPT to generate the subtitle and DALL-E to generate the photos. In the next chapters, we will see how to use more deeply generative AI tools for data storytelling.

5.4.1   *Creating a compelling subtitle*

We started a conversation with ChatGPT to extract a possible context that describes the situation where homeless individuals live. The context should be a short and engaging sentence. Figure 5.16 shows the steps involved in the conversation. *Q* refers to user questions (which will form the basis of our prompts), and *A* refers to ChatGPT's answers. You can read the complete conversation in 05/genAI/SubtitleConversation.txt.

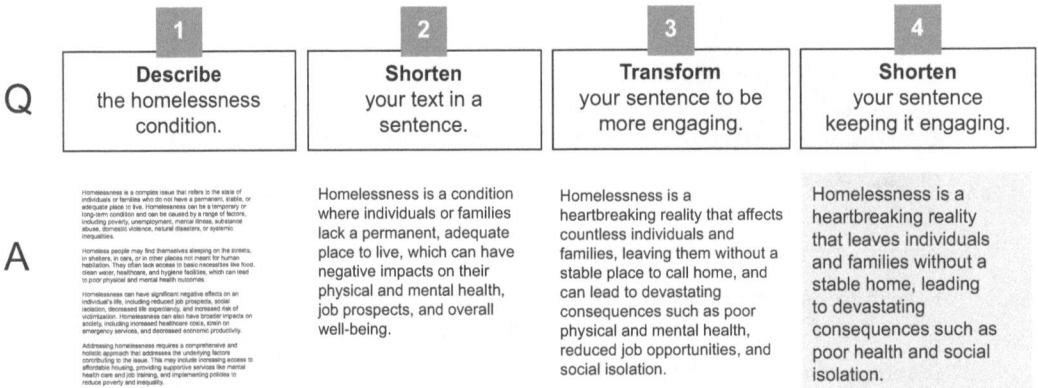

Figure 5.16   The steps to generate the text used for context

ChatGPT generated the text after four steps. We used the following strategy to make ChatGPT generate the desired text:

1 *Describe*—Ask ChatGPT to describe your problem in general. In this case, ask ChatGPT to describe the homelessness condition in general. As an answer, ChatGPT generates a long passage of text.
2 *Shorten*—Ask ChatGPT to write a summary of the generated text.
3 *Transform*—Ask ChatGPT to make the summary more engaging for the audience.
4 *Shorten*—If the text is still long, ask ChatGPT to reduce it.

Without realizing it, we applied the DIKW model to the use of ChatGPT. Starting from a long passage of text (data), we extracted the information (summary), and then converted it into knowledge and wisdom (engaging text). In other words, when you talk to ChatGPT to generate context, organize the conversation as a story (figure 5.17).

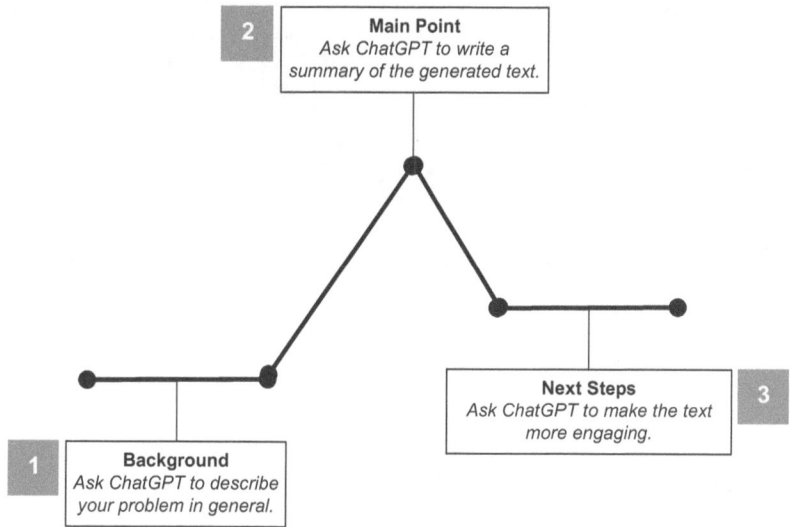

Figure 5.17   The mapping between the data story structure and the ChatGPT conversation to generate context

You can generalize the described procedure to generate a variety of possible subtitles and then choose the one that fits your needs. Additionally, you can add specific keywords that can help appeal to your target audience and increase the visibility of your text. We'll see how to add keywords to the ChatGPT conversation in chapter 6.

5.4.2   Generating images

Adding one or more images or photos to your chart helps give a face to the topic of your chart. However, images are often subjected to copyright, which means using them without permission from the owner can lead to legal consequences. We can use DALL-E and other AI tools for images to create unique and original images that are free from copyright restrictions.

In the chart about homelessness, we have added two photos representing homeless individuals. We have generated them using the following simple prompt to DALL-E: *a photo of a homeless individual.* DALL-E has generated four images, and we have chosen two (figure 5.18). In chapter 6, we will see more advanced techniques to generate images.

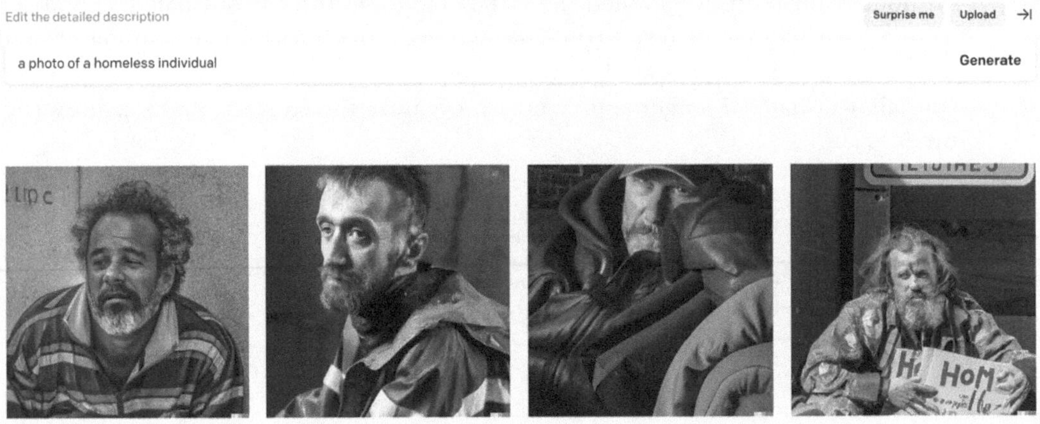

Edit the detailed description

Surprise me   Upload   →|

a photo of a homeless individual

Generate

Figure 5.18   The photos generated by DALL-E when prompted with *a photo of a homeless individual*

Now that you have learned how to use generative AI tools to transform your raw data visualization chart into a data story, let's look at another example, to consolidate the concepts.

5.5   Another example: Fake news

Imagine that LatestNews is an important website that publishes news from different contributors. At a given point, the editor-in-chief receives some complaints from different readers because they read a high percentage of fake news. The editor-in-chief contacts you to analyze the amount of fake news on the LatestNews website and advises the website editors to pay attention to the categories of news that have the highest probability of being fake. You already have collected data, and you have the dataset shown in table 5.2.

Table 5.2   An extract of the dataset of the example

Category	Number of Fake Articles	Number of Articles
Politics	1,235	1,300
Economy	1,456	1,678
Justice	300	570
Religion	30	100

The dataset shows the number of fake news stories and the total number of articles for each news category. You start by drawing a preliminary chart, paying attention to the percentage of fake articles for each category. Figure 5.19 shows the preliminary chart, and listing 5.7 shows the associated code. You can also find the code in the GitHub repository for the book under CaseStudies/fake-news/raw-chart.py.

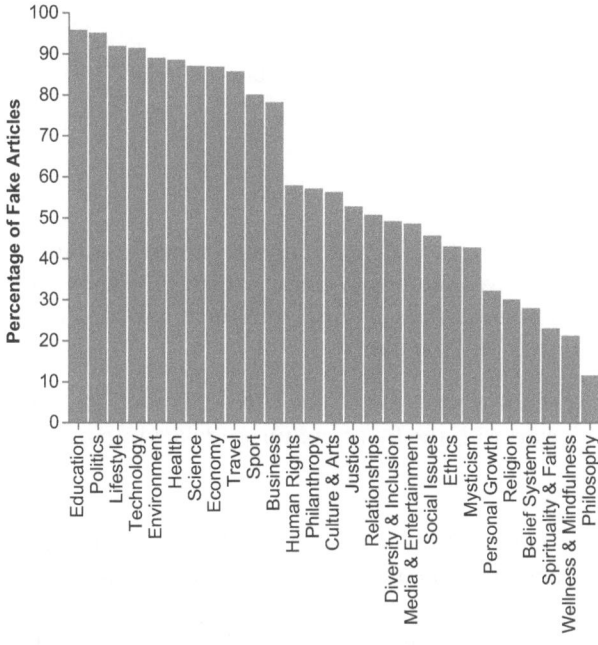

Figure 5.19 The raw chart showing the number of fake articles by category

Listing 5.7 Building the raw chart

```python
import pandas as pd
import altair as alt

# Load the dataset data/fakenews.csv.
df = pd.read_csv('data/fakenews.csv')

# Create a column chart of the number of articles per category:
# - Use the `Category` column for x channel.
# - Use the `Number of Articles` for y channel.

chart = alt.Chart(df).mark_bar(
    color='#81c01e'
).encode(
    x=alt.X('Category:N',
            sort='-y',
            title=None,
            axis=alt.Axis(labelFontSize=14)
            ),
    y=alt.Y('Percentage of Fake Articles:Q',
            axis=alt.Axis(labelFontSize=14, titleFontSize=14)
            )
).properties(
```

Uses Copilot comments to speed up the chart generation process

```
        width=400,
        height=300                      Uses configure_axis() to configure
).configure_axis(                       the axes' general properties
        grid=False
).configure_view(                       Removes the
        strokeWidth=0                   grid completely
).transform_calculate(
        'Percentage of Fake Articles', alt.datum['Number of Fake
        Articles']/alt.datum['Number of Articles']*100
)

chart.save('raw-chart.html')
```

NOTE Use `transform_calculate()` to dynamically add a new column to the DataFrame. This method receives the new column name as the first parameter (Percentage of Fake Articles, in the example) and the expression to calculate the new column as the second parameter.

In the remainder of the section, we will apply the DIKW pyramid to transform the chart into a data story. Let's start with the first step: turning data into information.

5.5.1 *From data to information*

Turning data into information means extracting some insights from data, an insight that is significant and helps your audience of editors understand which kind of news has the highest probability of being fake. Let's try the rotation strategy. This strategy, originally described by Berengueres in his previously quoted book, involves rotating the chart to obtain a pyramid and searching for some pyramid framework, such as moral, economic, and other similar frameworks. Figure 5.20 shows the rotation process.

Next, we can flip the chart horizontally and obtain the chart in figure 5.21, with corrected labels. You can also rewrite the Altair code directly to draw a column chart instead of a bar chart (the code is in the GitHub repository under CaseStudies/fake-news/bar-chart.py). We have preferred to show the rotation process instead of drawing the chart directly, to show how the rotation strategy works.

If you look at labels carefully, you may notice that at the bottom of the pyramid, there are categories related to material life (from Education to Business). In the middle of the pyramid, there are categories related to moral life (from Human Rights to Ethics). At the top of the pyramid, there are categories related to spiritual life (from Mysticism to Philosophy). This means that most fake news relates to material life (more than 70%) and moral life (more than 30% of fake news but less than 60%). You can highlight the model material–moral–spiritual life using different colors in the chart based on the different macro categories the news belongs to.

First, use Copilot to generate the list of macro categories, as shown in listing 5.8 and in the GitHub repository for the book under CaseStudies/fake-news/story-chart.py. The listing shows only how to generate the material life macro category, but you can apply the same strategy also for the other macro categories.

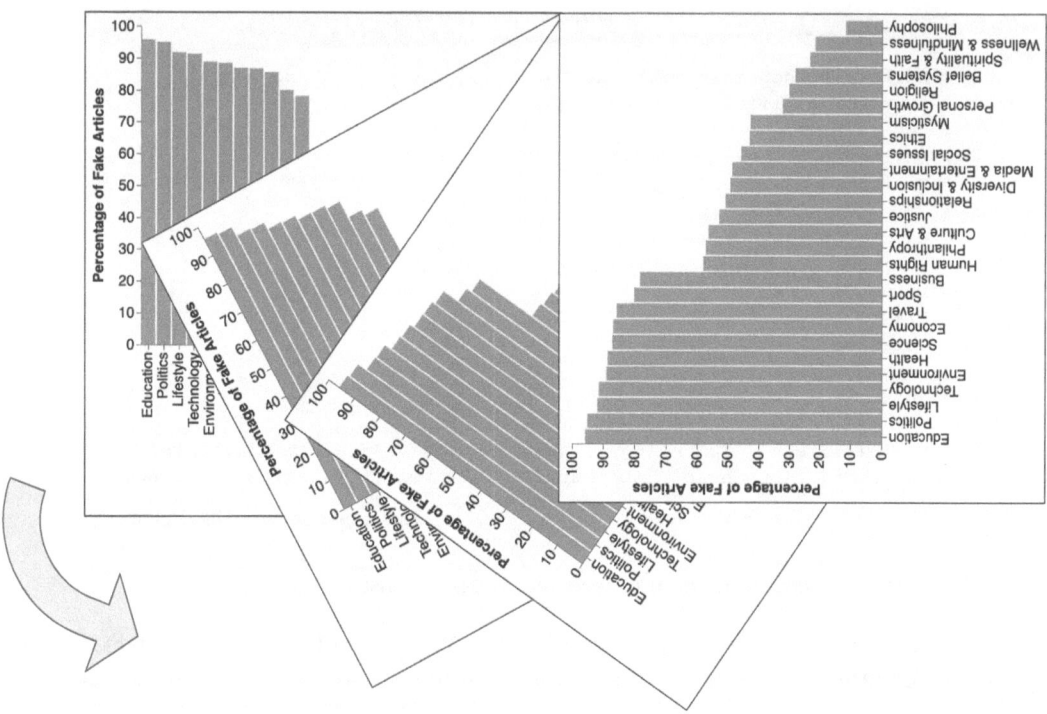

Figure 5.20   The rotation process of the chart in figure 5.19

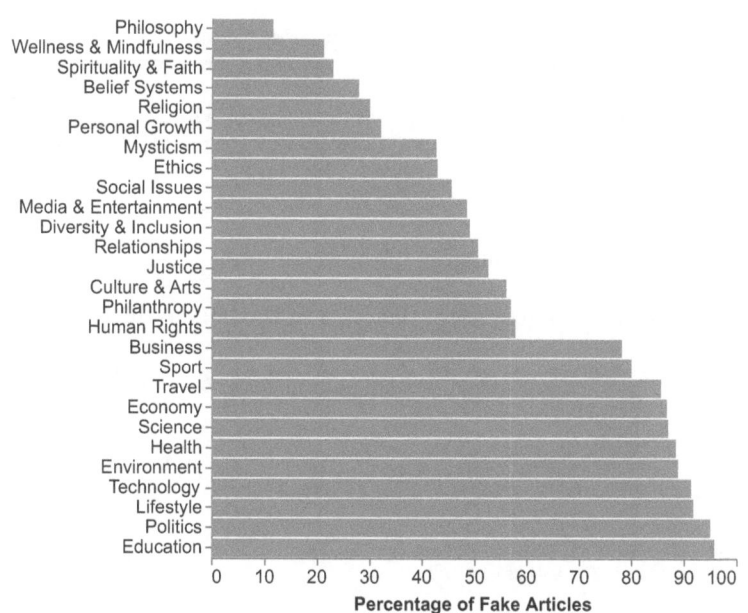

Figure 5.21   The rotation process, flipping the chart in figure 5.19 horizontally

---
Listing 5.8   Using Copilot to generate a list

```
# Build a Python list with the following categories and name it
    material_life:
# Technology
# Environment
# Health
# Science
# Education
# Business
# Lifestyle
# Travel
# Politics
# Economy
# Sport

material_life = ['Technology', 'Environment', 'Health', 'Science',
'Education', 'Business', 'Lifestyle', 'Travel', 'Politics', 'Economy',
'Sport']
```
⟵ **The output produced by Copilot**

NOTE   Describe the elements to include in the list.

Now, ask Copilot to generate the code to add a new column to the DataFrame, containing the macro category. The following listing shows the instructions for Copilot.

Listing 5.9   Using Copilot to add a new column to a DataFrame
---

```
# Add a new column to the dataframe df called Macro Category that contains
    the following values:
# - If the Category is in material_life, then the value is Material Life.
# - If the Category is in moral_life, then the value is Moral Life.
# - If the Category is in spiritual_life, then the value is Spiritual Life.

df['Macro Category'] = df['Category'].apply(lambda x: 'Material Life' if x in
material_life else ('Moral Life' if x in moral_life else 'Spiritual Life'))
```
The output produced by Copilot

NOTE   Describe how to build the new category of the DataFrame.

Next, use the new column Macro Category to set the color of the bars in the chart.

Listing 5.10   Using the new column for the bars' colors

```
chart = alt.Chart(df).mark_bar(
).encode(
    y=alt.Y('Category:N',
            sort='x',
            title=None,
            axis=alt.Axis(labelFontSize=14)
            ),
```

```
x=alt.X('Percentage of Fake Articles:Q',
        axis=alt.Axis(labelFontSize=14,
                      titleFontSize=14),
    ),
    color=alt.Color('Macro Category:N',
                    scale=alt.Scale(
                        range=['#991111', '#f38f8f','lightgray'],
                        domain=['Material Life', 'Moral Life', 'Spiritual Life']
                    ),
                    legend=None
        )
).properties(
    width=400,
    height=400
).transform_calculate(
    'Percentage of Fake Articles', alt.datum['Number of Fake
    Articles']/alt.datum['Number of Articles']*100
)
```

NOTE Add the color channel to set the color of the bars.

Figure 5.22 shows the resulting chart. We have used two tonalities of red to highlight the urgency of paying attention to material and moral life.

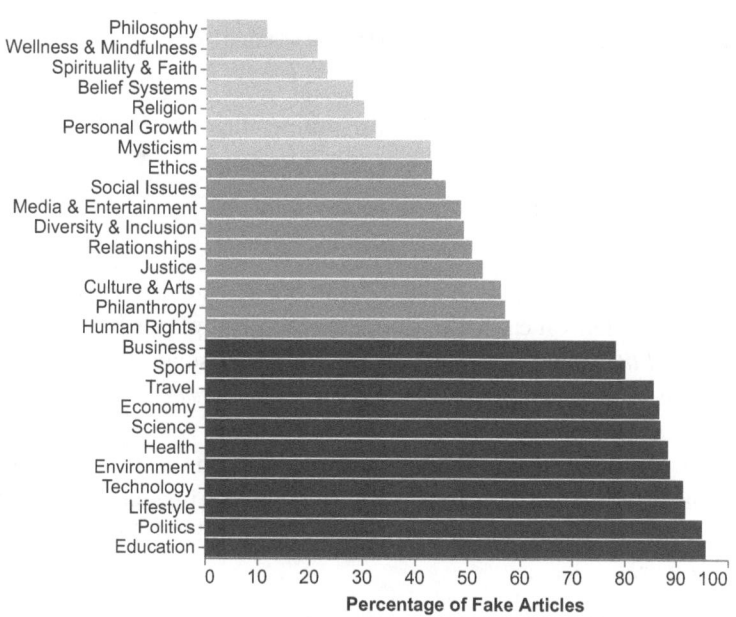

Figure 5.22 The chart of figure 5.21 with macro categories highlighted

We could improve the chart by simplifying it, for example, by grouping categories. However, our audience comprises experts (editors), who need very detailed information because they must know precisely which categories they must analyze more deeply. For this reason, we keep the chart very detailed. We have removed the legend from the chart because we want to replace it with some images, which also act as the context for our story. Let's see how to add these images in the next step: turning information into knowledge.

5.5.2 From information to knowledge

The idea is to add an image for each macro category, more specifically, an icon. We can use DALL-E to generate the three images. Use the following prompt to generate the spiritual life icon: a black-and-white icon with praying hands. DALL-E will generate four images, such as those shown in figure 5.23.

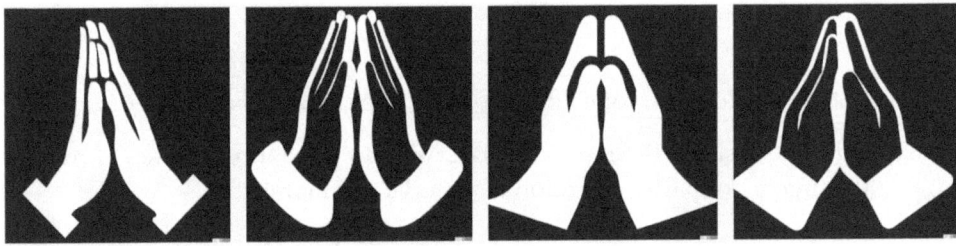

Figure 5.23   The images generated by DALL-E related to the spiritual life

Similarly, you can generate icons related to the other categories. We have used the following prompts:

- *A black and white icon with a balance symbol of moral life*
- *A black and white icon with a circle representing the world*

If you don't have any ideas on how to generate an image, you can use ChatGPT to get some ideas. For example, you can write the following prompt in ChatGPT: *How would you represent material life through an icon?* Among the other suggestions, ChatGPT suggests the following: *A circle to represent the world or universe, with various smaller icons or symbols placed within it to represent different facets of material life.*

Once you have generated the images, you can incorporate them into the chart. The following listing describes how to add the spiritual life image. You can adopt the same strategy also for the other images. Also, remember to load the generated HTML chart in a web server.

> **Listing 5.11   Adding an icon to the chart**

```
spiritual_image = alt.Chart(
    pd.DataFrame({'image_url': ['media/spiritual-life.png']})
).mark_image(
    width=80,
    height=80,
```

```
).encode(
    url='image_url',
    x=alt.value(270),   # pixels from left
    y=alt.value(50)
)                          ◁⎯⏌  Add an image to the chart.
```

> **NOTE** Calibrate the x and y positions manually, based on how the chart appears.

Near the icon, add a text describing the macro category.

Listing 5.12 Adding text to the chart

```
spiritual_text = alt.Chart(
    pd.DataFrame({'text': ['Spiritual Life']})
).mark_text(
    fontSize=30,
    color='black',
    align='center',
    baseline='middle',
    font='Monotype',
    fontStyle='italic'
).encode(
    x=alt.value(420),   # pixels from left
    y=alt.value(50),
    text='text'
)
```

> **NOTE** Calibrate the x and y positions manually, based on how the chart appears.

Finally, combine all the charts.

Listing 5.13 Combining the images, texts, and chart

```
chart = chart + spiritual_image  + spiritual_text + moral_image + moral_text
    + material_image + material_text

chart = chart.configure_axis(
    grid=False
).configure_view(
    strokeWidth=0
)
chart.save('story-chart.html')
```

> **NOTE** Use the + operator to combine all the elements of the chart

Figure 5.24 shows the resulting chart. The icons and the text act as the legend. In addition, they are the characters of our story.

The next step involves adding a textual background to our chart and setting the context. We can add it as the subtitle of our chart. Let's use ChatGPT to generate some ideas. Write the following prompt:

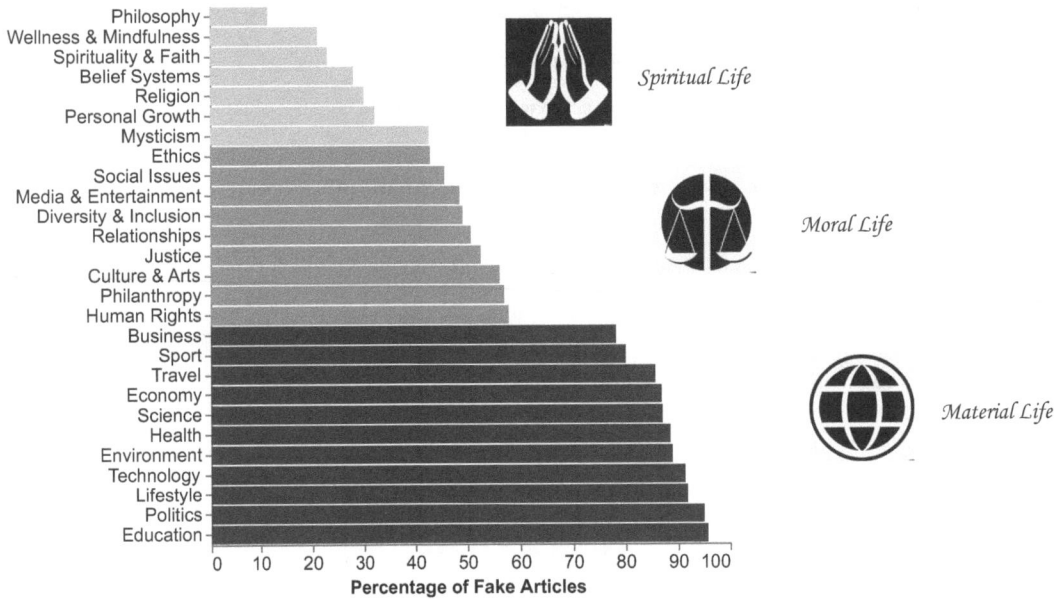

Figure 5.24   Adding the images related to spiritual life, moral life, and material life to the chart

Act as a storyteller. Describe the context of a visual chart showing that the highest number of fake news on the LatestNews website is related to material life, followed by moral life and finally by spiritual life. Use max 150 characters.

We have specified three main elements:

- The ChatGPT role (we'll see this in the next chapter in more detail)
- The scenario
- The maximum number of characters to use

ChatGPT may produce the following output: *In the intriguing chart, depicting fake news trends on the LatestNews website, material life claims the most prominent slice, trailed by moral life, and lastly, spiritual life.* We can use it as it is, we can modify it, or we can even ask for a new subtitle. In our case, we'll use this output as the context of our chart.

Let's add this to our chart, by simply setting the subtitle in the chart properties.

Listing 5.14   Adding a subtitle to the chart

```
chart = chart.properties(width=500,title=alt.TitleParams(
    subtitle=['The LatestNews website is a popular source of news and
      information, but it is also a source of fake news.'],
    subtitleFontSize=18,
    fontSize=30,
    offset=40
))
```

NOTE   Use `TitleParams()` to specify the title and subtitle parameters.

Now that we have turned information into knowledge, let's proceed with the last step: turning knowledge into wisdom.

5.5.3   *From knowledge to wisdom*

Wisdom involves adding a call to action to the story. In our case, we can simply say to the audience to pay attention to material and moral news because they have a high percentage of fake news. Let's use ChatGPT to generate an engaging title. Use the following prompt: *Transform the following text into an engaging title: pay attention to material and moral news because they have a high percentage of fake news. Generate 5 titles.* Figure 5.25 shows a possible output generated by ChatGPT.

1. "Navigating the News Landscape: Unmasking the Deceptive Veil in Material and Moral Stories"
2. "Decoding Truth: Why Vigilance Matters in Discerning Fact from Fiction in Material and Moral News"
3. "Separating Fact from Fiction: The Crucial Need to Scrutinize Material and Moral News"
4. "Truth Under Siege: Exploring the Surge of Fake News in Material and Moral Narratives"
5. "Guardians of Truth: Unveiling the Shadow of Fake News Lurking in Material and Moral Discourse"

Figure 5.25   A possible output generated by ChatGPT

The generated titles do not contain a call to action. To add it, let's improve the title by writing the following new prompt for ChatGPT in the same chat: *Transform the titles into a call to action that invites the audience to pay attention to news related to material and moral life.* Figure 5.26 shows a possible output produced by ChatGPT.

1. "Empower Your Judgement: Unmasking Deception in Material and Moral News – Take Action Today!"
2. "Elevate Your News Literacy: Act Now to Discern Truth in Material and Moral Stories!"
3. "Your Truth Guardian: Take a Stand Against Fake News in Material and Moral Narratives!"
4. "Defend Reality: Empower Yourself to Detect Fake News in Material and Moral Discourse!"
5. "Join the Truth Crusade: Be a Savvy Consumer of Material and Moral News – Start Now!"

Figure 5.26   A possible output generated by ChatGPT when asked to add a call to action

As an example, select the third title, *Your Truth Guardian: Take a Stand Against Fake News in Material and Moral Narratives*, and add it to the chart, simply setting the text parameter, as shown in the following listing. Figure 5.27 shows the final chart.

Listing 5.15   Adding a title to the chart

```
chart = chart.properties(width=500,title=alt.TitleParams(
    text=['Your Truth Guardian:', 'Take a Stand Against Fake News in Material
    and
➥ Moral Narratives'],
    subtitle=['The LatestNews website is a popular source of news and
    information, but it
➥ is also a source of fake news.'],
    subtitleFontSize=18,
    fontSize=30,
    offset=40)
)
```

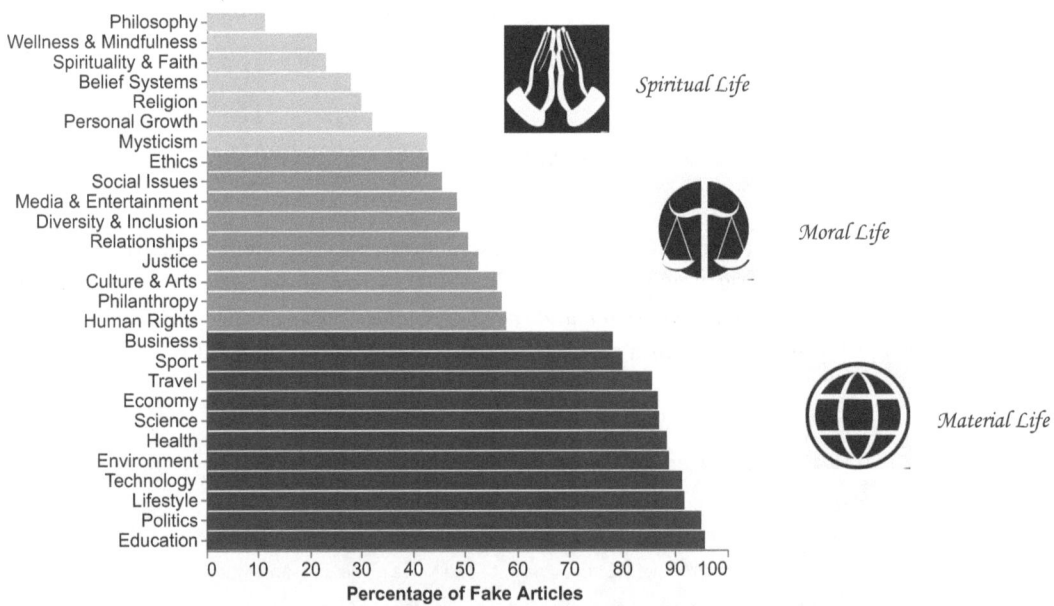

Your Truth Guardian:
Take a Stand Against Fake News in Material and Moral Narratives
The LatestNews website is a popular source of news and information, but it is also a source of fake news.

Figure 5.27   The final chart

NOTE   Use the text parameter to add a title to the chart.

In this chapter, you have learned the main concepts of building a data story using the DIKW pyramid and how to incorporate generative AI tools in the flow. We've had a high-level overview of implementing the DIKW pyramid as part of our data story as well as incorporating generative AI tools to enhance things further. In the next few chapters, we will look at traversing the DIKW pyramid in greater detail, starting by moving from *data* to *information.*

Summary

- A *story* is a narrative that engages its audience to communicate a message and move them to action.
- The *data storytelling arc* identifies five moments in a story: background, raising interest, main point, audience thinking, and next steps.
- Use the DIKW pyramid to implement the data storytelling arc practically.
- Use generative AI to assist you when scaling the DIKW pyramid, using prompts to produce what you want (text or images).

References

- Dykes, B. (2019). *Effective Data Storytelling: How to Drive Change with Data, Narrative and Visuals.* John Wiley & Sons.
- Cambridge Dictionary. (n.d.). *Story.* https://dictionary.cambridge.org/dictionary/english/story.

From data
to information:
Extracting insights

6

This chapter covers
- Introducing insights
- Choosing the characters of your story
- Choosing the right chart

An *insight* is something that is significant in data—something that you would want to communicate through a story. Extracting an insight from data is the most difficult task every data analyst must deal with. In the previous chapters, you learned that to turn data into information, you must extract insights from data. You also learned that an insight is something relevant you have found in your data. In this book, we will not focus on how to extract insights from data. Tons of books exist on the topic, so you can refer to them for a detailed description (see, e.g., Guja et al., 2024; De Mauro et al., 2021). In this chapter, we will follow a more intuition-based approach to extract an insight, describing the motivations guiding you while exploring data. I hope this approach will help you to broaden your point of view during your data analysis journey. Thus, the objective in this chapter is to understand that at the base of the DIKW pyramid, there is the extraction of an insight, its transformation into a character of a story, and its representation through a chart. First, we will focus on how to extract insights from data using an intuitive approach. Next, we'll describe how to choose the characters of a data story and how to choose the right chart. The story characters and the right chart help to communicate your insight to an audience. Finally, you'll learn how to implement some of the most popular data visualization charts in Altair. We'll exploit the power of Copilot to speed up the chart construction process and provide the code to implement decluttered charts, ready to be added to a data story.

6.1 *An intuitive approach to extract insights*

In the book *Journey to the Center of the Earth* by Jules Verne, Professor Otto Lidenbrock (1864) meticulously extracts information from a medieval Icelandic manuscript by employing his keen intellect and passion for exploration. With a deep understanding of languages, history, and geology, he deciphers cryptic clues, cross-references ancient texts, and applies scientific knowledge to unravel the secrets concealed within the pages. The manuscript provides details about the journey's route and the geological landmarks they would have encountered along the way. In other words, Professor Otto Lidenbrock turns the data contained in the old manuscript into information.

Similar to the story told by Jules Verne, you can start by turning data into information to tell any data story. In this section, we will describe some strategies that can help you transform your data into information.

In his book, *Seeing What Others Don't* (Klein, 2017), Gary Klein identifies four main strategies to help you to identify an insight:

- Connections
- Coincidences
- Curiosity
- Contradictions

Let's analyze each of the four proposed strategies separately, starting with the first: connections. Keep in mind that you can apply multiple strategies simultaneously.

6.1.1 *Connection strategy*

This strategy involves identifying the main points of your data and then connecting them to identify a story. In his book, Klein says that sometimes you extract an insight when you see a new way to combine different pieces of data. Other times, you can extract an insight simply by looking at the pieces of data you already have in a different way. To explain how the connection strategy works, consider the following scenario. Angelica is a data analyst working for an e-commerce website. One day, Angelica receives a report on customer reviews and ratings for each product on the website. Each product is rated on a scale of 1 to 5, and customers can leave text reviews. Table 6.1 shows the sales dataset Angelica must analyze.

Table 6.1   The sales dataset analyzed by Angelica

product_id	number_of_orders	product_rating	product_category	returns	number_of_reviews
P1001	30	4.5	Electronics	0	3
P1002	12	3.2	Home & Kitchen	6	6
...	...	...	...		...
P1006	24	4.2	Electronics	0	4
P1006	22	2.1	Electronics	18	20

Here is an explanation of the dataset columns:

- `product_id`—The unique identifiers for each product in the dataset.
- `number_of_orders`—The quantity of times a particular product has been ordered.
- `product_rating`—The rating assigned to each product by customers or users. Ratings could be on a scale (e.g., 1 to 5 stars) and reflect the satisfaction or perceived quality of the product.
- `product_category`—This column categorizes products into different groups or types.
- `returns`—The number of times a product was returned by customers. It provides insight into the rate of customer dissatisfaction or issues with products that lead to their return.
- `number_of_reviews`—The count of reviews received for each product.

Angelica starts by plotting the product ratings versus the number of reviews and the product ratings versus percentage of returns (figure 6.1). You can find this example in the GitHub repository for the book under 06/connections.

Angelica discovers the following main points:

- Some products have high ratings but relatively fewer reviews.
- Others have lower ratings but a substantial number of reviews.

Angelica uses her intuition to connect the points and discovers that negative reviews correlate with high return rates. This may hurt the overall customer experience and brand reputation.

The previous example demonstrates that once we have identified the "dots," extracting insights is just about connecting them. However, the main problem with this strategy is identifying the "non-dots," or the irrelevant messages. An example of a non-dot in the previous scenario is the product category versus the product rating.

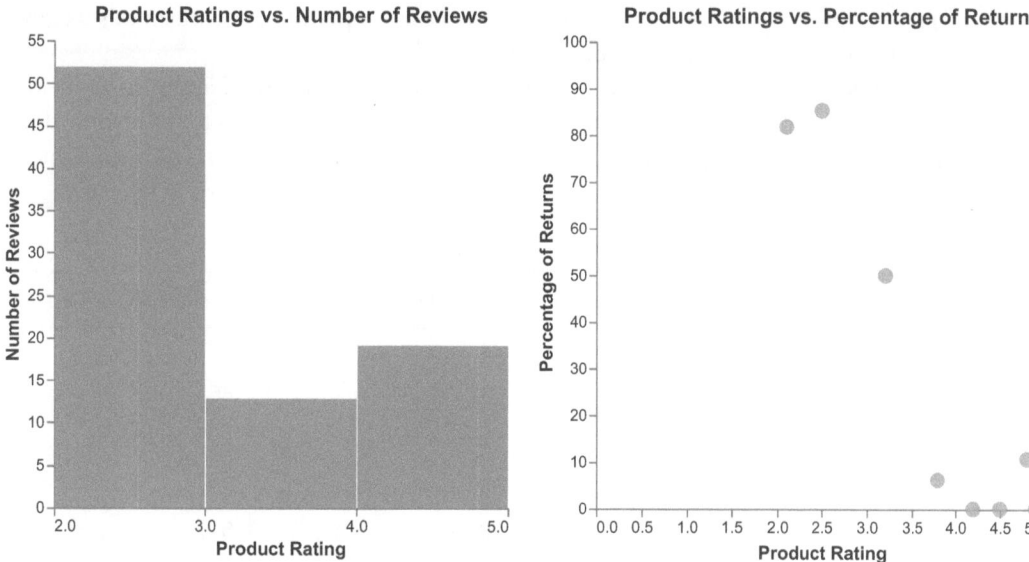

Figure 6.1   The charts drawn by Angelica

6.1.2   Coincidence strategy

A *coincidence* is an occurrence of events that happen at the same time due to randomness but which seem to have some connection—the unexpected convergence of two or more apparently significant situations, even though they may be disconnected. Coincidences can range from everyday events, like meeting someone you know in an unexpected place, to extraordinary events that seem incredibly unlikely. However, while coincidences are intriguing, they do not necessarily indicate anything beyond chance. They could be rare or unusually timed events capturing our attention due to their curious alignment.

In his book, *Seeing What Others Don't,* Gary Klein says, "Observing a coincidence means that we've spotted some events that seem related to each other even though they don't seem to have any obvious causal link." You should take a skeptical approach and consider something a coincidence (and, thus, ignore it) when it is an isolated incident with few data points to evidence a relationship between two variables. On the other hand, when events or trends repeat on multiple occasions, they should be considered less likely to be coincidental and, thus, worth investigating further.

To identify whether a repeated event is a coincidence or not, you could approach data with a deliberately skeptical mindset, ignoring isolated events, while looking for longer-term patterns that would suggest a relationship. The presence of correlated factors could indicate a potential causal relationship. For instance, by analyzing data over time, you might discover that certain variables precede the repeated event, hinting at a cause-and-effect dynamic. Through causality, you could distinguish between

mere coincidences and meaningful connections, explaining the mechanisms govern-
ing the observed phenomenon.

Let's consider the following example to explain how we could apply the coincidence
strategy. Imagine that Angelica works for a store selling accessories for electronic devices,
such as headphones, USB cables, smartphone covers, and so on. Also, imagine that
Angelica wants to monitor her orders. She is given the dataset shown in table 6.2.

Table 6.2   The sales dataset of the electronic store

Date	Headphones	USB Cables	Smartphone Covers
2023-01-01	14	34	32
2023-01-02	17	54	45
...	...	...	...
2023-01-30	16	34	34
2023-01-31	20	23	5

For simplicity, the dataset contains the number of orders for three products during
the month of January 2023: headphones, USB cables, and smartphone covers. Figure
6.2 shows a visual representation of the dataset. You can find the complete code in the
GitHub repository for the book under 06/coincidences.

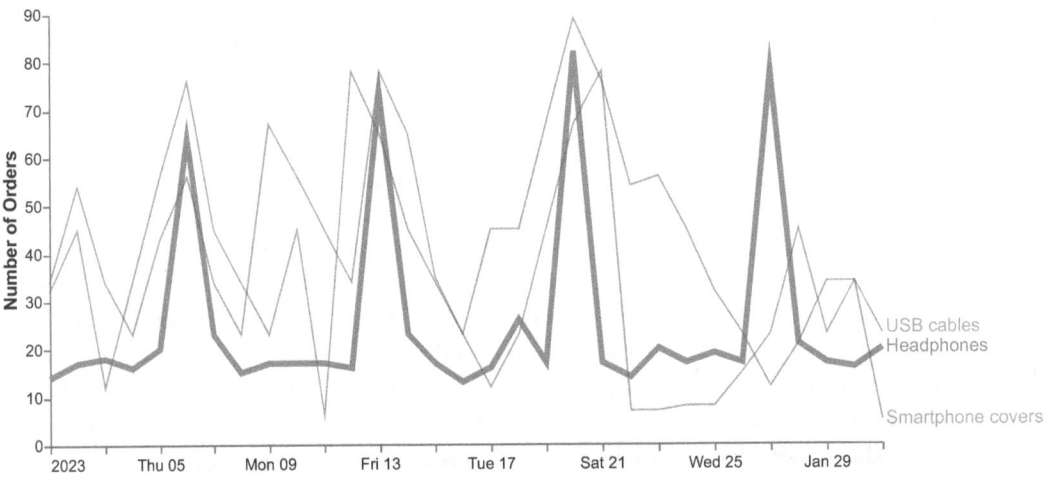

Figure 6.2   The electronic store orders, with a focus on headphones

The figure highlights the headphones orders. It's worth noting that headphone orders
peak every Friday. This could be a coincidence, but the repetitiveness of the event
should be considered. For example, Angelica might find that every Friday, there is an
exciting radio show that causes people to rush to buy new headphones. Or she might

find out that there's a law in her town on Fridays that doesn't allow speakers to be turned on. In any case, this repetitiveness of events requires further analysis.

Looking at the x-axis of figure 6.2, you can notice a strange labeling strategy: first, the year; then the weekdays; and, finally, the month. This is the default Altair labeling for dates. You can read more details on how to format dates in the Altair official documentation (https://mng.bz/1Mmy).

Coincidence insights differ from connection insights in the way we discover them. In connection insights, we focus on important details, while in coincidence insights, we focus on repeated events. In addition, connections are helpful when you have aggregated metrics, while coincidence is best applied to things like raw sales numbers over time.

6.1.3 *Curiosity*

In 1928, Sir Alexander Fleming, a Scottish bacteriologist, discovered penicillin while studying staphylococci bacteria. One day, while he was meticulously examining his petri dishes filled with these microorganisms, he noticed something peculiar. A petri dish left open by mistake had become contaminated with mold. Intrigued by this unexpected occurrence, Fleming examined the petri dish closely. To his astonishment, he observed that the bacteria surrounding the mold were gradually deteriorating. Something unleashed by the mold was defeating them. Driven by curiosity, Fleming embarked on a series of experiments to explore this phenomenon. He carefully isolated the mold and extracted the substance it produced, which he named penicillin. Curiosity prompted Fleming to investigate the problem further, leading him to the discovery of penicillin, the world's first antibiotic, saving countless lives and transforming medicine forever.

Curiosity is a driving force that sparks wonder, propels discovery, and fuels innovation, pushing us to question, seek answers, and expand the boundaries of knowledge. Curiosity is one strategy we can apply to extract insights. Curiosity about data can be triggered by a wide range of phenomena, including outliers, missing data, data gaps across time or space, sudden shifts or trends, unexpected patterns, and much more. When we note something curious in our data, we dig deeper, ask questions, and explore the underlying factors contributing to the observed patterns. This curiosity drives us to analyze the data from different angles, uncover hidden insights, and ultimately gain a deeper understanding of the phenomena.

Let's consider the following example to explain how we could apply the curiosity strategy. Consider again the example of the electronic devices in the previous section. Now, imagine that you have a different trend line for March, as shown in figure 6.3.

The figure shows a peak in sales on March 25. This event may raise some curiosity, which moves us to further investigation. Curiosity could help us during the brainstorming process. For example, it could lead us to search for various factors that could have contributed to this spike, such as marketing efforts, special promotions, or external events. As we proceed in the research, curiosity may spur us to add other factors, such as customer demographics during this period. This investigation will not only satisfy our curiosity but also help us to extract insights.

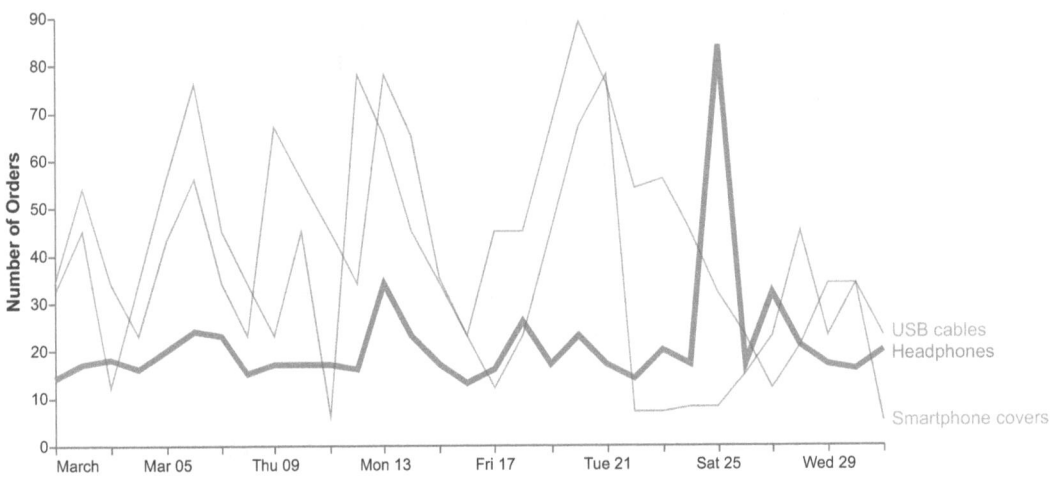

Figure 6.3   The electronic store orders in March, with a focus on headphones

6.1.4   Contradictions

A *contradiction* is a situation when two or more beliefs or pieces of evidence conflict, creating an inconsistency. According to Gary Klein, to extract insight from contradictions, you must approach the data with a skeptical mindset. Unlike a coincidence-based approach, where we look for repeating events, in the case of a contradiction-based strategy, we look for seemingly conflicting events. An insight extracted from contradictions can disrupt conventional thinking, challenge established notions, and provide alternative perspectives that foster creative problem solving. By embracing the inherent tension within contradictions, you can gain valuable insights, leading to breakthroughs, innovation, and growth.

Consider the chart in figure 6.4, showing product sales and prices over time. You can find the code of the produced chart in the GitHub repository for the book under 06/contradictions.

The chart shows that a price increase corresponds to an increase in sales. This event is quite contradictory because the general assumption is that a price increase should correspond to a decrease in sales. However, to understand this phenomenon, you could wear the skepticism lens and analyze this contradictory phenomenon more deeply. One possible motivation you could discover is the perceived value. When prices increase, customers may perceive your product as more exclusive, high quality, or desirable. This perception of increased value can make customers willing to pay more, leading to increased sales. An alternative explanation could be the sense of urgency among consumers who may fear missing out on the product, believing the higher price signifies scarcity or limited availability. Another reason may be inflation. In the context of inflation, when prices rise, people may buy more for fear that the prices may rise even more, making the product unattainable.

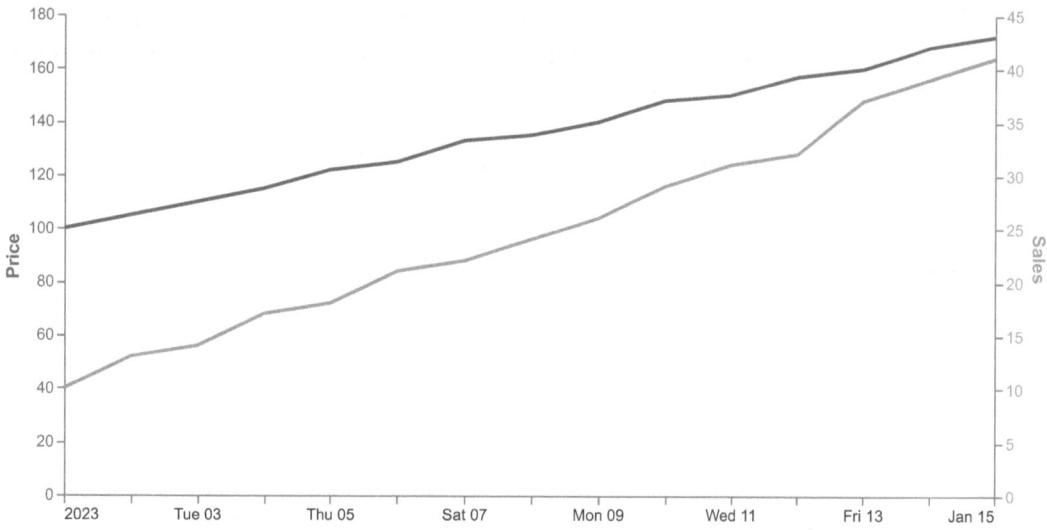

Figure 6.4   A chart showing product sales and price

All the proposed strategies, connections, coincidences, curiosity, and contradictions define some possible techniques to extract meaning. Regardless of which approach you take, always rely on your experience and a deep exploratory analysis phase to extract meaning from data. Now that you have learned how to extract insight from data, let's move on to how to choose the characters of your story.

6.2   Choosing the characters of your story

In the movie *Indiana Jones and the Raiders of the Lost Ark*, Indiana Jones, a brilliant archaeologist, embarks on a thrilling adventure pursuing an ancient artifact, the Ark of the Covenant. Set in the 1930s, Indiana races against time to find the sacred relic before the Nazis seize it for their wicked plans. This movie and the others of the Indiana Jones saga are examples of engaging stories, where the hero wants something.

Every story starts with a hero wanting something. Without any hero and something to achieve, there is no story. The same should be true in a data story. It might seem that the term *hero* is too much to define the subject of our story. However, in our case we really want to emphasize the similarity between a data story and a real story. In any story, there is always a hero who corresponds to the main character. Our idea is to see the subject to be represented just like the hero of our story, even if it is a simple product. Using this point of view, it will also be easier for us to construct a story:

- *Every data story should have a hero.* This could be a product, a customer, or whatever.
- *The hero must want something.* This may be a sales increase, increased satisfaction, or anything else.

Keep in mind that the hero of the story is something other than you. The hero is the main subject of your data story. It depends on the insight you have extracted. Examples of heroes include people, things, and places.

Once you have identified the hero of your story, the next step is asking the audience the following question: *Will the hero get what they want?* In his book, *Building a StoryBrand: Clarify Your Message So Customers Will Listen,* Donald Milligan (2017) says that "before knowing what the hero wants, the audience has little interest in her fate. This is why screenwriters have to define the character's ambition within the first nine or so minutes of a film getting started."

In addition to having a hero, your story should have at least two other characters: the guide and the adversary. The guide, often a wise and experienced mentor, helps the hero achieve their goal. The adversary can be

- A competitor who seeks to hinder or oppose the hero's progress
- A comparison character, which allows you to measure the hero's progress
- The same hero at a different time (e.g., comparing the hero's condition now to a year ago) or in a different space (e.g., comparing the hero here to another location)

Let's review all the examples we have seen in the previous chapters and identify the hero of each story. For convenience, figure 6.5 summarizes the examples analyzed. Try searching for the hero, the type of hero, what the hero wants, and the adversary in each.

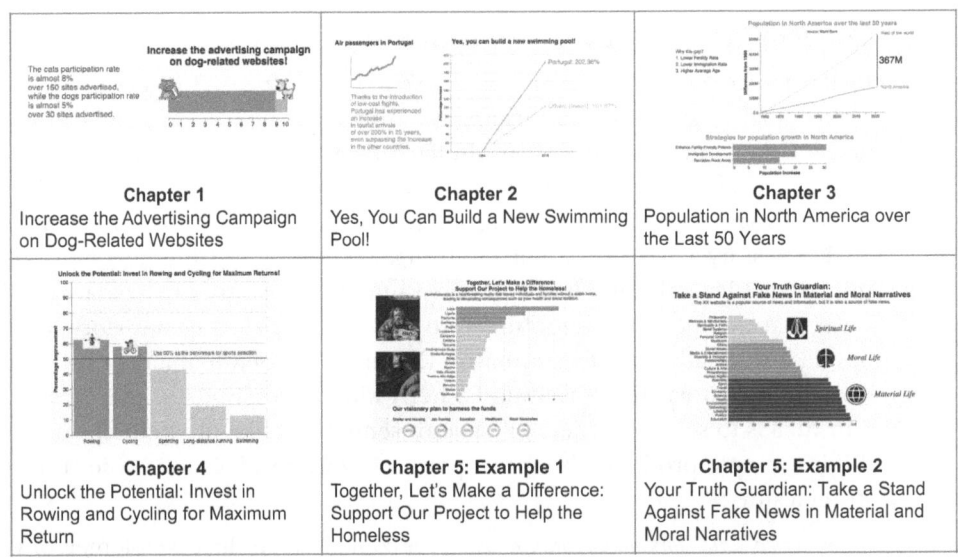

Figure 6.5   A summary of all the examples described in the previous chapters

Table 6.3 shows the hero and the adversary for each case study shown in figure 6.5. For each case study, the table also shows the type of hero and what the hero wants. In the example of chapter 2—*Yes, you can build a new swimming pool!*—there is the implicit assumption that an increase in tourist arrivals corresponds to the possibility of building a new swimming pool. This is, obviously, a simplistic assumption, but we simplified the example just to get started and show how the DIKW pyramid works.

Table 6.3   **The heroes and adversaries of the case studies analyzed in the previous chapters**

Case Study	Hero	Type of Hero	What the Hero Wants	Adversary
Chapter 1: Increase the Advertising Campaign on Dog-Related Websites	Dog-related websites	Thing	Increase the number of views	Cat-related websites
Chapter 2: Yes, You Can Build a New Swimming Pool!	The flow of tourist arrivals to Portugal	Process	An increase in tourist arrivals	The flow of tourist arrivals to Germany
Chapter 3: Population in North America Over the Last 50 Years	Population in North America	People	Explain the trend	The population in the rest of the world
Chapter 5, Example 1: Together, Let's Make a Difference: Support Our Project to Help the Homeless	Number of homeless people in Lazio, Piemonte, Liguria, and Sardegna	People	Reduce this number	Number of homeless people in the other Italian regions
Chapter 5, Example 2: Your Truth Guardian: Take a Stand Against Fake News in Material and Moral Narratives	News about the material and moral life	Thing	Establish whether they are genuine or fake	News about spiritual life

In all the examples shown in table 6.3, the adversary is always a character different from the hero, although it is of the same type as the hero. In some cases, the adversary can be the same hero in a previous condition, such as product sales in the past or within another geographical region. Now that you have learned how to choose the characters of your story, let's move on to choosing the right chart for your data story.

6.3   *Choosing the right chart*

The best chart depends on the insights you want to represent, the message you want to convey, and the audience you are targeting. Figure 6.6 shows how insight, message, and audience relate to each other in a chart. If you only consider the insight, the effect of your chart is to inform.

Figure 6.6   **The relationship between insight, audience, and message**

When telling a story, there are three levels:

- *Inform*—In this level, you describe the insight you have found in your data. No contact is made with the audience.
- *Communicate*—You involve the audience by providing them details about your data, but they are not motivated to do something after listening to your story.
- *Inspire*—Based on your data, you propose a message, which inspires the audience to do something. The audience is highly involved and will retell your story in other contexts.

In other words, if you only consider the insight, the effect of your chart is to inform. If you calibrate your insight to your audience, you can communicate it effectively. Finally, if you add a message to your chart, you can inspire your audience. We will see how to represent insights in the remainder of this chapter, how to calibrate the chart to your target audience in chapter 7, and how to add a message to a chart in chapter 9.

Let's focus on how to represent insights in a chart. The choice of a chart is driven by your specific goals. Different types of charts serve different purposes, based on the nature of the data and the message you want to convey. Table 6.4 describes which chart to use based on the information you want to convey. Consider that the list of described charts is not exhaustive, and you can use your imagination and expertise to build new charts. In addition, you can follow Dr. Abela's chart chooser (2020; https:// mng.bz/DdnE) to select the best chart for you.

Table 6.4   **Suggested charts based on the information to convey**

Information to Convey	Description	Suggested Charts
A single piece of information	A single number representing critical information	■ Big number (BAN) ■ Donut chart ■ 100% stacked bar chart ■ Waffle chart
Parts of a whole	The components contributing to the entirety of a system, object, or concept. The sum of all components must be 100%.	■ Pie chart ■ 100% stacked car chart ■ Multiple waffle chart ■ Donut chart
Comparison among entities	The similarities and differences between multiple entities to establish relationships and distinctions	■ Bar chart ■ Column chart ■ Slope graph ■ Dumbbell chart ■ Table

Table 6.4   Suggested charts based on the information to convey *(continued)*

Information to Convey	Description	Suggested Charts
Trend	The behavior of an entity over the time	■ Line chart ■ Small multiple line chart ■ Stacked area chart ■ Stacked column chart
Outcomes of a survey or a questionnaire	Answers to questions contained in a survey or a questionnaire	■ Stacked bar chart ■ Column chart ■ Multiple bar charts
Distribution	Spread of values across a dataset, indicating how frequently different values occur	■ Histogram ■ Pyramid ■ Box plot
Relationship	Association, connection, or correlation between different entities to identify patterns, trends, and dependencies	■ Scatterplot ■ Bubble chart ■ Heatmap
Spatial information	The behavior of an entity over the space	■ Choropleth map ■ Dot density map ■ Proportional symbol map ■ Heatmap
Flow	Represent a process	■ Sankey (not supported by Altair 5.0.1) ■ Chord (not supported by Altair 5.0.1)

As you can see from the table, you can use the same chart for different purposes. In addition, you can use many different charts for the same purpose. Your choice of chart depends on the audience, as we will see in chapter 7. In the remainder of this section, we'll describe how to implement some of the most important charts in Altair and Copilot, grouped by chart family.

We'll focus on the following chart families:

- Cooking charts
- Bar charts
- Line charts
- Geographical maps
- Dot charts

Let's start with the first family: cooking charts. You can find the code associated with each chart in the GitHub repository for the book under section 06. For each chart family, we will show only one representative chart. You can find more charts in appendix C.

6.3.1   The cooking charts family

The *cooking charts* family includes the pie chart and the donut chart. Use this category of charts only to represent numerical values. Lately, I have seen an unfounded opposition to these types of charts on social media and the web. The main argument is that these graphs are too generic and do not convey the results correctly. I was a victim of this sentiment, and for a while, I avoided this type of chart as well. Then, one fine day, on LinkedIn, I read a post (https://mng.bz/NRpD) by Brent Dykes reevaluating cooking charts, especially pie charts. In his post, Dykes pointed out that pie charts aren't great at representing the precise dimensions of the slices, but they do represent the parts of the whole. Dykes explicitly says, "I know everyone likes rules. 'Never use this chart type …,' 'Always use this chart type for…,' etc. However, it's not always so straightforward in data storytelling. It isn't just about how to display a certain type of data but how to best communicate a specific point to your audience."

Following Dykes's suggestions, you can use cooking charts in three main cases:

- You must represent a part of the whole. The total sum of the slices must be 100%.
- There are a maximum of two to three slices to represent (provided they are not very similar in size; otherwise, it is very difficult to determine which one is bigger than the others).
- You don't want to transmit details about your data but only a general overview. This could be the case for a general audience, not including technical experts.

In the remainder of this section, we will see how to implement a pie chart. For donut charts, refer to appendix C. A *pie chart* is a type of circular data visualization that displays data by dividing a circle into slices, each representing a proportion or percentage of a whole.

Suppose we want to use a pie chart to represent a value of 70%. First, create a pandas DataFrame containing your number and its complementary value (30%).

Listing 6.1   The DataFrame with the number

```
import pandas as pd
import altair as alt

data = {
    'percentage': [0.7,0.3],
    'label'     : ['70%','30%'],          ← The labels to show
    'color'     : ['#81c01e','lightgray']   ← Use a neutral color (light gray) for the complementary number.
}
df = pd.DataFrame(data)
```

NOTE   Create a pandas DataFrame with the number to represent and its complementary value.

Next, ask GitHub Copilot to draw the chart for you. A pie chart does not have the classical x and y channels to encode data. Instead, the pie chart uses the Theta and Color channels. The Theta channel specifies the angular position of data points in a polar plot. The Color channel refers to encoding data using different colors.

Listing 6.2 shows the starting prompt for GitHub Copilot.

Listing 6.2   How to generate a basic pie chart in Copilot

```
# Draw a pie chart in Altair with the following options:
# - Use the `percentage` column for theta channel
# - Use the `label` column for tooltip
# - Use the `color` column for color
# Save chart to `chart` variable.
# Save chart as 'pie-chart.html'.
```

NOTE   The basic instructions to draw a pie chart in Copilot.

As a result, Copilot will generate the code shown in the following listing. Figure 6.7 shows the resulting chart.

Listing 6.3   The code to generate a basic pie chart

```
chart = alt.Chart(df).mark_arc(
).encode(
    theta=alt.Theta('percentage', stack=True),       stack=True is not
    color=alt.Color('color', scale=None),            generated directly
    tooltip='label'                                  by Copilot.
).properties(
    width=300,
    height=300
)
chart.save('pie-chart.html')
```

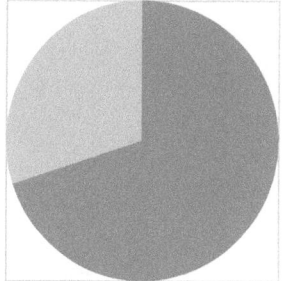

Figure 6.7   A basic pie chart in Altair

NOTE   Listing 6.3 is the code to generate a basic pie chart in Altair. Use the `mark_arc()` method to draw a pie chart. We add the `stack=True` property manually for the next steps. This property means that the individual slices within the chart will be stacked on each other instead of being displayed side by side.

We can improve the chart by adding the label value next to each slice. Ask Copilot to draw a text containing the labels, as shown in the following listing.

Listing 6.4   How to label to the chart in Copilot

```
# Add text near to each slice of the pie chart.
# - Use the `label` column for text channel.
# - Use the `color` column for color.

# Combine the pie chart and the text chart.
# - Use `+` operator to combine the charts.
# - Save the combined chart to `chart` variable.
```

NOTE   This is the prompt for Copilot to generate labels next to the pie chart.

As a result, Copilot will generate some partial code that does not show the desired output. Modify it as described in the following listing. Figure 6.8 shows the resulting chart.

Listing 6.5   The code to generate labels

```
text = chart.mark_text(
    size = 20,
    radius=180
).encode(
    text='label',
    color=alt.Color('color', scale=None)
).properties(
    width=300,
    height=300
)

chart = (chart + text
).configure_view(
    strokeWidth=0
)
```

Copilot uses alt.Chart(df) to draw
the chart. Use chart instead.

Add the size property
to set the font size.

Add radius to set the
distance from the slice.

Use strokeWidth=0
to remove borders
from the chart.

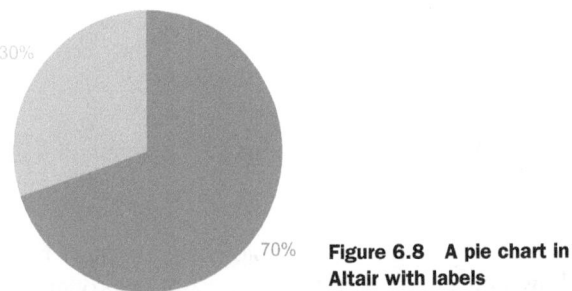

**Figure 6.8   A pie chart in
Altair with labels**

NOTE   Listing 6.5 is the code to generate the labels for the basic pie chart in Altair. Use the `mark_text()` method to write the text.

Now that you have learned how to implement a pie chart, let's move on to the next chart family. In the following section, we'll discuss bar charts.

6.3.2 The bar charts family

The family of *bar charts* includes various types of charts where each data series is visualized as rectangular bars. These bars can be oriented either vertically, known as *column charts*, or horizontally, known as *bar charts*. In addition to the classic bar and column charts, this family also includes stacked column/bar charts, pyramid charts, and histograms. Use these charts to represent categorical data or compare different categories by displaying their corresponding values along a horizontal or vertical axis.

In a way similar to the backlash against pie charts, lately, on the web and in social media, I have noticed an excessive abuse of bar charts. While these charts have the advantages of simplicity and effectiveness in conveying a message, using them repeatedly with the same audience could bore the audience and is likely a reason for the consternation directed at these charts.

You have already learned how to draw a bar chart and a column chart in the previous chapters. In this section, you'll learn how to improve bar chart readability, thanks to Copilot. For the other charts belonging to this family, refer to appendix C.

A bar chart is a type of chart that uses rectangular bars to represent data values, where the length of each bar corresponds to the quantity it represents. Use this chart to represent data across different categories. Consider the dataset shown in table 6.5, describing the number of likes for each type of meal.

Table 6.5 The dataset of the example

Meal Type	Number of Likes
Pizza	120
Burger	95
Pasta	80
Sushi	60
Salad	50
Steak	70
Tacos	90
Ice Cream	110
Curry	40
Sandwich	75

The following listing shows the code to generate a simple bar chart representing the previous dataset in Altair, and figure 6.9 shows the produced chart.

Listing 6.6 The code to create a bar chart

```
import pandas as pd
import altair as alt
```

```
df = pd.read_csv('data/meals.csv')

chart = alt.Chart(df).mark_bar(
    color='#81c01e'
).encode(
    y=alt.Y('Meal Type', sort='-x'),
    x='Number of Likes'
).properties(
    width=300,
    height=300
)

chart.save('bar-chart.html')
```

> **NOTE**   Use the `color` property to set the bar color and the `sort` property of the `y` channel to sort categories based on descending values of x.

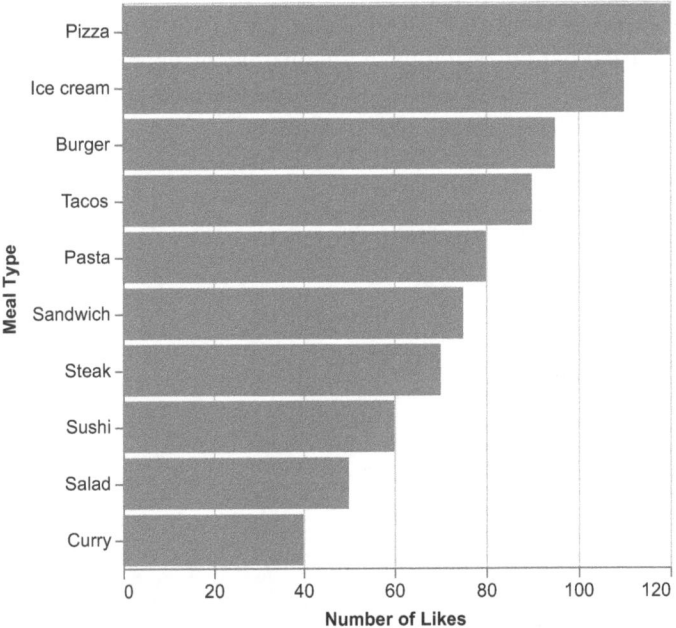

Figure 6.9   If you start writing, Copilot will suggest the next steps.

The chart needs to be decluttered, so we should remove unnecessary elements, such as the y-axis title (Meal Type), the grids, and so on. For example, we can add the value of each bar at its end and remove the x-axis labels entirely. Use Copilot to help you build the chart. As you use Copilot, you will find that it will suggest more and more things that fit your coding style. If you start writing something, Copilot will suggest the next steps based on what you have written previously. This means that you can use Copilot not only with predefined prompts but also to suggest the next steps in your prompts.

    To add the value of each bar at its end, start writing a prompt for Copilot, press Enter, and Copilot will suggest how to implement it, as shown in figure 6.10.

```
# Add text to the end of each bar
# - Use the `Number of Likes` column for x channel
```

Figure 6.10   An example of a Copilot suggestion

Simply press Enter, and Copilot will continue proposing how to add the text. The following listing shows the complete prompt proposed by Copilot.

Listing 6.7   The prompt generated by Copilot

```
# Add text to the end of each bar.
# - Use the `Number of Likes` column for text channel.
# - Use the `Meal Type` column for y channel and sort by the number of likes.
# - Set the color of the text to '#81c01'.
# - Set the text alignment to 'left'.
# - Set the text baseline to 'middle'.
# - Set font size to 14.
```

    NOTE Use Copilot to generate the prompt to build your chart, and then press Enter to let Copilot generate the code.

After the final prompt, press Enter, and Copilot will generate the code.

Listing 6.8   The code to generate labels

```
text = chart.mark_text(
    align='left',
    baseline='middle',
    dx=3,
    color='#81c01',
    fontSize=14
).encode(
    text='Number of Likes',
    y=alt.Y('Meal Type', sort='-x', title=None)
)
```

    NOTE Apply the `mark_text()` method to the chart to generate the labels. You can use the columns already set for the `chart` variable.

Now, add five vertical lines to the chart to ease the audience's comparison process. Ask Copilot to generate the code for you. Start writing the following prompt: `# Add five vertical lines to the chart`. Copilot will generate the prompt shown in the following listing.

Listing 6.9   The prompt to generate vertical lines

```
# Add five vertical lines to the chart.
# - Use alt.Chart(pd.DataFrame({'x': [20, 40, 60, 80, 100]})) to create a
    dataframe with six rows.
# - Use alt.Chart().mark_rule() to draw vertical lines.
# - Set the color of the lines to 'white'.
# - Set the line width to 1.
# - Set opacity to 0.5.
```

NOTE   Copilot generates the prompt to generate the code.

Press Enter, and Copilot will generate the code shown in the following listing.

Listing 6.10   The code to generate vertical lines

```
lines = alt.Chart(pd.DataFrame({'x': [20, 40, 60, 80, 100]})).mark_rule(
    color='white',
    strokeWidth=1,
    opacity=0.5
).encode(
    x='x:Q'
)
```

NOTE   Use the `mark_rule()` mark to add vertical lines to the chart.

Finally, remove the x-axis from the original chart in listing 6.6 (x=alt.X('Number of Likes',axis=None)), remove the title from the y-axis (y=alt.Y('Meal Type', sort='-x', title=None)), and combine the three charts, as shown in the following listing. Figure 6.11 shows the resulting chart.

Listing 6.11   Combining the charts to generate the final chart

```
chart = (chart + text + lines
).configure_view(
    strokeWidth=0
).configure_axis(
    grid=False
)

chart.save('bar-chart.html')
```

NOTE   Use the + operator.

Now you know how to draw a decluttered bar chart in Altair Copilot. Next, let's move on to the line chart family.

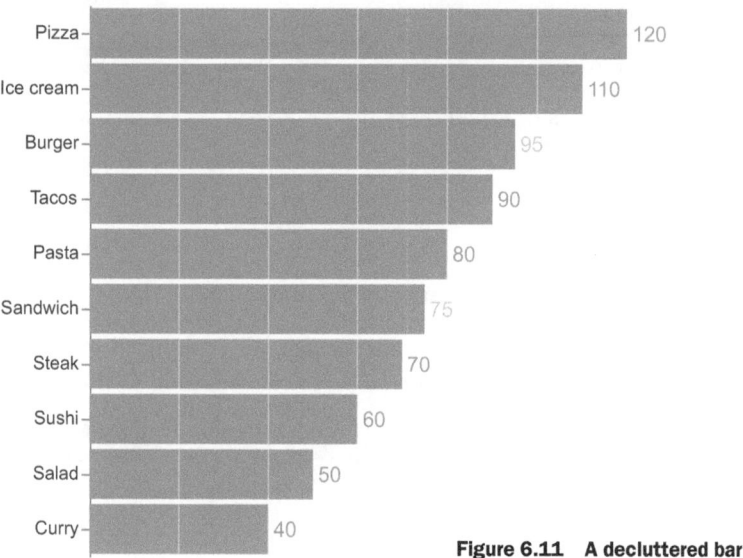

Figure 6.11   A decluttered bar chart

6.3.3   *The line charts family*

The *line charts* family includes all the charts that depict data using continuous lines to show the relationship or trends between different data points over a given time period. Common line chart types include basic line charts, area charts, slope charts, and dumbbell charts. In this section, we will analyze the line chart. Area charts, slope charts, and dumbbell chart are discussed in appendix C.

   Line charts, also known as *line plots* or *line graphs*, visualize data over time. They connect data points by straight lines, which makes it easy to see how the data has changed over time. Line charts are particularly useful for showing trends and patterns in data and comparing multiple data series.

   Imagine that you have a dataset showing the number of orders for pizza and spaghetti for each month of the year. The dataset contains three columns: Month, Meal Type, and Number of Orders. In her book, *Data Storytelling with Data* (Knaflic, 2015), Cole Nussbaumer Knaflic proposes to replace the legend from a line chart with labels near each line. To follow Knaflic's suggestion, we can build a line chart in Altair in three steps. First, we build a base chart, with the basic structure. It includes the general encodings and properties, as shown in the following listing.

Listing 6.12   How to build the base chart of a line chart

```
base = alt.Chart(df).encode(
    x=alt.X('Month',
            axis=alt.Axis(title=None,
                          labelAngle=0,
            ),
```

```
        sort=months
    ),
    y=alt.Y('Number of Orders'),
    color=alt.Color('Meal
     Type',scale=alt.Scale(range=['#81c01e','gray']),legend=None)
).properties(
    width=600,
    height=300
)
```

NOTE   To build a base chart, don't specify the mark.

Next, we draw the line chart by applying the `mark_line()` method to the base chart. Finally, we draw the labels by applying the `mark_text()` method to the base chart.

Listing 6.13   How to build the line chart and the labels

```
chart = base.mark_line()

text = base.mark_text(
    fontSize=14,
    baseline='middle',
    align='left',
    dx=10
).encode(
    text=alt.Text('Meal Type:N'),
).transform_filter(
    alt.datum['Month'] == 'December'
)

chart = chart + text              ◁──────┐   Combines the line
                                         │   chart and text mark
chart = chart.configure_view(            │   into a single chart
    strokeWidth=0
).configure_axis(
    grid=False
)
chart.save('line-chart.html')
```

NOTE   After defining the base chart, build the line chart by only specifying the `mark_line()` method. Use the `transform_filter()` method to select only some data. In our case, select only the last values (`alt.datum['Month'] == 'December'`) of the dataset to position the labels at the end of the lines.

Figure 6.12 shows the resulting chart. Now it's time to move to the next chart family: geographical maps.

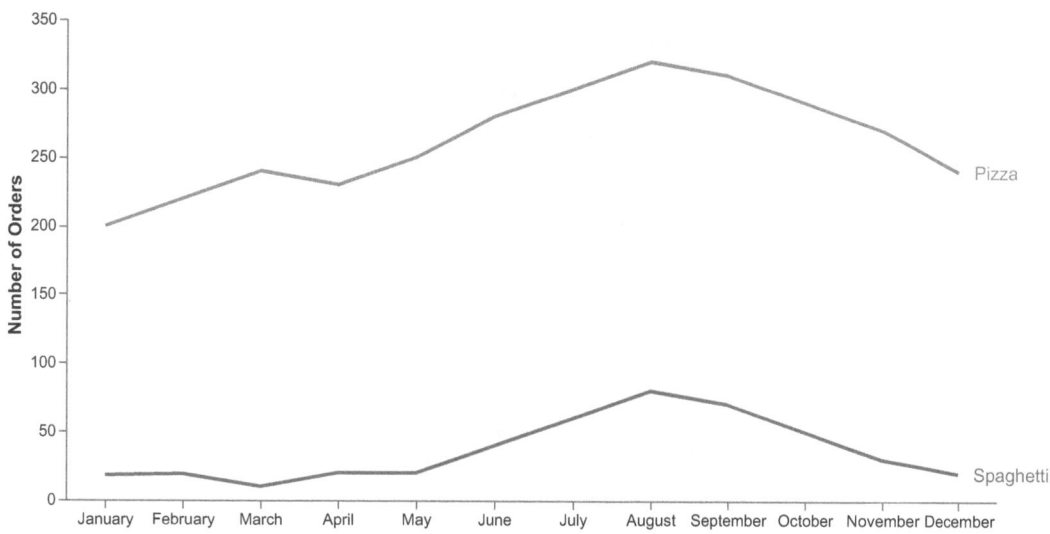

Figure 6.12 A line chart

6.3.4 *The geographical map family*

A geographical map shows the locations of different data points on the Earth's surface, such as countries, continents, oceans, and rivers. A geographical map also shows the distribution of resources and data on a given area of interest.

The most popular geographical maps include the following:

- *Choropleth maps*—These use color to show differences in values between different areas (figure 6.13).
- *Dot density maps*—These use dots to show the concentration of a certain value within an area (figure 6.14).
- *Proportional symbol maps*—These use symbols sized according to the value they represent (figure 6.15).

You can find the code associated with each type of geographical map in the GitHub repository for the book.

When you build a geographical map, you must set the projection to use. One of the most popular projection maps is the Mercator projection map. Although this map is very popular, it does not represent the world correctly. For example, on the map, Greenland appears to be nearly the same size as Africa, when in reality, Africa is about 14 times larger! This distortion occurs because it is difficult to accurately represent the curved surface of the Earth on a flat map. Now that you have learned some of the most popular geographical maps, let's move on to the dot charts family.

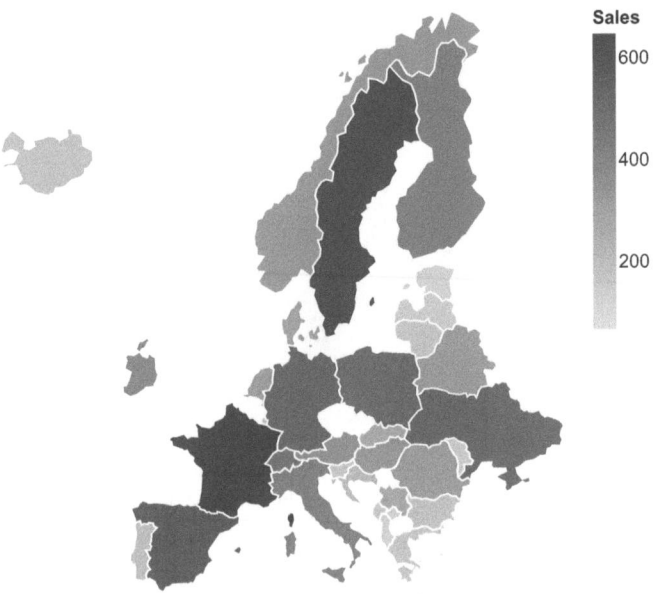

Figure 6.13   A choropleth map

Figure 6.14   A dot density map

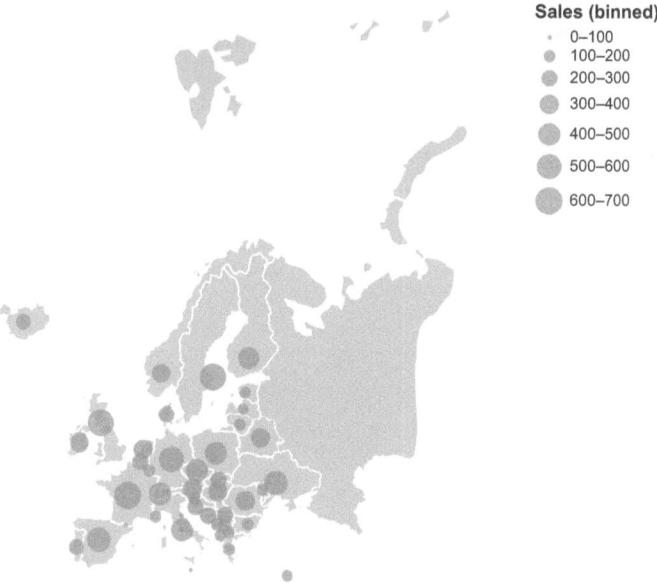

Figure 6.15   A proportional symbol map

6.3.5   Dot charts family

The *dot charts* family includes charts representing data points with dots along the x and y scales to show the relationship between two variables. Some of the most popular dot chart families include scatter plots and bubble charts. Use the `mark_point()` mark to draw a dot chart.

Scatter plots visualize the relationship between two numerical variables. Use them to identify patterns and trends in data and make predictions based on those trends. Figure 6.16 shows an example of a scatter plot. You can find the associated code in the GitHub repository for the book.

Bubble charts are scatter plots that use the size of the data points to encode an additional variable. In Altair, you can use the `size` channel to set the bubble size: `size=alt.Size('Ranking:Q',scale=alt.Scale(range=[1,200]),legend=None)`. Figure 6.17 shows an example of a bubble chart. Now that you have learned the basic chart families, let's move on to implementing a practical case study.

6.4   Case study: Salmon aquaculture

Imagine that you want to study safety in the salmon aquaculture industry in the United States. You have the aquaculture dataset (https://data.world/agriculture/aquaculture-data), provided by the US Department of Agriculture, Economic Research

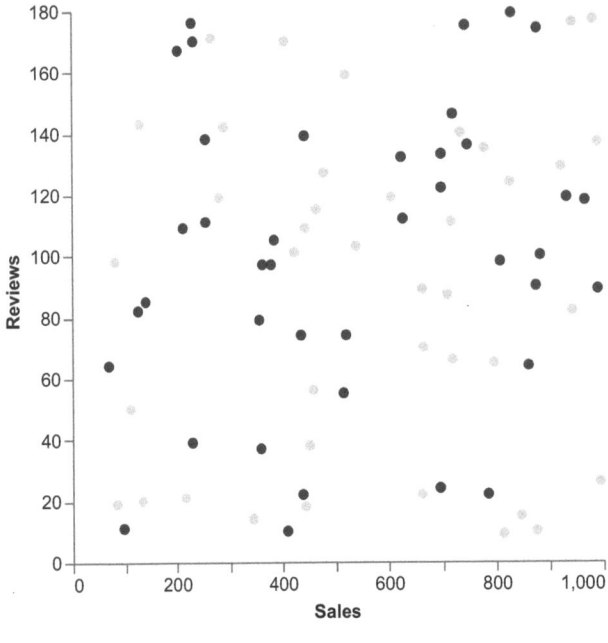

Figure 6.16   A scatter plot

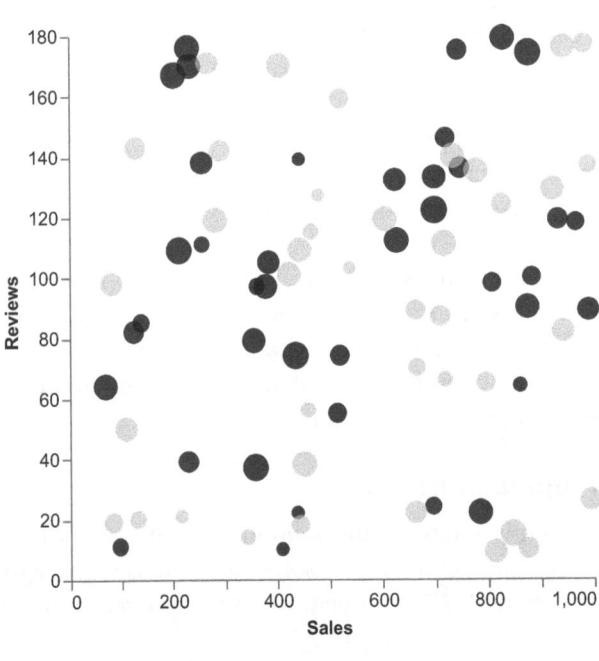

Figure 6.17   A bubble chart

Service (available under the CC-0 license). This dataset describes the exported kilograms and the dollars earned over time for each fish category, as shown in table 6.6.

Table 6.6   A snapshot of the aquaculture dataset

SOURCE_ID	HS_CODE	COMMODITY_DESC	GEOGRAPHY_CODE	GEOGRAPHY_DESC	ATTRIBUTE_DESC	UNIT_DESC	YEAR_ID	TIMEPERIOD_ID	AMOUNT
63	302110000	TROUT (SALMO TRUTTA, S. CLARKI ETC) FRESH, CHILLED	1	World	US Export, QTY	KG	1989	2	17,183
63	1604112000	SALMON, WHOLE/PIECES, IN OIL IN AIRTIGHT CONTAINER	2740	Trinidad and Tobago	US Export, VLU	U.S.$	1989	2	4,329

The dataset contains the following columns:

- `SOURCE_ID`—The source of the data
- `HS_CODE`—The Harmonized System code for the commodity
- `COMMODITY_DESC`—The description of the commodity
- `GEOGRAPHY_CODE`—The code for the country or region of origin or destination
- `GEOGRAPHY_DESC`—The name of the country or region of origin or destination
- `ATTRIBUTE_DESC`—The type of data, either US Export, QTY, or US Export, VLU
- `UNIT_DESC`—The unit of measurement, either KG or U.S.$
- `YEAR_ID`—The year of the data
- `TIMEPERIOD_ID`—The month of the data
- `AMOUNT`—The value of the data

The objective of your study is to establish whether the safety measures adopted so far in salmon aquaculture are sufficient or should be improved. For simplicity, you base your study on this dataset. However, in a real situation, you should also consider other aspects, such as the costs to maintain these measures and other similar analyses.

You decide to plot the salmon aquaculture sales trend line versus the other types of aquaculture. You can find the code associated with this case study in the GitHub repository for the book under CaseStudies/aquaculture/. Ask Copilot to generate the chart for you. Use the list of instructions described in the following listing to build the base chart.

Listing 6.14   The instructions for Copilot

```
# import required libraries
# load the dataset '../source/Aquaculture_Exports.csv' as a pandas dataframe
# apply the following filters to the dataframe:
#   - select only the rows where the 'GEOGRAPHY_DESC' column is 'World'
#   - select only the rows where the 'UNIT_DESC' column is 'U.S.$'
# add a new column to the dataframe called 'DATE' which is a date object
    build as follows:
#   - the year is the 'YEAR_ID' column
#   - the month is the 'TIMEPERIOD_ID' column
#   - the day is 1
# plot the dateframe using altair as follows:
#   - the x axis is the 'DATE' column
#   - the y axis is the 'AMOUNT' column
#   - the color is the 'COMMODITY_DESC' column
# save the plot as 'chart.html'
```

NOTE   After importing the required libraries, ask Copilot to import the dataset and apply some filters to select only the world exportations, in terms of dollars. Next, calculate the date from the YEAR_ID and the TIMEPEROID_ID column. Finally, plot the trend line in Altair.

The previous instructions help you build the basic chart. Work on it to produce the chart shown in figure 6.18.

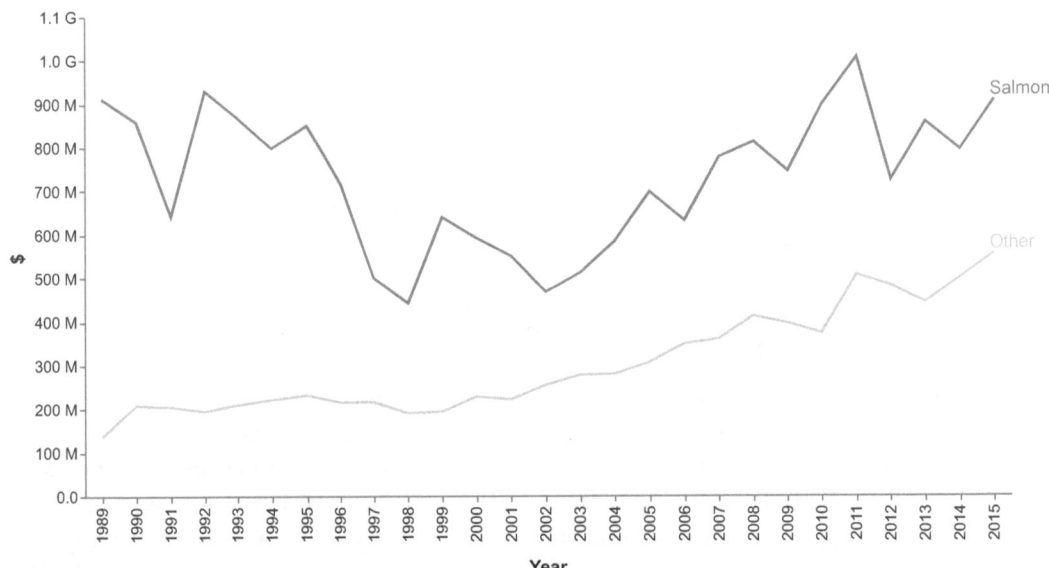

Figure 6.18   The aquaculture case study

You can find the resulting code in the GitHub repository for the book under Case-Studies/aquaculture/from-data-to-information/chart.py. As an insight, you discover that since 1998, there has been an increase in sales, following a period of decrease in sales from 1992 to 1998. Moved by curiosity on the decreasing period, you do some research and discover that the decreasing period was partially due to some health problems in the salmon aquaculture.

At the end of this first step, you have turned data into information. In the next chart, we will continue implementing this case study by turning the extracted information into knowledge.

This chapter has described how to turn data into information. First, we have described some techniques to extract insights. Next, we have seen how to select the characters of a story. Finally, we have learned how to use Copilot to implement some of the most popular charts in Altair. You can find other charts in the GitHub repository for the book under section 06/other-charts. In the next chapter, you will learn how to turn information into knowledge.

Summary

- Turning data into information means extracting meaning from data, an insight that is significant to you.
- Some of the techniques to extract insights include connections, coincidences, curiosity, and contradictions.
- Connections involve searching for details in your data and connecting them.
- Coincidences involve searching for repeated or random events that seem to have some connection.
- Curiosity is searching for strange events, such as missing values and anomalies.
- Contradiction is searching for events that apparently conflict with each other.
- Every story should have at least three characters: the hero, the guide, and the adversary.
- The right chart to tell your story depends on the information to convey.
- There are different families of charts, such as bar charts, line charts, geographical charts, and cooking charts.

References

- De Mauro, A., Marzoni, F., and Walter, A. J. (2021). *Data Analytics Made Easy.* Packt Ltd.
- Dorfer, T. A. (2022). *The Case Against the Pie Chart.* https://towardsdatascience.com/the-case-against-the-pie-chart-43f4c3fccc6.
- Guja, A., Siwiak, M., and Siwiak, M. (2024). *Generative AI for Data Analytics.* Manning Publications.
- Khalil, M. (2024). *Effective Data Analysis: Hard and Soft Skills.* Manning Publications.

- Klein, G. (2017). *Seeing What Others Don't: The Remarkable Ways We Gain Insights.* Nicholas Brealey Publishing.
- Miller, D. (2017). *Building a StoryBrand: Clarify Your Message So Customers Will Listen.* Thomas Nelson.
- Moses, B., Gavish, L., and Vorwerck, M. (2022). *Data Quality Fundamentals: A Practitioner's Guide to Building Trustworthy Data Pipelines.* O'Reilly Media.
- Nussbaumer Knaflic, C. (2015) *Storytelling with Data: A Data Visualization Guide for Business Professionals.* Wiley.

From information to knowledge: Building textual context

This chapter covers

- Introducing context
- Calibrating the story to the audience
- Using ChatGPT for commentaries and annotations
- Using large language models for textual context
- A case study: From information to knowledge (part 1)

Talking about knowledge in a computer science book might seem completely out of place. The word *knowledge* could inspire philosophical concepts or even intimidate. But in this chapter (and the next), we will not be talking about philosophical knowledge but, rather, the knowledge that helps the reader understand the context of a story. It is, therefore, knowledge applied to the context of our data story, rather than general knowledge. In these chapters, we will review the basic concepts behind context in a data story and how to adapt it based on the audience. First, we will focus on textual context in this chapter, and in the next one, we will cover images. We will introduce large language models (LLMs) and use ChatGPT as an example of LLM implementation for data storytelling. Finally, we will explore a practical example.

7.1   *Introducing context*

When I was a child, I often heard my parents discussing some topic and did not understand anything. Their words rang in my ears as meaningless until, eager to understand what they were talking about, I entered the conversation and asked for explanations. Then, my father or mother, very patiently, explained to me what they were talking about, adapting their adult reasoning to my child's mind so that I, too, could understand. Years later, I found myself in the same situation as a mother. My children often ask me to explain more complex speech *in words they can understand.* And the satisfaction is enormous when I see their faces light up and understand what I'm saying.

The examples described show us the need to adapt the words we use according to the audience we are addressing. If we ignore who will receive our story, we risk talking in a way that may make perfect sense to ourselves but which excludes our audience from the message we want to communicate.

In the previous chapter, we looked at how to extract and represent an insight through a chart. The next step is to enrich the chart with context (text and images), making reading easier for the reference audience. *Context* refers to the surrounding elements allowing the audience to understand the displayed information, such as texts, images, and symbols. Data context should prepare the scene of your data story and raise interest in your audience. In this chapter, we'll primarily be dealing with textual context, while in the next chapter, we'll look more at visual context.

Context depends on the type of audience you are addressing. For example, if you are talking with an adult about how much you paid for a product, you don't need to explain how money works. On the other hand, if you are talking to your kids about the same topic, you probably need to explain the denominations of the different banknotes and how the monetary system works.

You can use generative AI tools, such as ChatGPT for text and DALL-E for images, to ease context building. You have already learned the basic techniques for building context using generative AI tools. This chapter will focus on more advanced techniques to write an impactful context tailored to your audience.

We will consider the following types of context:

- *Commentary*—The text that precedes your insight. It includes the background that helps the audience to set the scene and understand the insight. In the example of the product cost explained to your kids, the commentary includes banknotes denominations, and how the monetary system works.
- *Annotation*—A short text that explains a detail of your chart, for example, an anomalous point or a trend line. Consider adding annotations only when necessary. Don't overload your chart with unnecessary annotations.
- *Image*—A picture enforcing the commentary or the annotation. In the example of the product cost, you could add banknote images to help your kids understand the different denominations.

- *Symbols*—Arrows, circles, lines, and so on, combined with annotations. They help the audience focus on particular points of your chart.

In the remainder of this chapter, we will use ChatGPT for commentaries and annotations. In the next chapter, we will focus on DALL-E for images and symbols. In addition, we will introduce LLMs and how to use them for commentaries and annotations. But first, let's describe how to calibrate the story to our audience.

7.2 Calibrating the story to the audience

A few years ago, I was invited to give a seminar to master's students. The seminar topic concerned the implementation of web applications for the construction of data journalism projects. Unfortunately, I found myself faced with a somewhat embarrassing situation. My seminar topic was very technical, even commenting on some pieces of code. As I began to speak, I realized that the audience couldn't follow me because they didn't have the technical skills required to properly understand. My presentation was technically correct, but having spoken too technically to a non-technical audience, the result of my talk was that the audience learned very little. The experience I gained from that episode taught me to always learn about the audience I will be addressing before communicating any message.

The *audience* is the person or the group of persons reading your data story. Understanding the target audience is crucial to building data stories that convey information effectively. In the previous chapter, we saw that you can use multiple types of charts to convey information (table 6.4). Once you've chosen the set of charts that answer your question, you can refine your choice, tailoring the chart to your audience.

In chapter 4, you learned that there are different types of audiences. For simplicity, in this chapter, we group them into three common types of audiences:

- General public
- Executives
- Professionals

Let's investigate each type of audience separately. To explain how you can calibrate the chart to the target audience, we will use the case study described in chapter 4. The objective of this case study was to understand which athletic disciplines our hypothetical team needed to continue to train to achieve the best possible results in the upcoming competitions. For convenience, figure 7.1 shows the complete data story we implemented: moving from data to wisdom.

7.2.1 General public

This audience includes individuals from various backgrounds and levels of knowledge. They may have little to no previous knowledge of your topic. When crafting data stories for the general public, use precise language, avoid overwhelming them with too much information, and focus on presenting the most relevant insights visually and engagingly. The general public could find the chart shown in figure 7.1 complex, with

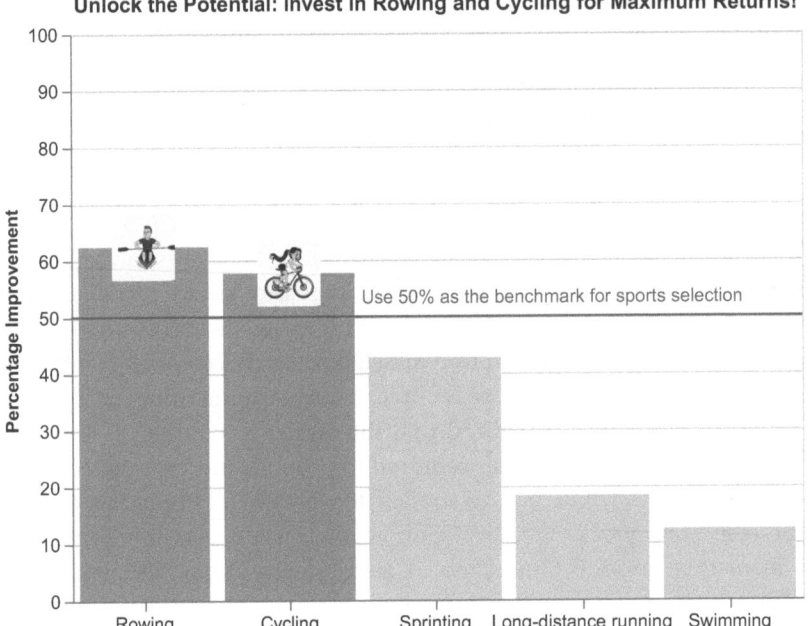

Figure 7.1   The use case described in chapter 4

an unnecessary baseline. As an alternative to the chart in figure 7.1, you could draw the chart shown in figure 7.2, which some audiences may find more appealing.

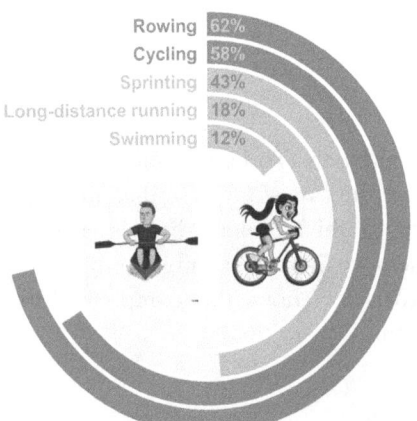

Figure 7.2   The use case adapted to the general public

This chart is called a multi-layer donut chart. We could have placed the images close to the relevant bars, but in this case, there wasn't enough space, so we placed them in the center of the chart. In other scenarios, you might consider placing images next to the bars. You can find the complete code to generate this chart in the GitHub repository for the book under 07/general-public.

7.2.2   Executives

Executives are typically high-level decision makers in organizations, who rely on data-driven insights to make essential business choices. They often have limited time and need concise and actionable information. When creating data stories for executives, it is essential to present key findings, trends, and recommendations up front.

Use visualizations highlighting the most critical data points and providing a straightforward narrative linking the data to strategic goals. It can also be helpful to provide additional context or industry benchmarks to support your analysis. The chart shown in figure 7.1 could be great for executives because it does not contain many details and describes why we chose some sports, thanks to its baseline of 50%.

7.2.3   Professionals

This audience consists of individuals with a specific domain expertise or professional background. They have a deeper understanding of data and require more analytical information. When creating data stories for professionals, explain the methodology, assumptions, and limitations of the data analysis. Consider including additional supporting data and references, allowing professionals to explore the data further.

As an alternative to the chart in figure 7.1, you could draw the chart shown in figure 7.3, which some audiences may understand easily. The figure shows only the chart, without any annotation or context. You can find the complete code to generate this chart in the GitHub repository for the book under 07/professionals.

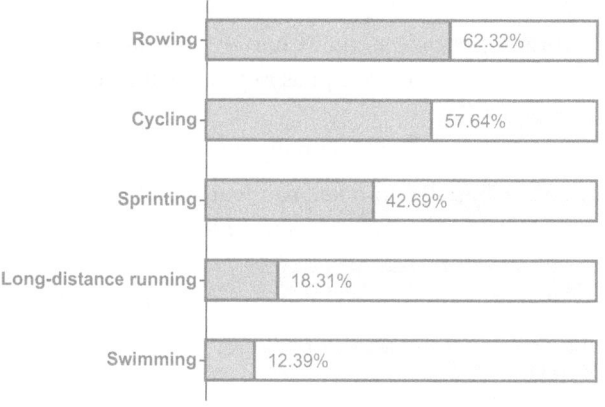

Figure 7.3   The use case adapted to professionals

Table 7.1 summarizes what to represent in a chart based on the audience type. Now that you have learned how to adapt your chart based on the audience type, let's move on to the next step: using ChatGPT for commentaries and annotations.

Table 7.1 **What to represent in a chart based on the audience type**

Audience Type	Requirements	What to Represent
General public	Understand data	An appealing overview of insights
Executives	High-level overview of data trends to aid strategic decision making	Highlight critical metrics and trends influencing business outcomes.
Professionals	Detailed insights to understand the phenomenon behind data	Add numbers, statistics, and useful information to understand insights deeply.

7.3   *Using ChatGPT for commentaries and annotations*

In his novella, *Metamorphosis*, Franz Kafka tells the story of Gregor Samsa, a traveling salesman who wakes up one morning transformed into a giant insect. Encased in this insect's guise, Samsa cannot interact with his family or communicate his thoughts. Gregor's family struggles to accept his transformation, leading to their relationship with Gregor deteriorating and Gregor becoming increasingly isolated. The novella unearths the fundamental isolation that emerges when one's inner world remains inaccessible to others. Data analysts could find themselves in a situation quite similar to that experienced by Gregor Samsa in Kafka's novella when they have to add text to a data visualization chart. The data analyst, by nature, is a technician and could encounter some difficulties in writing engaging text.

ChatGPT can assist you in adding textual context to your data visualization chart. You have already learned that a prompt's basic structure for ChatGPT comprises three main elements: role, audience, and task.

For example, you can write *Act as an entertainer* (role), *writing for decision makers* (audience). *Write 5 titles about <topic>* (task). The topic could be whatever you want. The main problem is structuring the topic so that ChatGPT produces the correct context. To include the topic in the ChatGPT prompt as well, we will generate context following the schema shown in figure 7.4.

In a prompt, we specify the following four main elements:

- *Role*—The role you want ChatGPT to take. You learned about many role types in chapter 4, including entertainer, educator, informer, inspirer, inviter to action, and relationship builder.
- *Audience*—The audience of your chart. There are different types of audiences, such as the general public, executives, and professionals.
- *Topic*—The subject of your chart.
- *Type*—The text type to generate, including annotations and commentaries.

The process of generating context is iterative, in the sense that you can generate the context multiple times if you are not satisfied with the produced result. For example, you can adjust one or more elements to make ChatGPT converge on the desired output.

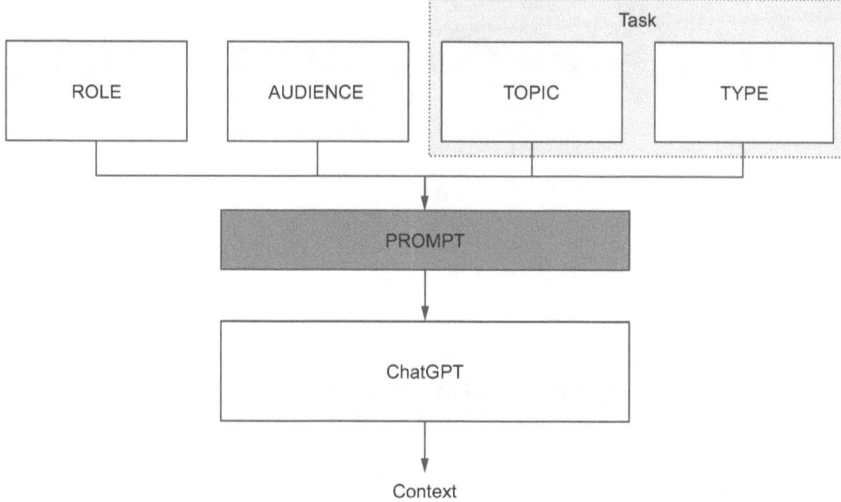

Figure 7.4   The schema used to generate context

In the remainder of this section, we will focus on how to write the topic and type elements of the schema while keeping the role and the audience simple. However, you can adapt the strategies described for the topic and the audience to the other elements.

As an example of how to build the context, we will focus on the case study described in chapter 4 and shown in figure 7.1. The following text summarizes the scenario for convenience: *Imagine you work in a sports company. You are training a team of young athletes in various disciplines. For each discipline, you have noted the world record and recorded the best time achieved by your team for comparison. Unfortunately, your company has limited investment funds available. Your boss asks you to understand which disciplines are worth training in, hoping to achieve good results in the upcoming competitions.*

7.3.1   Describing the topic

Describing the topic means composing simple words that precisely depict for ChatGPT what you have discovered and shown in your chart. The more precise you are, the better the output will be.

To describe the topic, focus on three aspects: scenario, data, and insight, as shown in figure 7.5. Let's go through each of those three aspects in a bit more detail.

SCENARIO

Describe an overview of your scenario, including the background and objective of the analysis. For the scenario in figure 7.1, we could write the following prompt for ChatGPT: *We are training a team of young athletes in various disciplines. For each discipline, we have calculated the percentage improvement of each discipline compared to the world record in that discipline. The objective is to search for the best two disciplines to fund.*

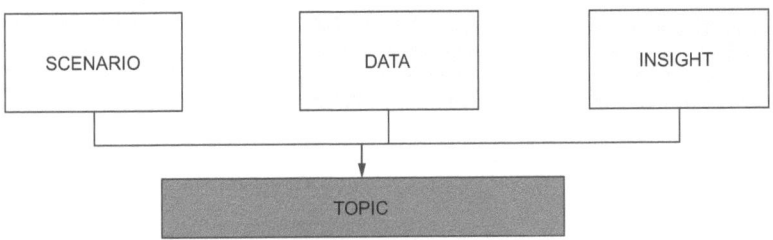

Figure 7.5   The elements used to describe the topic

DATA

Provide an overview of your data. This may include explaining the type of data, its source, and any manipulations you performed.

Describe the data using your own words, providing a more personalized description. By manually describing the data, you can highlight important patterns, trends, or correlations that may not be apparent through automated methods alone. Additionally, through manual descriptions, you can incorporate domain expertise observations.

For the scenario in figure 7.1, write the following prompt text: *There are five sports disciplines: Rowing (percentage improvement = 62.32%), Cycling (57.64%), Sprinting (42.69%), Long-distance running (18.31%), and Swimming (12.38%).* Now that you have learned how to describe the data, let's move on to the last step: describing insights.

INSIGHTS

Describe the central insights you have derived from the analysis, such as key patterns, trends, correlations, or relationships you have discovered. For the scenario in figure 7.2, include the following text: *Rowing and Cycling percentages are more significant than the baseline of 50%.* Before illustrating how we can build the audience description, let's test the prompt built so far on ChatGPT.

TEST

We write the following prompt: *Act as an inspirer. Write 5 titles for the following topic. There are five sports disciplines: Rowing (percentage improvement = 62.32%), Cycling (57.64%), Sprinting (42.69%), Long-distance running (18.31%), and Swimming (12.38%). Rowing and Cycling percentages are more significant than the baseline of 50%.* Figure 7.6 shows a possible output produced by ChatGPT.

As an alternative, you can ask ChatGPT to produce the context as follows: *Act as an inspirer. Write the context of a chart using 30 words for the following topic. There are five sports disciplines: Rowing (percentage improvement = 62.32%), Cycling (57.64%), Sprinting (42.69%), Long-distance running (18.31%), and Swimming (12.38%). Rowing and Cycling percentages are greater than the baseline of 50%.* Figure 7.7 shows a possible output produced by ChatGPT.

Now that you have learned how to describe the topic, try to generate the context for the case studies described in chapter 1: the pets scenario. For example, you can act

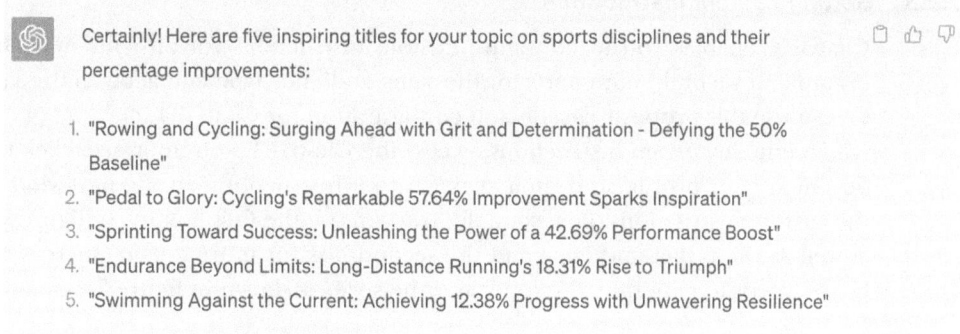

Figure 7.6   The five titles generated by ChatGPT

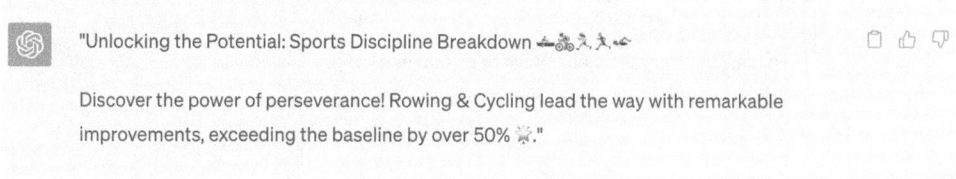

Figure 7.7   The context generated by ChatGPT

as an informer. For convenience, we summarize the scenario: *The organizers of an event dedicated to pets are collecting the type of pets that will participate. For each pet category, the organizers advertise the event on specific websites dedicated to that category. The organizers ask you to build a quick report about the current situation.*

For more details, please refer to chapter 1. You can find a prompt with the generated context here: https://mng.bz/EZvJ. For further practice, write the topic for the other scenarios described in the previous chapters. Now that you have learned how to describe the topic, let's move on to the next element: describing the type.

7.3.2   Describing the type

We consider the following types: commentary and annotations. In the previous sections, you have seen different ways to instruct ChatGPT to generate context types, such as writing the context of a chart using 30 words.

When describing the type, be as precise as possible, specifying the following aspects:

- The type (commentary, annotation, or general text)
- The maximum number of words to generate

My suggestion is to try different types and evaluate the results based on your needs.

7.3.3   Setting custom instructions

ChatGPT enables you to configure custom instructions for all your new chats. For example, if we build our charts for the same audience type and act with the same role, we can use this property as a default configuration for ChatGPT.

To enable custom instructions, access the ChatGPT web interface, click the three dots near your profile, and then click Custom Instructions. In the new window, write the custom instructions. For example, you can use the first box to configure your role as well as the target audience and the second box for more details, such as the number of words to generate, the tone, and the style, as shown in figure 7.8.

Custom instructions ⓘ

What would you like ChatGPT to know about you to provide better responses?

> - I'm a data analyst wanting to speak to an audience of professionals.

70/1500   Hide tips ⦸

How would you like ChatGPT to respond?

> - Generate an engaging response
> - Generate maximum 50 words
> - Be informal

73/1500

Enable for new chats 🔘   Cancel   Save

Figure 7.8   The Custom Instructions dialog box

Click the Save button to enable the custom instructions for new chats. If you want to disable this property, deselect the Enable for New Chats property in the dialog box. Then, click the Save button.

Let's try the custom instructions with the previous text (we have removed the number of words to generate, since we have configured them in the custom instructions): *Write the context of a chart for the following topic. There are five sports disciplines: Rowing (percentage improvement = 62.32%), Cycling (57.64%), Sprinting (42.69%), Long-distance running (18.31%), and Swimming (12.38%). Rowing and Cycling percentages are greater than the baseline of 50%.*

Figure 7.9 shows a possible output. Notice, for example, the informal tone we have set in the custom instructions.

 Sure thing! **ıl** Here's the context for your chart:

"Hey folks, check out this chart showcasing the percentage improvement in five sports disciplines. Rowing leads the pack with an impressive 62.32% improvement, followed closely by Cycling at 57.64%. Sprinting shows a solid 42.69% boost, while Long-distance running and Swimming have respectable gains of 18.31% and 12.38%, respectively. Notably, Rowing and Cycling exceed the 50% baseline, indicating their remarkable progress."

Figure 7.9   A possible output produced when configuring custom instructions

For optimal usage of custom instructions in data storytelling, I suggest using them to configure the role and the audience. In addition, you can configure other specific details correlated to your job or data, as specified in the ChatGPT documentation (https://mng.bz/8wAP).

Now that you have learned how to use ChatGPT for commentaries and annotations, let's move on to the next step: using large language models in a different way. So far, you have used the web interface provided by OpenAI to write your prompt for ChatGPT. Now, we will cover an advanced use of LLMs, based on APIs calls.

7.4   Using large language models for context

So far, you have used three generative AI tools: Copilot for code generation, ChatGPT for text generation, and DALL-E for image generation. All these tools are examples of application usage of a *large language model* (LLM). An LLM is a machine learning (ML) model aimed at predicting plausible language. LLMs have exploded in popularity since 2017, when Google researchers introduced the concept of transformers, a revolutionary architecture that allowed the training of large language models, such as generative pretrained transformers (GPTs), on which ChatGPT is based, and bidirectional encoder representations from transformers (BERT). Transformers allowed for the training of LLMs on massive datasets, resulting in models with incredible language generation capabilities.

In this book, we will not focus on how LLMs work. Instead, we aim to demonstrate how you can use them effectively for data storytelling. However, if you're interested in delving deeper into the technical aspects, a vast bibliography is available on the topic (Vaswani, 2017; Koenigstein, 2024).

Before you embark on using LLM to build your data-driven story, it's essential to ask yourself whether the model needs to know specific information related to your domain of work, as shown in figure 7.10. If the answer is no, then you can safely continue using ChatGPT. If, however, your answer is yes, then you can apply one of the following techniques:

- *Fine-tuning*—This technique adapts a pretrained LLM to a specific domain by updating its parameters on task-specific data, optimizing its performance for that domain.
- *Retrieval augmented generation*—This technique combines information retrieval and language generation, enabling LLMs to incorporate external knowledge sources during the generation process.

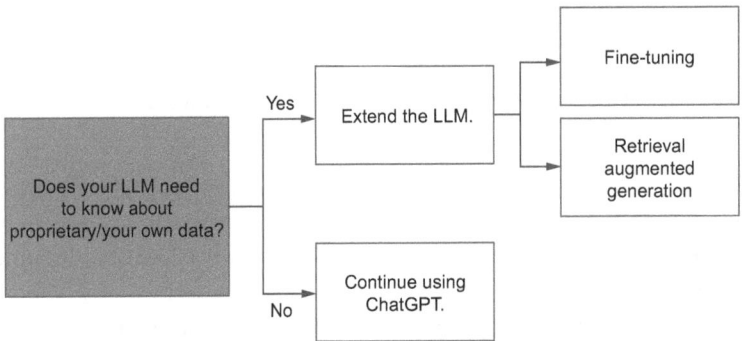

Figure 7.10   A criterion to establish whether to extend an LLM or not

In the remainder of this section, we assume that your answer is yes and that you must tailor your LLM to your specific domain. A practical case where fine-tuning is useful could be when you must generate different data stories for the same type of audience or even the same audience. In this case, you could build your database with the same structure of annotations so that all new annotations have the same structure as the previous ones. This may generate some familiarity for your audience when they read your data stories. In other cases, you may need to use RAG, for example, when you have a long document and you want to build a short annotation for your data story based on it. Using RAG could help you to build textual summaries. Now that you have learned the potential benefits of extending the LLM, let's start by analyzing the first strategy: fine-tuning.

7.4.1 Fine-tuning

GPT-3 was trained on 17 gigabytes of data, and GPT-4, the most recent model of OpenAI, has 45 gigabytes of training data. This means they contain a variety of information you can use in almost all cases. However, in some cases, fine-tuning your model could provide better results.

Fine-tuning is the process of further training a pretrained language model on a specific dataset that is more relevant to your specific domain. During fine-tuning, you use a smaller dataset, which typically contains examples and specific input–output pairs relevant to your task. In practice, the dataset is a collection of samples, each containing the prompt and the suggested completion.

When you apply fine-tuning to data storytelling, you can build a different dataset for each audience type, leading to better results. For example, you can build a dataset for the general public, one for professionals, and another for decision makers. You can even create a different dataset for each scenario you work with (products, topic, and so on) and for each type of text you want to generate (title, annotation, and commentary). The more specific your dataset is, the better your results will be.

Preparing the dataset is the most significant effort during the process of fine-tuning. In the remainder of this section, we will describe two strategies to prepare the dataset: manual building and building from sources. In both cases, we will use the OpenAI API. For more details on the installation, refer to appendix A.

MANUAL BUILDING

Manual building involves defining each pair (prompt, completion) manually. This solution enables you to obtain the best results since you specify the exact behavior of your model, given a specific input. Consider, for example, the following pair:

- *Prompt*—Generate a title for the general public about topic X.
- *Completion*—X revealed to you!

Now, imagine that you have fine-tuned your model with that pair and want to use your model to generate titles for the general public. If you give *the theory of relativity* topic as input to your fine-tuned model, it will probably generate a title similar to the following one: *The theory of relativity revealed to you!* The drawback of this strategy is that it is time consuming because you must write each pair manually.

To start, you can define a minimum number of curated samples covering all your possible cases. The OpenAI model requires you to represent at least 10 samples. Next, train the model. After that, proceed with the model evaluation by considering the original model (i.e., without fine-tuning) as a reference. Provide the same prompt to the two models, original and fine-tuned, and compare the produced outputs. Use your new model if your fine-tuned model performs better than the original one. Instead, if it performs worse than or has the same behavior as the original model, try to add new samples or improve the existing ones. Repeat this procedure until you reach a good result.

To show how manual building works, we will build a dataset tailored for the general public and generate a commentary as an output. In the previous chapter, you saw that you should use a different chart based on the information you want to convey. Here, we will build a different output based on the information to convey. We will build one or more (prompt–completion) pairs for each type of information to convey. Table 7.2 shows a possible dataset representing the described scenario. You can find the code described in this section in the GitHub repository for the book under 07/manual-chatgpt-fine-tuning.

Table 7.2   The samples based on the information to convey

Information to Convey	Prompt	Completion[a]
Parts of a whole	Percentage of participants in the conference by type (researchers 5%, students 30%, professors 65%)	More professors participated in the conference (65%). Researchers were not interested in the event (5%).
Comparison among entities	Comparison between red (80%), yellow, and green performance	Compared to yellow and green, red experienced an improvement in performance of 80%.
Trend	Sales changed in the last 12 months (–12%), due to fewer subscribers and video views.	Over the last 12 months, sales decreased by 12% based on various metrics, including subscribers and video views.
Outcomes of a survey or a questionnaire	Questionnaire: 3 positive answers, 7 negative answers	Three out of 10 people answered the questionnaire with a positive answer.
Distribution	Sales of product A (+30%) and product B over the last 12 months	Compared to product B, sales of product A have increased by 30% over the last 12 months.
Spatial information	Sales in North Europe (+23%) compared to South Europe	Compared to South Europe, sales in North Europe increased by 23%.
Relationship	The sales trend line from 1990 to 2020 increased by 120%.	Between 1990 and 2020, sales increased by 120%.
Comparison among entities	Top ingredients for our recipe: sugar and salt	The chosen ingredients for our recipe are sugar and salt.
Comparison among entities	Comparison between gold (30), silver (20), and bronze (40)	Bronze beats silver and gold with 40.
Distribution	Distribution of household chores (cooking 35%, cleaning 30%, laundry 20%, and yard work 15%)	Cooking takes up the most significant portion at 35%. Cleaning follows at 30%, while laundry and yard work account for 20% and 15%, respectively.

a. The word *completion* may be confusing, but it is used by the OpenAI API. *Completion* refers to the output produced by the model.

Once you have built the dataset, you must format it as a JSONL file. This file contains a list of messages. Consider each message as a separate chat where you can specify a general configuration, the user prompt, and the assistant (model) output, as shown in the following listing.

Listing 7.1   The structure of a JSONL file

```
{
  "messages": [
    {
   "role": "system",
 "content": "You are a data analyst showing data to the general public."
    },
    {
        "role": "user",
        "content": "Distribution of household chores
                              (Cooking 35% Cleaning 30% Laundry 20%, Yard
    work 15%)"
    },
    {
        "role": "assistant",
        "content": "Cooking takes up the largest portion at 35%.
                    Cleaning follows at 30% while laundry
                    and yard work accounts for 20% and 15% respectively."

    }
  ]
}
```

NOTE Use the keyword messages to define the list of samples. Imagine each sample as a separate chat, where you can specify the model role: system, for general configuration; user, for user prompt; and assistant, for model output.

If your dataset is saved as a CSV file, use the code shown in the following listing, which is also available in prepare-data.py, to convert it into JSONL.

Listing 7.2   How to convert the CSV file into JSONL

```
import pandas as pd
import json

df = pd.read_csv('general-public.csv')

json_list = []

for index, row in df.iterrows():
    json_object = {
        "messages": [
            {
                "role": "system",
                "content": "You are a data analyst showing data to the
    general public."
            },
            {
                "role": "user",
                "content": row['prompt']
            },
```

```
            {
                "role": "assistant",
                "content": row['completion']
            }
        ]
    }
    json_list.append(json_object)

with open('general-public.jsonl', 'w') as outfile:
    for json_object in json_list:
        json.dump(json_object, outfile)
        outfile.write('\n')
```

NOTE First, load the dataset as a pandas DataFrame. Next, format it in the JSONL format, as described in listing 7.1. Finally, save the generated JSONL file.

Now, we are ready to fine-tune our model. We need an OPENAI_API_KEY, as specified in appendix A. If you are transitioning from a free to a for-fee plan, you might need to generate a new API key because the initial key does not work after the switch to a for-fee plan. Open a terminal, and export your OPENAI_API_KEY as an environment variable (export OPENAI_API_KEY='my key'). Next, upload the produced file to the OpenAI server, and when the uploading process is complete, create a job for fine-tuning. The following listing shows the code to perform these operations. Alternatively, read the tune-model.py script in the GitHub repository for the book. Remember that this option is exclusively available with the for-fee version.

Listing 7.3   How to fine-tune the model

```
import os
import openai
import time                                      An alternative way to get
                                                 your key: openai.api_key
openai.api_key = os.getenv('OPENAI_API_KEY')  ◄  = 'MY_KEY'

dataset = openai.File.create(file=open('general-public.jsonl',
[CA]'rb'), purpose='fine-tune')                  Creates a new dataset
print('Uploaded file id', dataset.id)            and uploads it to the
                                                 OpenAI server
while True:
    print('Waiting while file is processed...')
    file_handle = openai.File.retrieve(id=dataset.id)
    if len(file_handle) and file_handle.status == 'processed':
        print('File processed')
        break
    time.sleep(3)

job = openai.FineTuningJob.create(training_file=dataset.id, model="gpt-3.5-
    turbo")                                  Create a new
                                             fine-tuning job
Enters into a loop until
the model is fine-tuned
```

```
while True:
    print('Waiting while fine-tuning is completed...')
    job_handle = openai.FineTuningJob.retrieve(id=job.id)
    if job_handle.status == 'succeeded':
        print('Fine-tuning complete')
        print('Fine-tuned model info', job_handle)
        print('Model id', job_handle.fine_tuned_model)        ◁──┐  Prints the
        break                                                     │  model ID
    time.sleep(3)
```

NOTE First, use the `openai.File.create()` method to create a new dataset and upload it to the OpenAI server. Next, use the `openai.FineTuning-Job.create()` method to create a fine-tuning job using GPT-3.5 Turbo. Wait until the job is completed. This could take a long time, depending on the dataset size. Once the model is trained, use the `fine_tuned_model` variable to print the information associated with the fine-tuned model.

The following listing shows an example of information printed after the execution of the fine-tune-model.py script. This fine-tuning would cost around $0.05.

Listing 7.4   An example of information associated with a fine-tuned model

```
fine-tuned model info {
  "object": "fine_tuning.job",
  "id": "your model id",
  "model": "gpt-3.5-turbo-0613",
  "created_at": 1693347869,
  "finished_at": 1693348340,
  "fine_tuned_model": "ft:gpt-3.5-turbo-0613:personal::7t1Xuct5",
  "organization_id": "org-jWkYw8hPpaNwkesXezsWOwK8",
  "result_files": [
    "file-ro0BoeariIjOl7NSGRC80v8r"
  ],
  "status": "succeeded",
  "validation_file": null,
  "training_file": "file-InGnigMTto3YLrsiLuIUr7ty",
  "hyperparameters": {
    "n_epochs": 10
  },
  "trained_tokens": 5930
}
```

NOTE Some helpful information is provided, including the model type, the model ID, the hyperparameters used, and more.

Now, we can use the fine-tuned model to generate new commentaries tailored to the general public. Use the value corresponding to the `fine_tuned_model` key of the previous listing to refer to your model (`"ft:gpt-3.5-turbo-0613:personal::7t1Xuct5"` in the example).

To generate a new commentary, start a new chat session by using the `openai.Chat-Completion.create()` method, as shown in the following listing and in the generate-description.py script of the GitHub repository for the book. As a use case, consider again the example of figure 7.1.

Listing 7.5 How to generate a new commentary

```
import os
import openai

openai.api_key = os.getenv("OPENAI_API_KEY")

model_id = "ft:gpt-3.5-turbo-0613:personal::7t1Xuct5"

completion = openai.ChatCompletion.create(
    model=model_id,
    messages=[
        {
            'role': 'system',
            'content': 'You are a data analyst showing data to the general
    public.',
        },
        {
            'role': 'user',
            'content': 'Top sports: rowing (62%) and cycling (58%)'
        },
    ],
)

print(completion.choices[0].message)
```

NOTE Create a new `ChatCompletion` instance by specifying the model ID and the list of messages. The example defines only one message with the same system role as the fine-tuning dataset and the user role with a short description of our scenario.

The following listing shows an example output.

Listing 7.6 An example output

```
{
  "role": "assistant",
  "content": " \"The most popular sports are rowing and cycling with 62% and
    58% of people practicing them respectively.\""
}
```

NOTE The output contains the role (assistant) and the content.

Incorporate the produced content into your chart as a commentary. This example has demonstrated how you can perform model fine-tuning using a manual dataset. In the example, the output is straightforward. If you want your model to produce more complex outputs, you must complicate your fine-tuning dataset—for example, by adding new pairs (prompt–completion) specifically designed for your audience or your topic.

For comparison with the fine-tuned model, figure 7.11 shows the output produced by ChatGPT (without fine-tuning) with a similar input. Now that you have learned how to build your dataset manually, let's move on to the next strategy, building from sources.

 You are a data analyst showing data to the general public. Write a commentary of max 50 words for the following text: Top sports: rowing (62%) and cycling (58%)

 Rowing dominates as the most popular sport, with a remarkable 62% participation rate, showcasing its enduring appeal. Meanwhile, cycling follows closely at 58%, highlighting the strong sporting culture and the growing interest in outdoor activities among the general public.

Figure 7.11 The output produced by ChatGPT without fine-tuning

BUILDING FROM SOURCES

This strategy involves building your dataset from external sources, such as your company website or a domain-specific blog. For example, if you work in the health field, you could download the title and abstracts of scientific papers about health. This enables you to build a dataset with a very domain-specific language. Or if you work in the ICT field, you can download the titles and subtitles of blog articles from feeds to build your technical dataset. Anyway, you must pay attention to the data license in all cases. If the license explicitly prohibits their usage, you cannot use those sources, and you must search for other data sources. In some cases, contacting the data author directly could be sufficient—for example, if you want to download their data.

In the remainder of this section, we will build a dataset tailored to a technical audience by extracting feeds from my Medium blog. The objective is to build a chart title corresponding to the blog title, providing the blog subheading as an input. You can find the example in the GitHub repository for the book under 07/from-source-chatgpt-fine-tuning.

First, download the data. Ask Copilot to generate the code for you. The following listing shows the instructions for Copilot.

Listing 7.7 The instructions for Copilot

```
# import required libraries
# extract the title and link from the following rss/feed url:
    https://alod83.medium.com/feed
# for each extracted link, extract the subheading from the article
```

```
# create a dataframe with the following columns: 'prompt', 'completion'
# save the dataframe to a csv file called 'medium-articles.csv'
```

NOTE Specify the feed URL and the information to extract for each item. Also, ask Copilot to generate the code to save the extracted items into a CSV file.

Copilot will generate an output similar to that shown in listing 7.8. Set the prompt to the subheading and the completion to the title. Save the script, and run it. You can find the code generated by Copilot in the GitHub repository for the book in the download-raw-data.py script. You should see the medium-articles.csv file in your working directory.

Listing 7.8 How to extract data from feeds

```
import feedparser
import pandas as pd
import requests
from bs4 import BeautifulSoup

url = 'https://alod83.medium.com/feed'
feed = feedparser.parse(url)

titles = []
links = []
subheadings = []

for entry in feed.entries:
    titles.append(entry.title)
    links.append(entry.link)
    print(entry.link)
    response = requests.get(entry.link)
    soup = BeautifulSoup(response.content, 'html.parser')
    subheading = soup.find('h2', attrs={'class': 'pw-subtitle-
      paragraph'}).text
    subheadings.append(subheading)

df = pd.DataFrame({'prompt': subheadings,'completion': titles})
df.to_csv('medium-articles.csv', index=False)
```

NOTE Use the `feedparser`, `requests`, and `bs4` libraries. If you don't have them in your environment, install them using the pip package manager.

Once you have built the dataset, follow the procedure described in section 7.4.1 to fine-tune the dataset (listings 7.2–7.6). You can find the complete example in the GitHub repository for the book.

To test the fine-tuned model, provide the following prompt as input: *A chart on selecting the best sport to fund*. The model generates an output similar to the following: *How to Choose the Best Sport to Fund: A Data-Driven Approach*. Try a similar prompt with ChatGPT. Figure 7.12 shows a possible output. Since ChatGPT is not fine-tuned, you must specify more details in your prompt, as previously seen. Instead, for your fine-tuned model, describing the content in your prompt is sufficient. Now that you have learned how to perform fine-tuning for data storytelling, let's move on to the next strategy for adapting your model to your specific context: retrieval augmented generation.

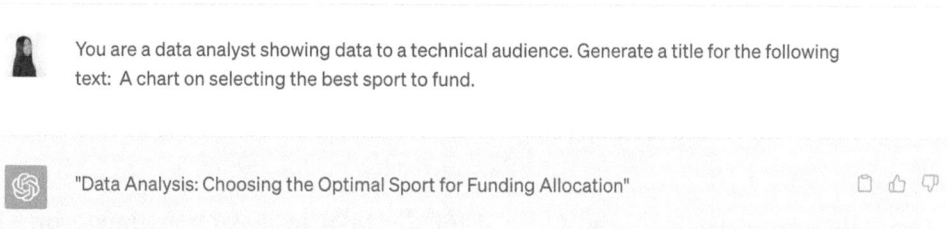

Figure 7.12   The output produced by ChatGPT

7.4.2   *Retrieval augmented generation*

So far, you have seen how to adapt an LLM to a context by building an ad hoc dataset. The effort, in this case, consists in preparing the dataset. Imagine how nice it would be to pass a text directly to the LLM without converting it to a specific format. Well, the good news is that this is possible, thanks to retrieval augmented generation (RAG).

RAG is an advanced natural language processing (NLP) technique that combines elements of information retrieval and text generation. First, RAG performs a retrieval step, which queries an external knowledge source, such as a vast text corpus or a structured database. Next, RAG uses this knowledge source to enhance its response generation. RAG integrates the retrieved facts into its generated text.

In the data storytelling domain, you can use RAG to adapt your LLM to your topic, such as a product, real-time data, customer reviews, or other relevant information. For instance, by querying the knowledge base for specific product details, you can generate ad hoc commentaries and annotations.

Imagine you want to build a RAG-based system that retrieves information about a product from your website company. Figure 7.13 shows the architecture of the RAG system we will implement.

First, we will download the text from a specified URL, split it, and represent it as vectors we store in a vector database. We will provide the vector database as an input to an LLM application, which can answer queries by querying the vector database. We will implement an example that generates commentaries for a specific smartphone, based on its description contained in an HTML page. In practice, we will load the HTML page into the vector database, and then we will implement an LLM application to query it. We will use LangChain to implement the LLM application, Chroma for

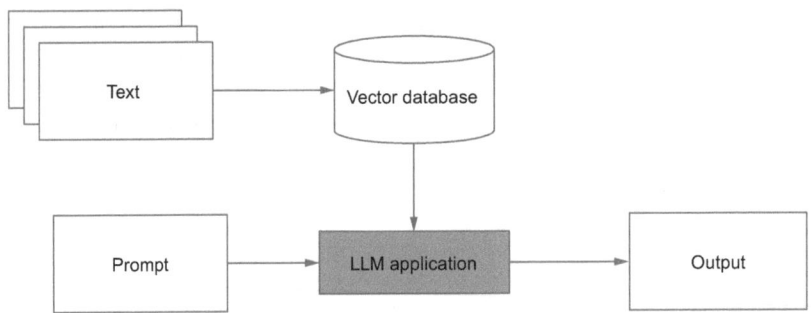

Figure 7.13   A RAG-based system

the vector database, and OpenAI for the LLM to make everything work. For more details on how to install these tools, refer to appendix A.

In the remainder of this section, you will learn how to implement the described system. We will start by introducing LangChain. Next, we will see how to store data in Chroma. Finally, you will learn how to query the built system.

INTRODUCING LANGCHAIN

LangChain (https://www.langchain.com/) is a framework that enables you to create applications that connect an LLM to other sources. LangChain supports several providers, including OpenAI, Google, Microsoft, Hugging Face, and many more. In this book, we will focus on the models provided by OpenAI.

The core idea behind LangChain is the concept of a chain, which consists of several components from different modules. There are three main components:

- *LLM wrappers*—Wrappers for LLMs provided by external providers, such as OpenAI and Hugging Face
- *Prompt templates*—Templates for different prompts, such as chatbot and question answering
- *Indexes*—External structures you can use to provide additional context to an LLM

The LangChain-based applications are *context aware* because they connect LLM to external sources. Additionally, such applications are useful because they can answer questions based on the provided context, what actions to take, and so on.

The most straightforward chain consists of just one LLM chained with a prompt that enables you to query the model. In the remainder of this section, we will implement a LangChain composed of the components shown in figure 7.14: the vector database (Chroma), the prompt template, the LLM model (GPT-3.5 Turbo by OpenAI), and the retrieval interface.

You can find the full code described in this example in the GitHub repository of this book under 07/rag. We need an OPENAI_API_KEY, as specified in appendix A.

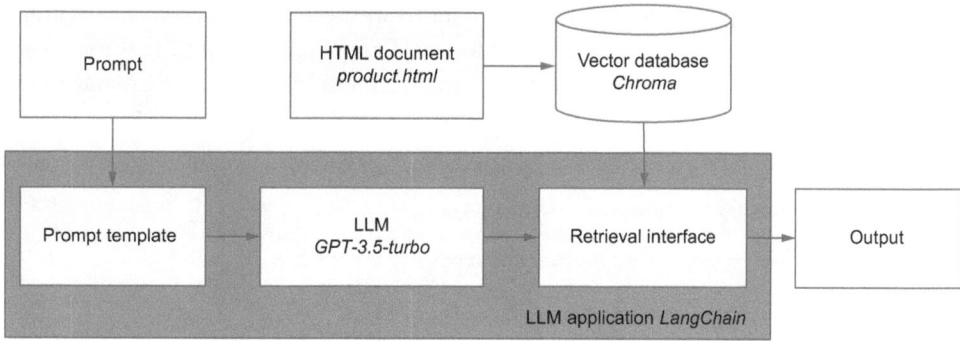

Figure 7.14   The implemented architecture

Open a terminal, and export your OPENAI_API_KEY as an environment variable (export OPENAI_API_KEY='my key').

Chroma (https://www.trychroma.com/) is an embedding database you can use as an indexer for your LangChain. To install and configure Chroma, refer to appendix A. An *embedding* is a numerical representation of data that is easy to index and retrieve, often for real-time tasks (Lane and Dyshel, 2024). Before storing a text in Chroma, we must convert it into vector embeddings. For more details about embeddings, refer to the references section of this chapter.

Consider the product description available on a hypothetical website, as shown in figure 7.15. The objective of our task is to store the product description shown in figure 7.15 in Chroma.

Introducing the New SmartX 2023

Key Features:

- 5.7-inch Super AMOLED Display
- Octa-core Processor for Lightning-Fast Performance
- Up to 256GB of Storage Space
- Triple-Lens 48MP Main Camera
- Ultra-Fast 5G Connectivity
- Long-Lasting 4000mAh Battery
- Wireless Charging Support

Description:

Get ready to experience the future with the all-new SmartX 2023. This cutting-edge smartphone combines sleek design, powerful performance, and innovative features to redefine your mobile experience.

The SmartX 2023 features a stunning 5.7-inch Super AMOLED display that delivers vibrant colors and sharp contrast, making your photos, videos, and games come to life like never before.

Under the hood, the octa-core processor ensures lightning-fast performance, allowing you to multitask effortlessly and run the latest apps and games with ease. With up to 256GB of storage space, you'll have ample room for all your photos, videos, and apps.

Capture stunning photos and videos with the triple-lens 48MP main camera. Whether you're taking breathtaking landscapes or detailed close-ups, the SmartX 2023's camera delivers exceptional results every time.

Stay connected at blazing speeds with 5G connectivity, and enjoy all-day usage with the long-lasting 4000mAh battery. Plus, wireless charging support means you can power up without the hassle of cables.

Price and Availability:

The SmartX 2023 is available now with prices starting at $699.99. Get yours today and experience the future of smartphones!

© 2023 SmartX Technologies. All rights reserved.

Figure 7.15   The HTML page with the product description

The first step involves loading the data from the URL, as shown in the following listing. Since Chroma is fully integrated with LangChain, we will use it to accomplish our task. LangChain supports multiple formats, including PDFs, URLs, and more.

Listing 7.9   How to load the HTML document in LangChain

```
from langchain_community.document_loaders.html import UnstructuredHTMLLoader
from langchain.text_splitter import RecursiveCharacterTextSplitter
from langchain_community.vectorstores import Chroma
from langchain_openai import OpenAIEmbeddings
from langchain.chains import RetrievalQA
from langchain.prompts import PromptTemplate
from langchain_openai import ChatOpenAI

loader = UnstructuredHTMLLoader('product.html')          ◁── Loads data
data = loader.load()
```

> **NOTE**   To load an HTML document in LangChain, build an `Unstructured-HTMLLoader()` object.

Next, split the data into chunks of 20, as shown in the following listing. We could have chosen any number smaller than the total text size for the chunk size.

Listing 7.10   How to split the text into chunks

```
text_splitter = RecursiveCharacterTextSplitter(
    chunk_size = 100,
    chunk_overlap  = 20,
    length_function = len,
    is_separator_regex = False,
)

splitted_data = text_splitter.split_documents(data)
```

> **NOTE**   Create a `RecursiveCharacterTextSplitter()` object to split the text into chunks.

After that, convert the split text into embeddings and store them in Chroma.

Listing 7.11   How to generate embeddings in Chroma

```
embeddings = OpenAIEmbeddings()

store = Chroma.from_documents(
    splitted_data,
    embeddings,
    ids = [f"{item.metadata['source']}-{index}" for index, item in
     enumerate(splitted_data)],
    collection_name='Product-Info',
persist_directory='db',
)
store.persist()
```

NOTE   First, create a new `OpenAIEmbeddings()` object. Next, create a Chroma store with the split data and the embeddings and associate it with the collection `Product-Info`. Finally, store the Chroma store on the filesystem, using the `persist()` method.

Now, our vector store is ready, so we can move on to the next step: defining a prompt template.

DEFINING A PROMPT TEMPLATE

A prompt template is a predefined text used for generating prompts for LLMs. A prompt template may include instructions, examples, context, and questions appropriate for your task. The following listing shows an example of a prompt we can provide as an input to our system.

Listing 7.12   How to structure a prompt template

```
template = """You are a bot that answers questions about the product New
    SmartX 2023, using only the context provided.
If you don't know the answer, simply state that you don't know.

{context}

Question: {question}"""

prompt = PromptTemplate(
    template=template, input_variables=['context', 'question']
)
```

NOTE   First, define the structure of your template. Use brackets to define input variables. In the example, there are two variables: `context` and `question`. Next, create a new `PromptTemplate()` object, and pass it the template and the input variables as parameters.

Once we have built the prompt template, we are ready to proceed with the last step: retrieval and query.

RETRIEVAL INTERFACE

A *retrieval interface* is an interface that enables us to combine the data stored in the Chroma database and the OpenAI LLM. We can use a retrieval to query our system and generate commentaries and annotations to incorporate in our charts. The following listing shows an example of the usage of a retrieval.

Listing 7.13   How to build a retrieval

```
llm = ChatOpenAI(temperature=0, model='gpt-3.5-turbo')

qa = RetrievalQA.from_chain_type(
    llm=llm,
    chain_type='stuff',
    retriever=store.as_retriever(),
```

```
        chain_type_kwargs={'prompt': prompt, },
        return_source_documents=True,
)

print(
    qa.invoke({"query": 'Describe the product New SmartX 2023 using 30 words'})
)
```

> **NOTE**   First, create an LLM instance using `ChatOpenAI()`. Set the tempera-
> ture to 0 for conservative output. The temperature spans from 0 (low creativ-
> ity) to 1 (high creativity). Set the model to `GPT-3.5-turbo`. Next, create a
> retrieval interface using `RetrievalQA()` by specifying the LLM, the vector
> store (`retriever`), the prompt, and other parameters. Set the `chain_type` to
> `stuff`, a prepackaged document chain that takes a list of documents and
> inserts them into the prompt, which is then passed to the LLM. Finally, ask
> the question.

The following listing shows the produced output. You can insert the produced text (in
bold) in your chart.

Listing 7.14   The produced output

```
{'query': 'Describe the product New SmartX 2023 using 30 words',
    'result': 'The New SmartX 2023 is a cutting-edge smartphone with a 5.7-
        inch Super AMOLED display and a high-quality camera that captures breath-
        taking landscapes and detailed close-ups.',
    'source_documents':
    [Document(page_content='© 2023 SmartX Technologies. All rights reserved.',
        metadata={'source': 'product.html'}),
        Document(page_content='Get ready to experience the future with the
    all-new SmartX 2023. This cutting-edge smartphone',
        metadata={'source': 'product.html'}),
        Document(page_content='Introducing the New SmartX 2023\n\nKey
    Features:\n\n5.7-inch Super AMOLED Display',
            metadata={'source': 'product.html'}),
        Document(page_content="you're taking breathtaking landscapes or
    detailed close-ups, the SmartX 2023's camera delivers",
            metadata={'source': 'product.html'})
    ]
}
```

> **NOTE**   The output contains the text to insert in the chart (in bold) and other
> useful information, such as the original query and the source documents.

Now that you have learned how to apply LLMs to build context in data storytelling,
let's move on to a practical example.

7.5   Case study: From information to knowledge (part 1)

In the previous chapter, we analyzed how to turn data into information in the aquaculture case study. As a quick reminder, the case study involved building a story around the problem of safety in the salmon aquaculture in the United States. We decided to plot the salmon aquaculture sales trend line versus the other types of aquaculture. As an insight, we discovered that since 1998, there had been an increase in sales, following a period of decrease in sales from 1992 to 1998. We discovered that the decreasing period was partially due to some health problems in the salmon aquaculture. Figure 7.16 shows the chart produced at the end of the first step of the DIKW pyramid: from data to information.

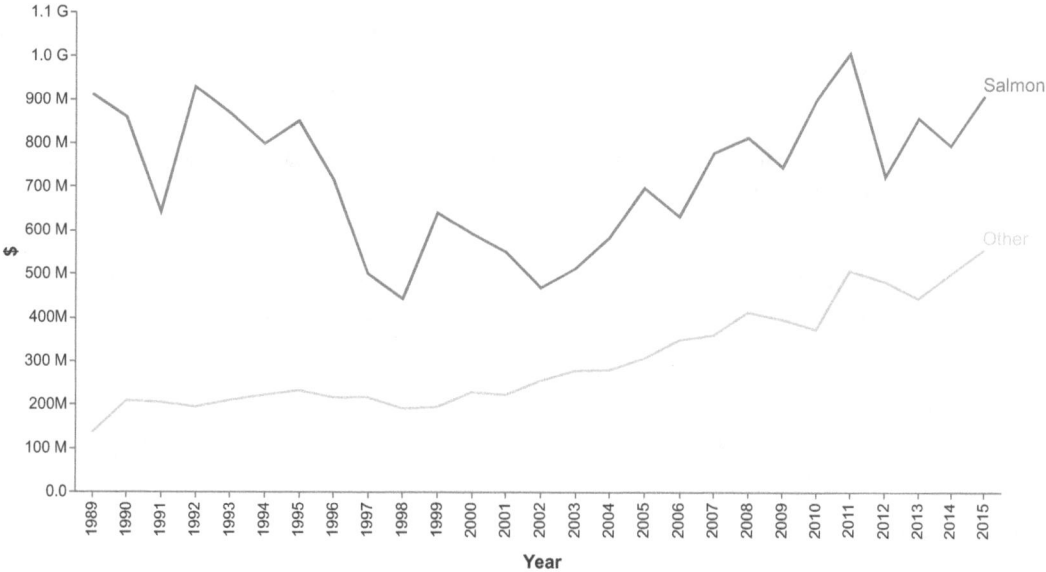

Figure 7.16   The chart produced at the end of the data-to-information phase

To transform the chart into a data story, the next step involves turning information into knowledge. We will accomplish this by doing the following:

- Some design considerations to tailor the chart to the audience
- Adding a commentary describing the general situation regarding safety in aquaculture
- Adding an annotation and a symbol to highlight the period of decrease in sales

Let's start with the first point: tailoring the chart to the audience.

7.5.1   *Tailoring the chart to the audience*

The scenario required us to present the data story to an audience of executives, which meant we needed a chart easy enough to understand that they could quickly make decisions based on its information. In general, executives are familiar with trend lines, so we do not need to modify the chart. In addition, the chart is neither too detailed nor too simple. It contains the right level of detail to allow the audience not to be overwhelmed by information. Additionally, the chart doesn't give the impression of being sparse. Therefore, we can conclude that the chart type is perfect for our audience.

   We also suppose that our audience is familiar with the $ symbol on the y-axis and the Years label on the x-axis, so we do not need to add any further specifications. We can leave the comparison between the salmon trend line and the others because it is useful for our audience to understand how the salmon sales behave compared to the other categories. It is not necessary to add further details.

> **Challenge: How could you tailor the chart to the general public or to an audience of professionals?**
> For the general public, you could consider simplifying the chart—for example, by reducing the number of years. You may also need to better explain the meaning of the y-axis. For professionals, you could add more details, such as points with values for each year, or you could even show the other aquaculture categories.

Now that we have discussed some design considerations to tailor the chart to the audience, let's move on to the next step: adding a commentary. We will use RAG to generate the commentary.

7.5.2   *Using RAG to add a commentary*

We will add a commentary to the chart immediately under the title. Our commentary should explain how safety works in US aquaculture. We will base the commentary on "Aquacultured Seafood" (https://mng.bz/WEW0), an official FDA fact sheet. This document describes, among other topics, the safety levels of aquaculture seafood.

   You can implement a RAG-based system that builds required commentary using the code implemented in section 7.4.2. You only need to provide this prompt: *Describe the safety of aquaculture seafood in the U.S.* The code of the implemented RAG system is also available in the GitHub repository for the book under CaseStudies/aquaculture/from-information-to-knowledge/rag.py. The following listing shows the produced output, containing the required commentary.

Listing 7.15 The produced output

```
{'query': 'Describe Safety of Aquaculture Seafood in the U.S.', 'result':
'Aquaculture seafood in the U.S. is regulated by the FDA to ensure safety.
Strict standards are in place to monitor water quality, feed, and disease
control. Regular inspections and testing are conducted to minimize risks and
protect consumers.', 'source_documents': [Document(page_content='Safety of
Aquaculture Seafood', metadata={'source': 'aquaculture.html'}),
Document(page_content='Regulatory Requirements for Aquacultured Seafood',
metadata={'source': 'aquaculture.html'}), Document(page_content='Domestic
Aquaculture Seafood', metadata={'source': 'aquaculture.html'}),
Document(page_content='for additional information on how the FDA ensures the
safety of imported seafood products.', metadata={'source':
'aquaculture.html'})]}
```

NOTE Use the produced output as a commentary for the chart.

Now, we can add this text as a commentary for our chart. Listing 7.16 shows only the modifications to our original chart including the commentary. You can find the complete code in the GitHub repository for the book under CaseStudies/aquaculture/from-information-to-knowledge/chart.py.

Listing 7.16 The commentary

```
commentary = ['Aquaculture seafood in the U.S. is regulated by the FDA to
    ensure safety. Strict standards are in place to monitor water quality,
    feed, and disease control.',
'Regular inspections and testing are conducted to minimize risks and protect
    consumers. (Source: U.S. Food and Drug Administration)'
]

base = alt.Chart(df).encode(
    x=alt.X('YEAR_ID:O', title=''),
    y=alt.Y('AMOUNT', title='$',axis=alt.Axis(format='.2s')),
    color=alt.Color('CATEGORY',
        legend=None,
        scale=alt.Scale(range=range, domain=domain)
    )
).properties(
    width=800,
    height=400,
    title=alt.TitleParams(
        text='Aquaculture Exports of Salmon in the U.S.',
        subtitle=commentary,
        fontSize=20,
        subtitleFontSize=14,
        align='left',
        anchor='start',
        offset=20,
        color=color,
        subtitleColor='black'
    )
)
```

> **NOTE**   Use the `title` property to add commentary to the chart, immediately before the title. Also, add a provisory title to the chart.

The next step involves highlighting the period of decrease in sales. So let's proceed.

7.5.3   *Highlighting the period of decrease in sales*

The period of decrease in sales ranges from 1992 to 1998. We want to highlight it to let the audience know that during this period, there were health problems in the salmon aquaculture. This will prepare the audience to consider respecting the safety rules to avoid the same problems in the future. We will add two elements to highlight this decreasing period:

- A light-gray rectangle covering the decreasing period
- A textual annotation describing the health problems

The following listing shows the code to build the rectangle.

Listing 7.17   **The rectangle**

```
N = 100000000
y = df['AMOUNT'].max() + N          ◁——┐  A magic number to set the
                                        │  upper part of the chart
rect_df = pd.DataFrame({'x': [1992],
            'x2': [1998],
            'y' : [0],
            'y2': [y]
        })

rect = alt.Chart(rect_df).mark_rect(
    color='lightgrey',
    opacity=0.5
).encode(
    x='x:O',
    x2='x2:O',
    y= 'y:Q',
    y2= 'y2:Q'
)
```

> **NOTE**   First, build a DataFrame with the rectangle's coordinates. Next, draw the rectangle using `mark_rect()`.

> **Challenge: Which instructions could you write for Copilot to speed up the coding process?**
>
> You could try adding the following instruction to generate the rectangle: `# Add a rectangle starting from 1993 to 2000`. What output would you obtain?

The following listing shows the code to add the annotation. The decline in sales was partially due to fish health issues.

Listing 7.18   The annotation

```
ann_df = pd.DataFrame({'x': [1992, 1992, 1992],
            'y': [y, y-N/3*2, y-N/3*4],
            'text': ['The decline in sales was',
            'partially due to fish',
            'health issues']
            })

annotation = alt.Chart(ann_df
).mark_text(
    align='left',
    baseline='middle',
    fontSize=14,
    dx=5,
    dy=10
).encode(
    x='x:O',
    y='y:Q',
    text='text:N'
)

chart = (chart + text + rect + annotation
).configure_axis(
    labelFontSize=14,
    titleFontSize=16,
    grid=False
).configure_view(
    strokeWidth=0
)
chart.save('chart.html')
```

NOTE   First, build a DataFrame with the annotation text and its position information. Next, draw the annotation using `mark_text()`. Finally, plot and save the chart.

Figure 7.17 shows the final chart, after adding context; we have turned information into knowledge. In the next chapter, we will further enrich context by adding some images, and in chapter 9, we will complete the story by adding the next step: wisdom.

You have now implemented a practical example of turning information into knowledge. Before moving to the next chapter, let's further solidify the concept by completing a practical exercise.

Aquaculture Exports of Salmon in the U.S.
Aquaculture seafood in the U.S. is regulated by the FDA to ensure safety. Strict standards are in place to monitor water quality, feed, and disease control. Regular inspections and testing are conducted to minimize risks and protect consumers. (Source: U.S. Food and Drug Administration)

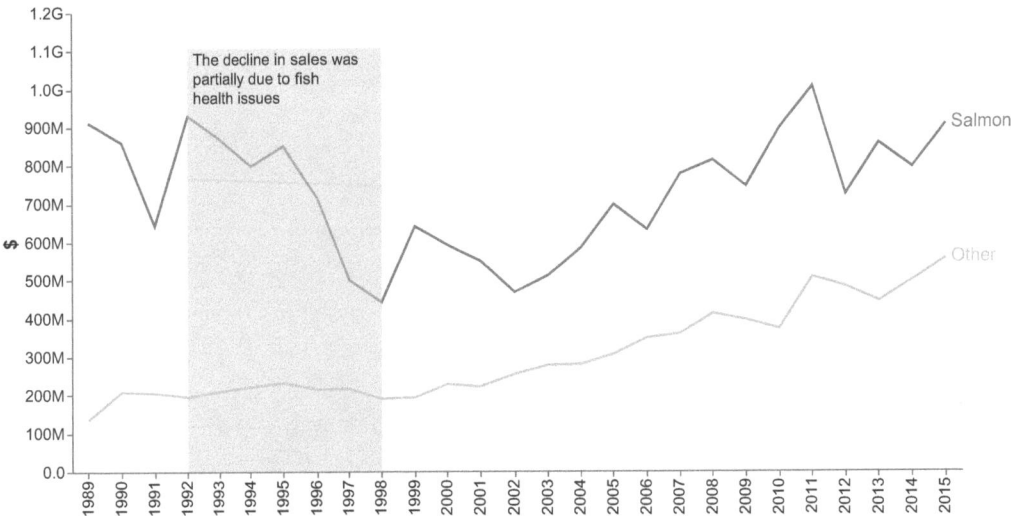

Figure 7.17   **The chart produced at the end of the information-to-knowledge phase**

7.5.4   *Exercise*

Modify the previous chart as follows:

1 Tailor the chart to an audience of professionals.

   a Add points to the salmon line chart. Suggestion: Use `point=True` as a parameter of `mark_line()`.

   b Add values for each point. Suggestion: Use `mark_text()` to add values for each point.

2 Implement a RAG-based system to extract an annotation for the decreasing period from the *Governor's Task Force on the Planning and Development of Marine Aquaculture in Maine Report and Recommendations* (pp. 28–32, https://mng.bz/ jXqx). Suggestion: Use `PDFMinerLoader()` to extract data from PDF. You may need to install some additional Python packages, including pdf2image, pdfminer, and pdfminer.six. You can find the solution in the GitHub repository for the book under CaseStudies/aquaculture/from-information-to-knowledge/rag-annotation.py.

In the first part of this chapter, you learned how to turn information into knowledge by adding context to your data visualization chart. You saw that context depends on the audience reading your chart. For example, if your chart will be read by the general public, avoid technical details and use an appealing visualization. On the other hand, if your chart will be read by technical experts, add as many details as you can, while keeping the chart easy to read. In the second part of the chapter, you saw how to use generative AI tools as assistants to build your context. Finally, you learned where to put the textual context in your chart. In the next chapter, you will see how to add images to your chart to enrich context.

Summary

- Adding context to your data visualization is crucial for turning information into knowledge. Textual context includes all the relevant facts and events useful for the audience to understand data.
- When you build a chart, tailor it to the audience. In general, there are three types of audiences: the general public, professionals, and executives.
- Use generative AI tools as assistants to help create context for your data. In particular, use ChatGPT to generate commentaries and annotations.
- If ChatGPT needs to know custom data or topics, extend your LLM with fine-tuning or RAG.
- Fine-tuning enables you to optimize a pretrained LLM based on a dataset of prompt–completion pairs.
- Retrieval augmented generation uses an external database, called a *vector database*, to extend the LLM knowledge with domain-specific topics.

References

Embeddings

- Lane, H. and Dyshel, M. (2024). *Natural Language Processing in Action* (2nd ed.). Manning Publications.
- OpenAI. (n.d.). *Embeddings.* https://platform.openai.com/docs/guides/embeddings.

Fine-tuning

- Bantilan, N. (2023). *Fine Tuning vs. Prompt Engineering Large Language Models.* https://mlops.community/fine-tuning-vs-prompt-engineering-llms/.
- Jolley, E. (2023). *Introduction to Retrieval Augmented Generation.* https://arize.com/blog-course/introduction-to-retrieval-augmented-generation/.
- Marcelo, X. (2023). *How to Fine-Tune OpenAI GPT.* https://medium.com/@marceloax.br/how-to-fine-tune-openai-gpt-3-d06741f915f4.
- OpenAI. (n.d.). *Fine-Tuning.* https://platform.openai.com/docs/guides/fine-tuning.

LangChain

- Biswas, A. (2023). *How to Work with LangChain Python Modules.* https://www.packtpub.com/article-hub/how-to-work-with-langchain-python-modules.

- Geeks for Geeks. (2024). *Introduction to LangChain.* https://www.geeksforgeeks .org/introduction-to-langchain/.
- Pinecone. (n.d.). *LangChain: Introduction and Getting Started.* https://www.pinecone .io/learn/series/langchain/langchain-intro/.

LLM

- De Angelis, L., Baglivo, F., Arzilli, G., Privitera, G. P., Ferragina, P., Tozzi, A. E., and Rizzo, C. (2023). *ChatGPT and the Rise of Large Language Models: The New AI-Driven Infodemic Threat in Public Health. Frontiers in Public Health, 11,* 1166120. https://doi.org/10.3389/fpubh.2023.1166120.
- Google Developers. (n.d.). *Introduction to Large Language Models.* https:// developers.google.com/machine-learning/resources/intro-llms?hl=en.

RAG

- Jolley, E. (2023). *Introduction to Retrieval Augmented Generation.* https://arize .com/blog-course/introduction-to-retrieval-augmented-generation.
- Needham, M. (2023). *Learn Data with Mark.* https://github.com/mneedham/ LearnDataWithMark/tree/main.
- ———. (2023). *Retrieval Augmented Generation with OpenAI/GPT and Chroma.* https://www.youtube.com/watch?v=Cim1lNXvCzY.
- Routu, V. (2023). *Answering with OpenAI and LangChain: Harnessing the Potential of Retrieval Augmented Generation (RAG).* https://www.linkedin.com/pulse/ transforming-question-answering-openai-langchain-harnessing-routu/.
- Schwaber-Cohen, R. (2023). *What Is a Vector Database?* https://www.pinecone.io/ learn/vector-database.

Thinking for the audience

- Bettes, S. (2019). *Technical and Professional Writing Genres.* https://open.library .okstate.edu/technicalandprofessionalwriting/chapter/chapter-2/.
- Emery, A. K. (2021). *Why "Know Your Audience" Is Terrible Dataviz Advice—And What to Do Instead.* https://depictdatastudio.com/why-know-your-audience-is -terrible-dataviz-advice-what-to-do-instead/.
- QuantHub. (2023). *How to Identify Your Audience for Impactful Data Storytelling.* https://www.quanthub.com/how-to-identify-your-audience-for-impactful-data -storytelling/.
- LinkedIn Community with AI. (n.d.). *How Do You Engage and Nurture Your Technical Audience and Build Trust and Authority?* https://www.linkedin.com/advice/ 0/how-do-you-engage-nurture-your-technical-audience.
- WirelessLAN Professionals. (n.d.). *How to Present to a Technical Audience.* https:// wlanprofessionals.com/how-to-present-to-a-technical-audience/.

Transformers

- Koenigstein, N. (2024). *Transformers in Action.* Manning Publications.
- Vaswani, A., Shazeer, N., Parmar, N., Uszkoreit, J., Jones, L., Gomez, A. N., Kaiser, Ł., and Polosukhin, I. (2017). "Attention Is All You Need." *Advances in Neural Information Processing Systems, 30.*

From information
to knowledge: Building
the visual context

Generally, looking at something interesting arouses an *emotion*, a set of feelings that strike us. Of course, written texts can arouse emotions, and the power of visual elements to generate emotions is likely even greater. So why not exploit emotions to involve the audience more in our data stories? In this chapter, we will review the basic concepts behind visual context in a data story with a focus on emotions. We will describe how to set color, size, and interactivity in a data story. Next, we will deepen the usage of DALL-E for building images. Afterward, we will focus on how to place the context in your chart. Finally, we will focus on a practical case study implementing the described concepts.

8.1   Emotions: The foundations of visual context

I went to the Natural History Museum with my family some time ago. There were many animals, reconstructions of archaeological excavations, some primitive men, and much more. At the end of the visit, the children were so enthusiastic that they talked about nothing else. Again, after a few days, my youngest told me he wanted to build an archaeological excavation in his grandmother's garden!

The same type of thing should happen with our stories: our audience should still be talking about them after some time. They should be memorable. One way to achieve this, while directly involving our audience, is by eliciting *emotions*. Emotions are central to perceiving, interpreting, and remembering visual information. Emotions bridge the gap between raw data and human comprehension. Visual stimuli trigger emotional responses, influencing our perception and engagement with what we see. The *Collins Dictionary* (https://mng.bz/9dZl) defines an emotion as "a feeling such as happiness, love, fear, anger, or hatred, which can be caused by the situation that you are in or the people you are with."

Emotions play a crucial role in the effectiveness of visual communication. The visual context, including colors, shapes, and layout, can evoke specific emotional responses in the audience. For instance, warm colors like red and orange might evoke feelings of excitement or urgency, while cooler tones like blue and green can convey calmness or trust. Intentionally using these emotional cues in visual elements can significantly impact how information is perceived and retained by the audience.

In their book, *Communicate to Influence: How to Inspire Your Audience to Action*, Ben and Kelly Decker explain that we can understand emotions through the dual dimensions of energy and mood, which help categorize various emotional states based on the level of positivity or negativity of the emotion, as shown in figure 8.1 (Decker, 2015).

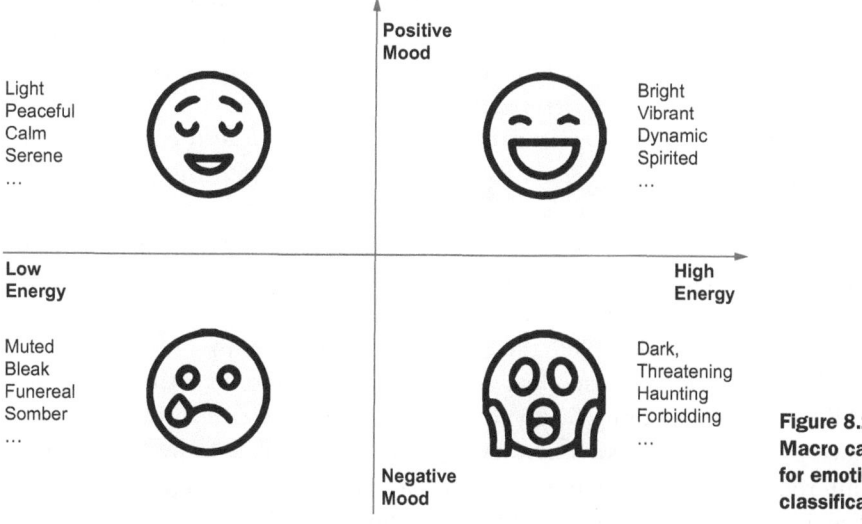

Figure 8.1 Macro categories for emotion classification

Emotions can vary widely along these dimensions, and individuals may experience and express emotions in unique ways, influenced by their personal and cultural backgrounds.

Considering the energy axis (x-axis), we have two types of emotions:

- *High-energy emotions*—These are often associated with heightened physiological arousal. They typically involve activating the body's "fight-or-flight" response. Examples of high-energy emotions include excitement, anger, and fear. These emotions are characterized by increased heart rate, heightened alertness, and a sense of urgency. High-energy emotions are often expressed through animated body language, enthusiastic gestures, and rapid speech.
- *Low-energy emotions*—These involve reduced physiological arousal and are characterized by a sense of calm and relaxation. Examples of low-energy emotions include contentment, tranquility, and sadness. People experiencing low-energy emotions may exhibit slower movements, a quieter demeanor, and reduced physical tension. Low-energy emotions often lead to contemplation and introspection.

Considering the mood axis (y-axis), we have two types of emotions:

- *High-mood emotions*—These are associated with positive valence, which evokes happiness, joy, and positivity. These emotions create an optimistic mood and often lead to smiles, laughter, and general well being. Examples include happiness, excitement, and love.
- *Low-mood emotions*—These are characterized by negative valence, leading to feelings of sadness, anger, or anxiety. These emotions create a more somber mood. Low-mood emotions may result in expressions of sadness, frowns, and a general sense of distress. Examples include sadness, fear, and anger.

We can combine mood and energy to generate different emotions, as shown by the emoticons in figure 8.1.

When choosing colors, you should also consider making your content accessible and understandable to people with disabilities. For more details, refer to the Web Content Accessibility Guidelines (WCAG, https://wcag.com/) website and WebAIM (https://webaim.org/articles/contrast/).

Incorporating emotions in data stories is not a mere aesthetic choice but *a strategic approach to conveying knowledge*. By aligning emotions with the intended message, we can transform the audience's passive reception of information into an active quest for knowledge and understanding. For example, high-energy emotions can capture immediate attention and incite curiosity, driving the viewer to explore the visual content more deeply. Conversely, low-energy emotions can encourage reflection, allowing the audience to contemplate the content more profoundly.

The quickest and most effective way to arouse emotions is through *sight*. Just as my children at the Natural History Museum were impressed by what they saw, our audience should be impressed by what they see in our visual representations. There are different ways to capture the audience's attention through sight, such as chart colors, size, interactivity, and images. There are tons of books and resources dealing with all these types of sights; thus, this book will give only a general overview of them with a focus on how to configure them in Altair. You can refer to the bibliography for more details. In addition, this chapter will focus on how to use generative AI, specifically DALL-E, to generate images with specific emotions. Let's start with a general overview of colors and how to implement them in Altair.

8.2 Color

Color helps communicate a message by setting the scene and the tone. It creates an emotional connection between the data and the audience. Color comprises three main components:

- *Hue*—This refers to the dominant color wavelength. It identifies color family or color name (such as red, green, or purple).
- *Saturation*—This component describes the intensity of that color. It defines how pure a color is.
- *Brightness*—This measures how light or dark a color appears.

In the previous chapters, we saw that every data story has a main character and an adversary. When drawing a chart, choose a primary color to highlight your main character, and always use it when referring to them. Use an alternative color for your adversary, but if you want to move the adversary to the background, use a gray tonality for them. You can choose colors using one of the following palettes (see the e-book for full-color pictures):

- *Sequential color palettes organize quantitative data from high to low using a single color in various gradients or saturations.* Use this palette to show a progression rather than a contrast (figure 8.2).

Figure 8.2 A sequential color palette

- *Diverging color palettes highlight the ranges of quantitative data by using two contrasting hues on the extremes and a lighter-tinted mixture to highlight the middle range.* A diverging palette shows where data are on a spectrum, such as cold to hot (figure 8.3). Cold could represent lower quantitative data values, typically depicted with a cooler hue, while hot could refer to higher values, often represented with a warmer hue.

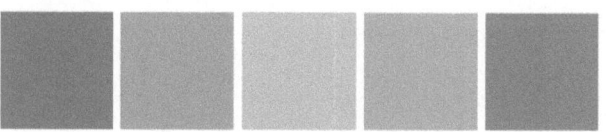

Figure 8.3 A diverging color palette

- *Qualitative color palettes highlight qualitative categories.* This palette type uses a different hue for each concept (figure 8.4).

Figure 8.4   A qualitative color palette

You can use colors to focus attention and highlight important information the chart wants to show. In addition, you can use color to create associations, such as brown or green for nature and shades of yellow for sunshine or bees. You can even use colors to evoke emotions, such as red for energy and blue for peace. Do not use overlapping colors to represent different concepts or too many colors.

> **Challenge**
>
> Consider the chapter 5 case study on the topic of homelessness. We used a tonality of green to color bars in our chart, which should evoke hope in the audience. The objective of the color was to demonstrate that our project would help solve the homeless problem. What happens if we use another color, such as #6F4E37 (coffee)? The message transmitted will probably be less intense than if green is used.

8.2.1   Setting colors in Altair

In the previous chapters, we have seen how to set colors in Altair using the color channel. In this section, we will provide more details. The following listing shows a quick reminder of how to use colors in Altair.

Listing 8.1   Setting colors in Altair

```
import pandas as pd
import altair as alt

df = pd.DataFrame({
    'Category' : ['A', 'A', 'B', 'B', 'C', 'C', 'C'],
    'Cost' : [12,45,64,23,45,78,72],
    'Sales' : [34,56,45,34,44,23,23],
    'Product' : ['P1','P2','P3','P4','P5','P6','P7']
})

chart = alt.Chart(df).mark_bar().encode(
        x=alt.X('Product:N'),
        y=alt.Y('Cost:Q'),
        color=alt.Color('Category:N')
    )

chart.save('chart.html')
```

NOTE   Use the alt.Color() channel to set the color.

The color channel may receive a specific color palette using the scheme parameter as input. Refer to the Altair documentation (https://vega.github.io/vega/docs/schemes/) for the supported palettes. In addition, you can use the scale parameter to specify the range and the domain of your colors:

- *The range parameter specifies the output range of the scale.* For example, when mapping a quantitative variable to color, use a color scale ranging from blue to red. Specify the range as a list of colors or as one of Altair's built-in color schemes.
- *The domain parameter specifies the input domain of the scale.* For example, if you have a quantitative variable with values between 0 and 100, set the domain of a color scale to [0, 100].

8.2.2   *Exercise: Setting colors*

Consider the chapter 3 case study on the population in North America over the last 50 years. You can find the complete code of this case study in the GitHub repository for the book under CaseStudies/population/population.ipynb. The final data story compared the population in North America with that in the rest of the world and showed a gap of 367 million people. We used a tonality of green to represent the North American trend line.

Now, take the following steps:

1 Change the color of the main chart (i.e., the chart showing the difference from 1960) to #963232 (a tonality of red), while keeping the green tonality in the next step chart (i.e., the chart entitled Strategies for Population Growth in North America).
2 Compare the obtained results with the chart implemented in chapter 3. Did the use of different colors change the message in any way? You will probably notice a negative emphasis on the fact that population growth in North America is slower than in the rest of the world. However, the next steps still remain optimistic.

A POSSIBLE SOLUTION

There are different ways to obtain the same result. A possible solution is to set the colors locally for each chart. For example, you can set the colors for the next step charts.

Listing 8.2   Setting colors in the next step chart

```
color='#80C11E'

df_cta = pd.DataFrame({
    'Strategy': ['Immigration Development', 'Enhance Family-Friendly Policies',
     'Revitalize Rural Areas'],
    'Population Increase': [20, 30, 15]  # Sample population increase percentages
})
```

```
# Creating the stacked column chart
cta = alt.Chart(df_cta).mark_bar(color=color).encode(
    x='Population Increase:Q',
    y=alt.Y('Strategy:N', sort='-x', title=None),
    tooltip=['Strategy', 'Population Increase'],
).properties(
  title=alt.TitleParams(
      text='Strategies for population growth in North America',
      color=color)
)
```

> **NOTE** Add a local color to a chart. Using this method, when you configure all the layered charts, this chart will not be affected by the global configuration.

This example shows a basic example of conveying emotions using colors. For more details, refer to specific books about the topic, such as those in the references section at the end of the chapter. Now that you have learned the basic concepts of conveying emotions using colors, let's move on to the next step: using size.

8.3 Size

When creating a chart, consider the *size*, as it can significantly affect how the audience perceives your data. Size impacts your visual representation's readability as well as overall impact and effectiveness. Choosing an appropriate size for your chart depends on multiple factors, such as where you will display it, what kind of content you want to show, and who will view it. The best way to determine the correct size is through experimentation; try different sizes until you land on one that looks good enough.

Moreover, choosing a smaller or larger-sized chart affects how much information can fit into one frame without cluttering or squishing text together. It also influences how easily readers can read all data points without straining their eyes or getting lost in too many elements. You can configure two main types of sizes in a chart: font size and chart size.

8.3.1 Font size

Font size directly affects the readability and comprehensibility of the data. The font size should be set to make it easy for the reader to understand and interpret the information presented in the chart.

When setting up a chart, you can specify different font sizes for different chart components, like titles, axis labels, or tick labels. A good practice is to keep them consistent throughout your plot so that they look visually appealing and organized.

By default, Altair sets the font size to 12 px. However, you can adjust the font size to make your charts more legible or to match the style.

To change the font size in Altair, identify which part of your visualization you want to modify. Once identified, use Altair's `configure_*()` method along with the `axis`, `title`, `text`, or `legend` property, depending on what component you want to modify. You can then pass parameters like `fontSize` or `labelFontSize` along with their desired value.

It is essential not to go overboard with increasing font sizes, as larger fonts might lead to cluttered plots that are hard on the eyes. On the other hand, smaller fonts may result in difficult-to-read charts. So always pick a reasonable font size that complements your charts well without negatively affecting readability.

8.3.2 Chart size

The final aspect to consider when setting the size of your Altair chart is the actual chart size. You can adjust this using the width and height arguments in your property. Both arguments are expressed in pixels.

There is no one-size-fits-all approach when determining the appropriate chart size. It will depend on factors such as where you plan on displaying the chart, what type of data it represents, and how much detail you want to include.

However, you can follow these general guidelines:

- If you want to create a simple chart to include in an email or presentation slide, use a smaller size.
- If you want to create a complex chart, use a larger size.

To set the chart size in Altair, use one of the following:

- `properties(width=my_width, height=my_height)` for single charts
- `configure_view(width=my_width, height=my_height)` for single or compound charts.

8.3.3 Exercise: Setting size

Consider the case study implemented in chapter 1—Increase the Advertising Campaign on Dog-Related Websites (https://mng.bz/KZaK):

1 Change the size of the x-axis label only for the value 9, as shown in figure 8.5.
2 Compare the obtained result with the original chart. Is the chart clearer?

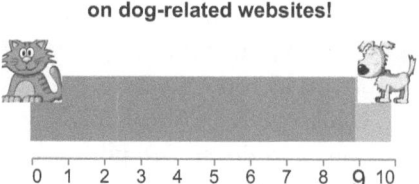

Figure 8.5   The case study from chapter 1 with increased font size for the value 9

SOLUTION
The following listing shows how to encode the x channel to increase the font size for just one axis label.

Listing 8.3   Setting label size

```
x=alt.X('x:Q',
        scale=alt.Scale(domain=[0, 10]),
        axis=alt.Axis(tickMinStep = 1,
            grid=False,
            title=None,
            orient='bottom',
            labelFontSize=alt.condition(alt.datum.value == 9,
                alt.value(25),
                alt.value(12)))
        ),
```

NOTE   Use the `alt.condition()` method to set the label size based on a condition.

Now that you have learned how to configure size in Altair, let's move on to the next step: interactivity.

8.4   *Interaction*

So far, we have implemented static charts that present a static story to the audience. The benefit of this chart type is that you decide the story and guide the audience through it. In other words, you don't allow the audience to explore data autonomously.

Adding interactive elements to a chart, such as buttons, menus, and zoom options, transforms a static chart into a dynamic one. With a dynamic chart, you build dynamic data stories, where the audience builds their own story. With a dynamic chart, you lose control over the audience and the possibility of having a focused and precise message. The audience is free to explore data and extract their insights, potentially leading to other calls to action. For this reason, it's important to design in advance which parts of the charts you want to make browsable by the audience based on the message to communicate.

Use interactive charts when you want the audience to explore data dynamically or when dealing with large datasets. For example, for financial data or in an educational context, an interactive chart allows the audience to zoom in on specific time-frames or hover over data points for details. Conversely, you may prefer static charts for simple presentations or reports where the data doesn't require exploration. For instance, a static chart could be ideal for a printed annual report that presents trends or comparisons and doesn't require user interaction. Additionally, in situations where the message needs to be straightforward and consistent, static charts may be more effective.

Consider, for example, the variant of the scenario described in chapter 3. The chart contains a drop-down menu enabling you to select the reference country to show in the chart. This chart type is okay for a data exploratory phase, but it won't work to build a data story simply because it has no focus. Instead, if you focus on North America and leave the audience the possibility to compare it with another country, you are still telling the story about the population in North America while giving the audience the freedom to compare it with other countries.

There are two main types of interactivity:

- *Passive interactivity*—This refers to features that allow the audience to interact with the chart without changing its state, such as tooltips or hover effects.
- *Active interactivity*—This involves changing the chart's state in response to user input.

In the remainder of this section, we will see how to add a tooltip, a slider, and a drop-down menu to an Altair chart.

8.4.1   Tooltip

A *tooltip* is a tiny pop-up box that appears when you hover over certain chart elements, such as data points or bars. To add a tooltip to your Altair chart, specify which element(s) you want it to appear for. In Altair, you can do it through the tooltip encoding channel using square braces [], as shown in the following listing. The example uses the population dataset of chapter 3, focusing on a completely different insight: the population growth in 2018 by continent.

Listing 8.4   Setting a tooltip

```
import pandas as pd
import altair as alt

df = pd.read_csv('data/population.csv')

# [...]

df = df[df['Country Name'].isin(continents)]

color = '#80C11E'

chart = alt.Chart(df).mark_bar(
    color=color
).encode(
        y=alt.Y('Country Name:O', sort='-x', title=''),
        x=alt.X('Population:Q'),
        tooltip=['Country Name', 'Population']
).transform_filter(
    alt.datum.Year == 2018
)                                        ◁──┐ Creates
                                             │ visualization
# [...]

chart.save('tooltip.html')
```

NOTE   Use the `tooltip` channel to set a tooltip in Altair.

You can find the complete code of the example in the GitHub repository for the book under 08/tooltip.py. Figure 8.6 shows a snapshot of a tooltip produced when hovering over a bar. Now that you have learned how to generate a tooltip in Altair, let's move on to the next interactive element: the slider.

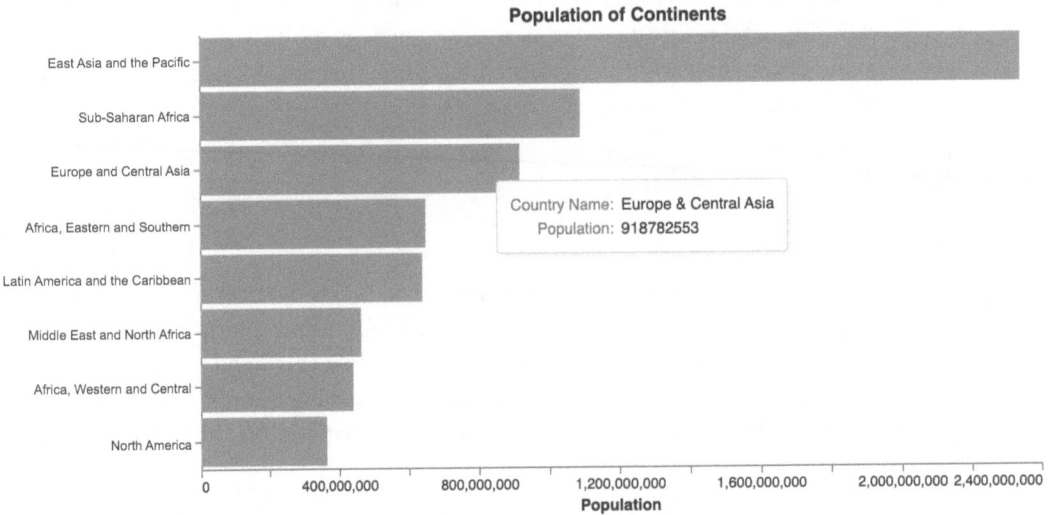

Figure 8.6   An example of a tooltip

8.4.2   Slider

A *slider* is a graphical element that allows users to input or adjust a value within a specified range by moving a visual indicator along a track or bar. To build a slider in Altair, follow these steps:

1 Create a `binding_range()`, an object that defines the range of values to which the slider can be set.
2 Create a selection using the `selection_point()` method. The selection is an object that defines how the slider will interact with the chart.
3 Add the selection to the chart using the `add_params()` method.
4 Optionally, apply the `transform_filter()` method to the chart to show only the selection.

Listing 8.5 shows an example of the usage of sliders. The example still uses the population dataset from chapter 3, focusing on the population growth over the years by continent. You can find the complete code in the GitHub repository for the book under 08/slider.py. The listing shows only the relevant code to implement a slider. For more details, read the code in the GitHub repository.

Listing 8.5   Setting a slider

```
import pandas as pd
import altair as alt

df = pd.read_csv('data/population.csv')

# [...]

color = '#80C11E'
slider = alt.binding_range(min=1960, max=2021,step=1)

select_year = alt.selection_point(name="Select", fields=['Year'],
                                    bind=slider)

chart = alt.Chart(df).mark_bar(
    color=color
).encode(
        y=alt.Y('Country Name:O', sort='-x', title=''),
        x=alt.X('Population:Q'),
).add_params(
    select_year
).transform_filter(
    select_year
)

# [...]

chart.save('slider.html')
```

> **NOTE**   After importing the required libraries, use the `binding_range()` function to define a binding range for the slider with a minimum of 1960, maximum of 2021, and step size of 1. Next, use the `selection_point()` function to create a selection named Select, and specify the field it should bind to (in this case, Year) using the slider. After that, use the `add_params()` method to add the Select_year selection to the chart and the `transform_filter()` method to filter the data based on the selected year from the slider.

Figure 8.7 shows a static representation of the resulting chart. Now that you have learned how to generate a tooltip in Altair, let's move on to the next interactive element: the dropdown menu.

8.4.3   *Drop-down menu*

A *drop-down menu* is a graphical element that typically appears as a small rectangular box with an arrow or triangle icon, indicating that additional options or choices are available. When a user clicks or hovers over the drop-down menu, a list of selectable items or options "drops down," or expands, below or above the menu. The user can

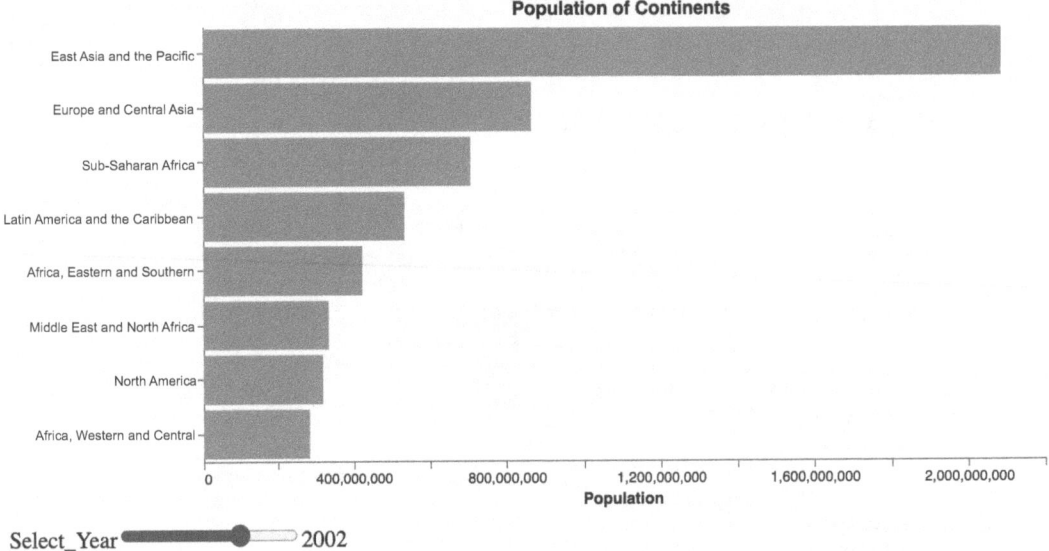

Figure 8.7   An example of a slider

then choose from the available options by clicking them. To create a drop-down menu in Altair, follow these sequential steps:

1 Begin by generating a `binding_select()` object. This object defines the list of values assigned to the drop-down menu.
2 Create a selection using the `selection_point()` method. This selection object governs the interaction between the drop-down menu and the chart.
3 Integrate the selection into the chart using the `add_params()` method.
4 Apply the `transform_filter()` method to display only the selected data.

Listing 8.6 shows an example of the usage of a drop-down menu. The example uses the population dataset of chapter 3, focusing on the population growth over the years by continent. The drop-down menu enables you to select a specific continent. You can find the complete code in the GitHub repository for the book under 08/dropdown.py. The following listing shows only the relevant code to implement a slide. For more details, read the code in the GitHub repository.

Listing 8.6   Setting a drop-down menu

```
import pandas as pd
import altair as alt

df = pd.read_csv('data/population.csv')

# [...]
```

```
color = '#80C11E'
input_dropdown = alt.binding_select(options=df['Country Name'].unique())
select_country = alt.selection_point(name='Select',fields=['Country Name'],
    bind=input_dropdown, value=[{'Country Name': 'Africa Eastern and
    Southern'}])


# Create visualization
chart = alt.Chart(df).mark_line(
    color=color
).encode(
        y=alt.Y('Population:Q', title=''),
        x=alt.X('Year:O'),
).add_params(
    select_country
).transform_filter(
    select_country
)


# [...]
chart.save('dropdown.html')
```

NOTE   After importing the required libraries, use the `binding_select()` function to define a binding selection for the drop-down menu. Next, proceed as you do for sliders.

Figure 8.8 shows a static representation of the resulting chart.

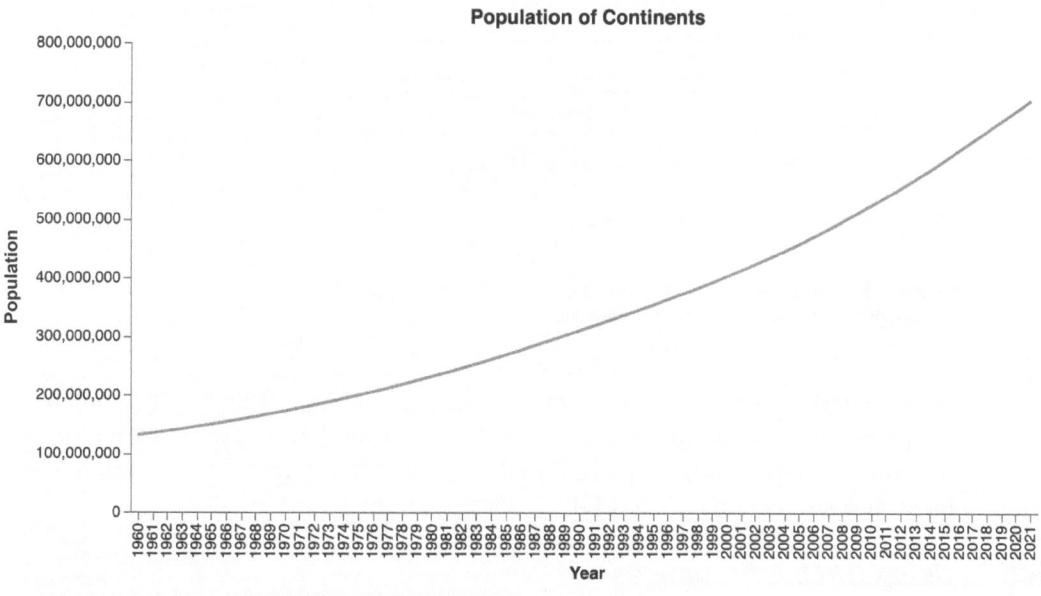

Figure 8.8   An example of a drop-down menu

Now, it's time to practice. So let's proceed with an exercise.

8.4.4   *Exercise: Setting interactivity*

Consider the case study implemented in chapter 3: Population in North America Over the last 50 Years. You can find the complete code of this case study in the GitHub repository for the book under CaseStudies/population/population.ipynb:

1 Remove the Others line, and add a dynamic line showing a selectable continent from a drop-down menu, as shown in figure 8.9. Leave the North America line static.
2 Compare the obtained result with the original chart. For which type of audience is this dynamic graph most suitable?
3 Optionally, adapt all the graph elements to the animations.

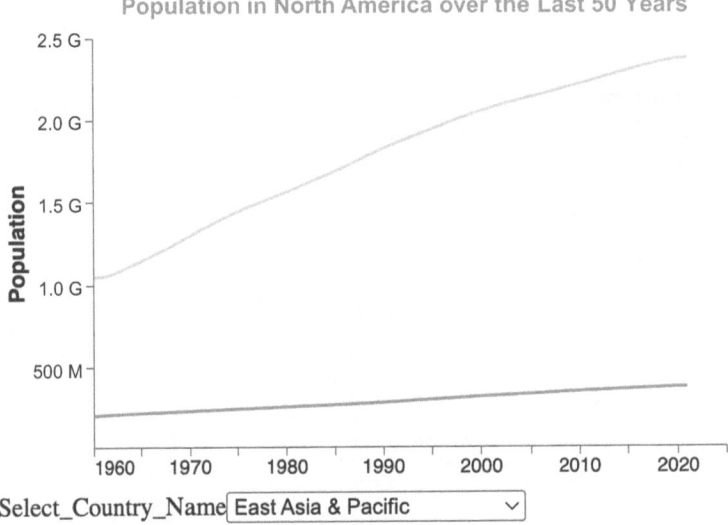

Figure 8.9   The case study of chapter 3 with a dynamic drop-down menu for the continent to compare with North America

You can find the solution to the exercise in the GitHub repository for the book under 08/population-exercise.py. Now that you have learned how to introduce interactivity in Altair through tooltips, sliders, and drop-down menus, let's move to the next element, images, focusing on DALL-E.

8.5   *Using DALL-E for images*

Imagine reading a text that describes the following scene: *a beautiful little girl with black hair and dark eyes, wearing a yellow dress, running in a green meadow full of yellow flowers.*

You can use your creativity to set up the scene in your mind, but this could require a little effort from your brain. Now, look at figure 8.10 (realized with DALL-E), showing the same scene visually. You see exactly what we described before, but surely, you depicted the scene differently in your mind.

Figure 8.10   An image of a little girl wearing a yellow dress

The same situation occurs when you describe the subject of your data story. If you don't give a face to your subject and give only a general description, your audience is free to imagine the subject in various ways, depending on their background. Instead, if you add one or more images to your data story, you help the audience have a clear representation of the story's subject. You will bring the audience up to the intended emotional level through images, especially photos, and they will be more open to listening to your data story. Adding images to a data story reinforces the context and prepares the audience to accept the data story message.

In chapter 4, you learned how to draw images in DALL-E and insert them into a chart. In this section, you'll see how to use DALL-E-generated images to evoke emotions. We'll also look at how to maintain consistency in the images we're generating in DALL-E. Let's start with emotions.

8.5.1   Adding emotions

The DALL-E book of prompts suggests some keywords to add emotions to images. Figure 8.1 shows some possible adjectives associated with each emotion category:

- *High energy, positive mood*—Bright, vibrant, dynamic, spirited
- *High energy, negative mood*—Dark, threatening, haunting, foreboding

- *Low energy, positive mood*—Light, peaceful, calm, serene
- *Low energy, negative mood*—Muted, bleak, funereal, somber

You can read the complete list of keywords in the DALL-E book of prompts (https://mng.bz/0GNx).

Using emotional keywords enables you to obtain different results. Figure 8.11 shows a possible output produced by DALL-E for the following input: *a calm woman, playing in a garden, photograph.*

Figure 8.11   An image of a calm woman in a garden

Figure 8.12 shows a totally divergent result, produced by simply changing the keyword *calm* to *muted.*

Figure 8.12   An image of a muted woman in a garden

Incorporating this type of image into a data story helps the audience give your main character a face. The main character of your story can communicate directly with the audience through their emotions and expressiveness. Now that you have learned how to add emotion to images generated by DALL-E, let's move on to the next step: generating consistent images.

8.5.2 Generating consistent images

Consistent images are a set of images sharing common characteristics, such as colors, style, and so on. It is important to build consistent images when you want to add more than one image to your data story. At the time of writing this book, DALL-E does not provide any explicit way to build consistent strategies. However, in this section, you will see a trick proposed by Tyler Taggart in his YouTube video (Taggart, 2023) to build consistent images in DALL-E. To do it, we will use the Editor tool, provided by DALL-E and introduced in chapter 4.

To show how the proposed strategy works, we will redraw the images related to rowing and cycling used in the chapter 4 case study: Unlock the Potential: Invest in Rowing and Cycling for Maximum Returns! The idea is to insert real photos in the data story, tailored to executives to capture their attention better.

Start by writing the following prompt in DALL-E: *a photo showing the effort of a rowing athlete. Show off the sweat and muscles. The athlete wears a green jacket.* Figure 8.13 shows a possible output produced by DALL-E.

Figure 8.13　A possible output produced by DALL-E

Select a photo among the four available, and click Edit. Now, click Add a Generation Frame, and insert it near the image, as shown in figure 8.14.

Now, edit your prompt as follows: *a photo showing the effort of a cycling athlete. Show off the sweat and muscles. The athlete wears a green jacket.* You have simply changed the word *rowing* to *cycling*. DALL-E will generate four images. Select the one that best fits your requirements, or generate new images. Figure 8.15 shows a possible output. The two parts of the photos (left and right) are very similar. For example, they use the same green tonality.

As you can see, the two photos are linked to each other. The result looks a bit strange, but you can split them (producing two normal-looking photos) by downloading them and using any image editing tool.

Generation frame: 1024 x 1024

Figure 8.14   Where to put the generation frame

Figure 8.15   The output produced by DALL-E

Now that you have obtained photos of the two best-performing sports, you can replace them in the chart from the case study in chapter 4, as shown in figure 8.16. We have also added the subtitle produced with fine-tuning to the chart. The photos in the story highlight the effort made by the athletes to achieve their records, thus inviting the executives to continue financing them.

Now, it's time to practice. Let's move on to an exercise.

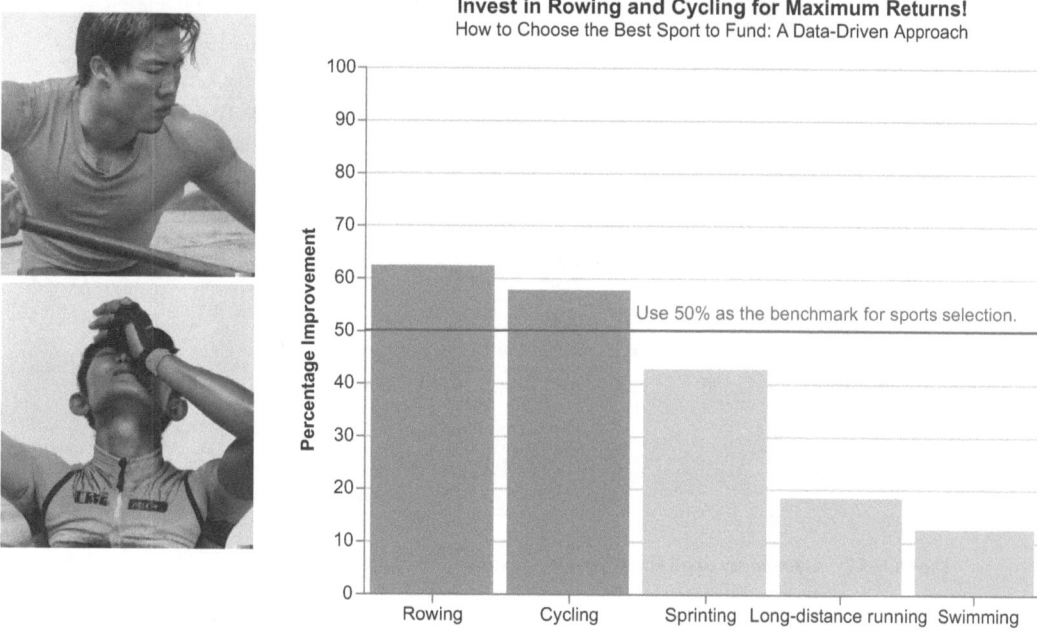

Figure 8.16   **The final chart tailored to an audience of executives**

8.5.3   *Exercise: Generating images*

Consider the case study in chapter 1: Increase the Advertising Campaign on Dog-Related Websites:

- Generate two consistent images for a dog and a cat, and replace the original ones in the chart with the generated images.
- Generate a new image for a dog (or a cat) with a specific emotion (e.g. angry, calm, relaxed). Is DALL-E able to generate emotional images for animals?

We have just completed our journey toward using DALL-E to generate images for data stories. As a last step in this chapter, we will consider tips on positioning your chart in your data story.

8.6   *Strategic placement of context*

Context should precede the main point of your data story. For this reason, I suggest you place your context in one of the following three main positions in your data visualization chart: on the top, on the left, or within the chart. In the remainder of this section, we will investigate each context position separately. As an example of using each context position, we will focus on the case studies described in chapters 1 through 5. For convenience, figure 8.17 summarizes the case studies analyzed.

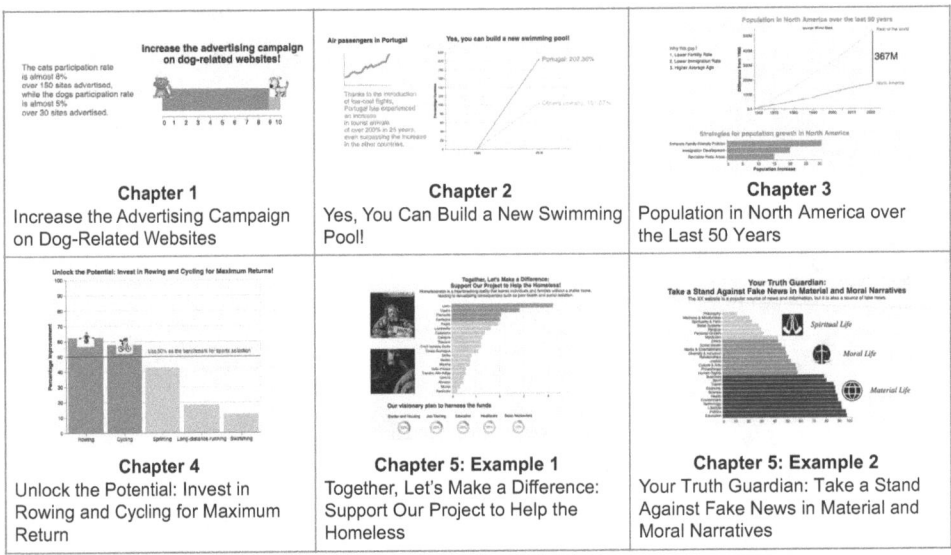

Figure 8.17   A summary of all the examples described in chapters 1 through 5

8.6.1   Top placement

Adding context at the top of a chart means adding a textual description immediately under the title, as shown in figure 8.18.

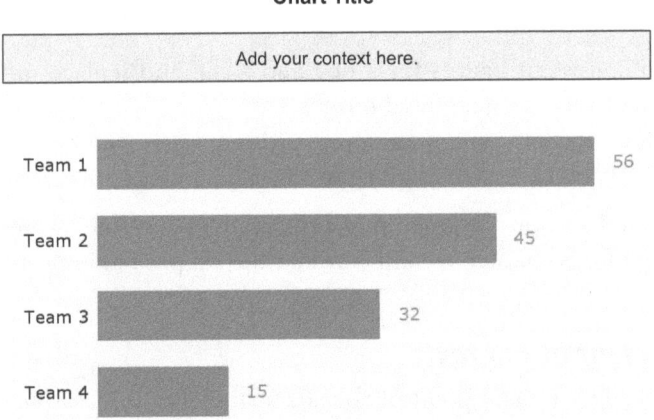

Figure 8.18   The context position at the top of the chart

Use the top of the chart to describe an overview of the topic or data sources. You can also use the top part of the chart to add the data source. Adding data sources means acknowledging the source of information used in a chart. This gives credibility to the information presented. When the audience sees that we have correctly credited the data source, they are more likely to trust the information presented.

Consider the case studies in figure 8.17. Table 8.1 shows the case studies using the top context and the type of context used.

Table 8.1   Case studies using the top context position

Name	Type of Top Context	Text
Chapter 3: Population in North America Over the Last 50 Years	Data source	Source: World Bank
Chapter 5—Example 1: Together, Let's Make a Difference: Support Our Project to Help the Homeless	Textual description	Homelessness is a heartbreaking reality that leaves individuals and families without a stable home, leading to devastating consequences, such as poor health and social isolation.
Chapter 5—Example 2: Your Truth Guardian: Take a Stand Against Fake News in Material and Moral Narratives	Textual description	The LatestNews website is a popular source of news and information, but it is also a source of fake news.

8.6.2   Left placement

Adding context on the left of a chart involves adding one or more annotations immediately before the title, as shown in figure 8.19.

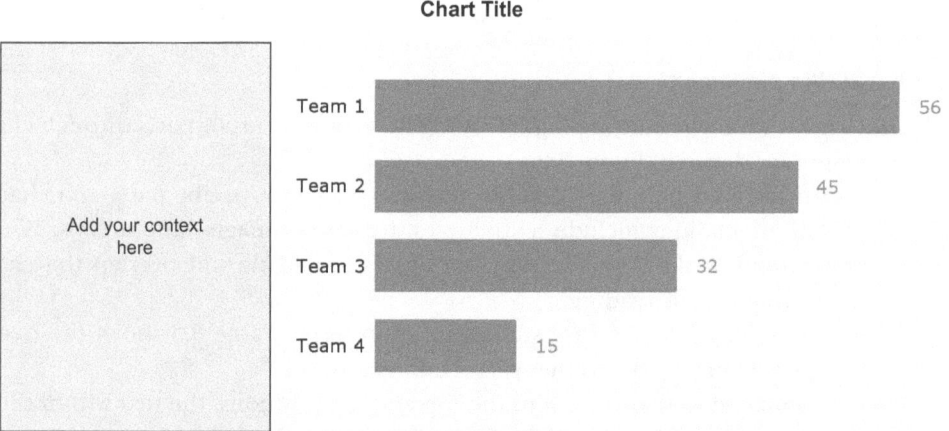

Figure 8.19   The context position on the left of the chart

Use the left part of the chart to add preliminary charts or deeper details, such as textual descriptions, images, or photos that affect audiences' emotions. Consider again the case studies in figure 8.17. Table 8.2 shows the case studies using left context and showing the type of context used.

Table 8.2   Case studies using the left context position

Name	Type of Left Context	Text
Chapter 1: Increase the Advertising Campaign on Dog-Related Websites	Textual description	The cat participation rate is almost 8% over 150 sites advertised, while the dog participation rate is almost 5% over 30 sites advertised.
Chapter 2: Yes, You Can Build a New Swimming Pool!	Secondary chart Textual description	Thanks to the introduction of low-cost flights, Portugal has experienced an increase in tourist arrivals of over 200% in 25 years, even surpassing the increase in the other countries.
Chapter 3: Population in North America Over the Last 50 Years	Textual description	Why this gap? 1. Lower fertility rate 2. Lower immigration rate 3. Higher average age
Chapter 5—Example 1: Together, Let's Make a Difference: Support Our Project to Help the Homeless	Images	–

Compared to that written in the top position, the text in the left position contains additional details, sometimes extracted from other sources.

8.6.3 *Within placement*

Context within a chart highlights some interesting points or parts through chart annotations, as shown in figure 8.20.

Annotations help the audience to focus on some specific parts of a chart. Examples of annotations include text, circles, baselines, images, and arrows. When using annotations, make sure they are not excessive and do not overlap the chart itself, interfering with its reading.

Consider again the case studies in figure 8.17. Table 8.3 shows the case studies using the left context and the type of context used.

Compared to that written in the top and left positions, the text within the annotation contains concise text, highlighting data details. Now that you have learned how to add a visual context to your data story, let's implement a practical case study.

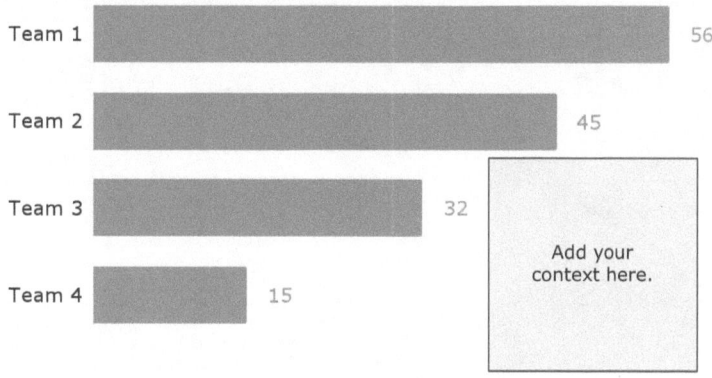

Figure 8.20   The context position within the chart

Table 8.3   Case studies using the within context position

Name	Type of Within Context	Text
Chapter 1: Increase the Advertising Campaign On Dog-Related Websites	Image	–
Chapter 3: Population in North America Over the Last 50 Years	Text Vertical line	367M
Chapter 4: Unlock the Potential: Invest in Rowing and Cycling for Maximum Return	Text Baseline Images	Use 50% as the benchmark for sports selection.
Chapter 5—Example 2: Your Truth Guardian: Take a Stand Against Fake News in Material and Moral Narratives	Images Text	Material life Moral life Spiritual life

8.7   Case study: From information to knowledge (part 2)

In the previous chapter, we analyzed how to turn information into knowledge through textual context in the aquaculture case study. Just for a quick reminder, the case study involved building a story around the safety problem in salmon aquaculture in the United States. We decided to plot the salmon aquaculture sales trend line versus the other types of aquaculture. As an insight, we discovered that since 1998, there has been an increase in sales, following a period of decrease in sales from 1992 to 1998. We discovered that the decreasing period was partially due to some health problems in the salmon aquaculture. Figure 8.21 shows the chart produced when adding textual context to turn information into knowledge.

Aquaculture Exports of Salmon in the U.S.
Aquaculture seafood in the U.S. is regulated by the FDA to ensure safety. Strict standards are in place to monitor water quality, feed, and disease control. Regular inspections and testing are conducted to minimize risks and protect consumers. (Source: U.S. Food and Drug Administration)

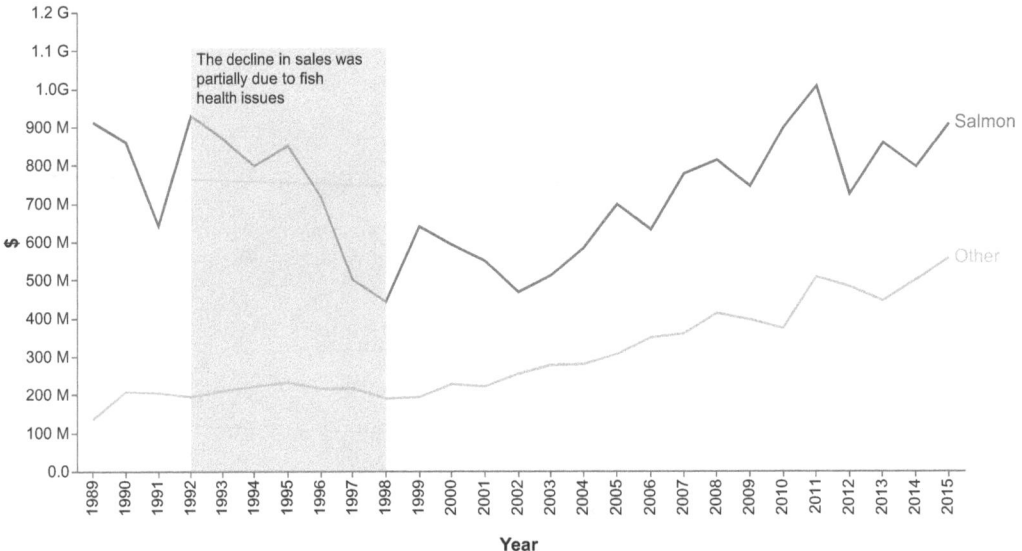

Figure 8.21 The chart produced when adding textual context

The next step involves adding visual context to the story. We will follow two different scenarios:

- Communicating a negative mood, with a focus on the negative years (1992–1998)
- Communicating a positive mood, with a focus on the years where the trend is increasing (from 1998).

In both cases, we will generate images, set colors, and pick a size related to the specific mood. Let's start with the first point: communicating a negative mood.

8.7.1 Setting a negative mood

In this scenario, the focus is on the negative years. The aim is to show the dreadfulness of this period to raise awareness among the audience of the urgency of ensuring that it never happens again. We will generate shocking images and use dark colors.

IMAGES

Generate shocking images (high energy, negative mood) about salmon. Let's use DALL-E with the following prompt: *a shocking photo of a dead salmon from a disease.* Figure 8.22 shows a possible output of images generated by DALL-E.

Let's choose photo number 2 and add it to the chart. Listing 8.7 shows the instructions for Copilot to generate the image in Altair.

Figure 8.22   The output produced by DALL-E with the following input: *a shocking photo of a dead salmon from a disease*

Listing 8.7   Instructions for Copilot

```
# build a DataFrame named img_df with the following columns:
#   - 'url' with the values '../images/deadsalmon.png'
#   - 'x' with the value 1993
#   - 'x2' with the value 1997
#   - 'y' with the value 0

# draw a image with the img_df DataFrame:
# - set x to 'x'
# - set y to 'y'
# - set x2 to 'x2'
```

> **1 is a symbolic number.**

> **NOTE**   First, instruct Copilot to generate a DataFrame containing all the relevant channels. Next, specify how to draw the image using Altair. We have asked Copilot to set y = 0. This is an initial value. You can adapt it manually by looking at the image.

You can find the complete code in the GitHub repository for the book under Case-Studies/from-information-to-knowledge-visual/chart-negative.py.

COLOR AND SIZE

In addition to the photo, we can use dark colors to emphasize the negative scenario. For example, we can use a dark red tonality (#460805), which suggests blood. Next, we can increase the stroke width in correspondence to the years 1992–1998. We generate another line for those years, overlapping with the original chart. The following listing shows the code to generate the line.

Listing 8.8   Adding a focused line

```
base = alt.Chart(df
).encode(...)

chart = base.mark_line()

chart_line = base.mark_line(
    strokeWidth=8
```

```
).transform_filter(
    (alt.datum.CATEGORY == 'Salmon') &
    (alt.datum.YEAR_ID >= 1992) &
    (alt.datum.YEAR_ID <= 1998)
)
```

NOTE   First, build a base chart with all the encoding options. After drawing the main chart, draw another chart focusing on a selection. Use the `transform_filter()` method to select only some data.

Again, you can find the complete code in the GitHub repository for the book under CaseStudies/from-information-to-knowledge-visual/chart-negative.py. Figure 8.23 shows the resulting chart.

Aquaculture Exports of Salmon in the U.S.
Aquaculture seafood in the U.S. is regulated by the FDA to ensure safety. Strict standards are in place to monitor water quality, feed, and disease control. Regular inspections and testing are conducted to minimize risks and protect consumers. (Source: U.S. Food and Drug Administration)

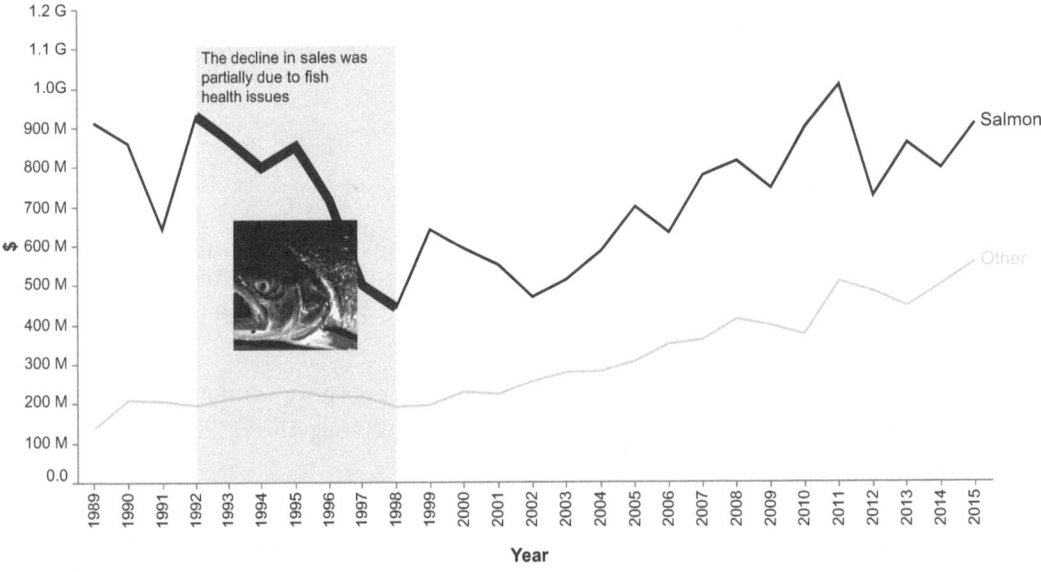

Figure 8.23   **The chart after adding a negative visual context**

Now that you have implemented the negative-mood scenario, let's move to the opposite one: the positive mood.

8.7.2   *Setting a positive mood*

In this scenario, the focus is on the positive trend line, beginning in 1998. The objective is to encourage the audience to continue to adopt the current strategy for the sake of the salmon population. Instead of focusing on the negative effects, here, we can focus on the positive scenario, mentioning the negative aspects only as a possibility. We will use relaxing photos and colors.

IMAGES

The procedure to generate the relaxing images is similar to the negative scenario, with the difference of using words falling in the top-left clock face of figure 8.1 (low energy, positive mood). Use the following prompt for DALL-E to generate the required images: *a calm aquaculture of safe salmon*. Figure 8.24 shows a possible output. Let's choose photo number 1.

Figure 8.24   The output produced by DALL-E with the following input: a calm aquaculture of safe salmon

COLOR AND SIZE

Use a peaceful color (e.g., #105473) to represent the salmon line. In addition, highlight the trend line since 1998. Figure 8.25 shows the resulting chart. You can find the complete code in the GitHub repository for the book under CaseStudies/from-information-to-knowledge-visual/chart-positive.py.

We have generated two types of charts—negative and positive—but not both together (say, with both images in the same graph). This is because we don't want conflicting emotional cues, potentially confusing the audience or diluting the intended message we aim to convey. Now that you have learned how to add a positive or a negative mood to a data story, let's move on to an additional exercise: adding interactivity.

8.7.3   Exercise: Making the chart interactive

Modify the previous chart to make it interactive as follows: define a drop-down menu with two values: Positive and Negative. When the user selects Negative, the chart highlights the salmon trend line between 1992 and 1998. When the user selects Positive, the charts highlights the trend line since 1998. You can find the solution to the exercise in the GitHub repository for the book under CaseStudies/from-information-to-knowledge-visual/chart-dropdown.py.

In the first part of this chapter, you learned how to turn information into knowledge by adding visual context to your data visualization chart. You saw how to add emotions to your charts in terms of colors, size, and interactivity. You also learned how

Aquaculture Exports of Salmon in the U.S.

Aquaculture seafood in the U.S. is regulated by the FDA to ensure safety. Strict standards are in place to monitor water quality, feed, and disease control. Regular inspections and testing are conducted to minimize risks and protect consumers. (Source: U.S. Food and Drug Administration)

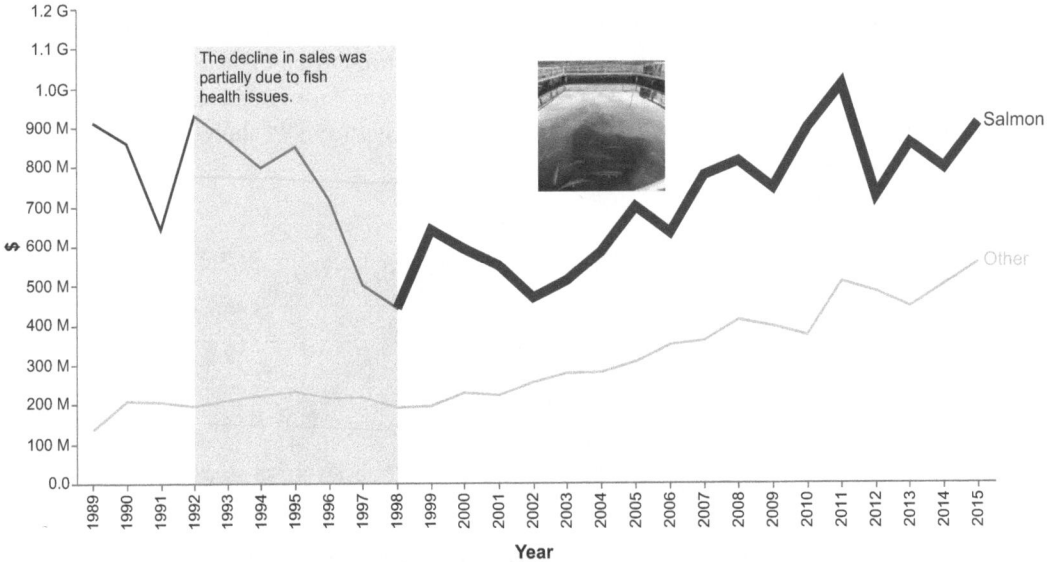

Figure 8.25   The chart after adding a positive visual context

to generate mood-based and consistent images in DALL-E. Next, you learned where to put the context in your chart. Finally, you implemented a practical case study. In the next chapter, you will see how to turn knowledge into wisdom by adding a call to action and possible next steps to your chart.

Summary

- Adding visual context to your data visualization is crucial for raising audience emotions.
- Incorporating colors, size, and interactivity helps you to better focus your story and tailor it to your audience.
- When you draw your images in DALL-E, keep them consistent using the editing tool.
- Consider positioning the context strategically within your chart to enhance its effectiveness.

References

Using emotions for communication

- Decker, B. and Decker, K. (2015). *Communicate to Influence: How to Inspire Your Audience to Action.* McGraw Hill.

- Abbott, D. (2023). *Everyday Data Visualization.* Manning Publications.
- Christiansen, J. (2022). *Building Science Graphics: An Illustrated Guide to Communicating Science through Diagrams and Visualizations.* A K Peters/CRC Press.
- Strachnyi, K. (2022). *ColorWise: A Data Storyteller's Guide to the Intentional Use of Color.* O'Reilly Media.

Images

- QuantHub. (2023). *Photos and Illustrations in a Visual Narrative.* https://www.quanthub.com/photos-and-illustrations-in-a-visual-narrative/.
- *The DALL·E 2 Prompt Book.* (n.d.). https://dallery.gallery/wp-content/uploads/2022/07/The-DALL%C2%B7E-2-prompt-book-v1.02.pdf.
- Gemignani, Z. (2021). *Tips for Using Photos in Data Storytelling.* https://www.juiceanalytics.com/writing/tips-for-using-photos-in-data-stories.
- Hang, I. (2023). *How to Create Consistent Characters in Midjourney: EASY Step by Step AI Tutorial for Amazon KDP.* https://www.youtube.com/watch?v=zdSIPkbvsek.
- Nussbaumer Knaflic, C. (2018). *Using Images.* https://www.storytellingwithdata.com/blog/2018/3/26/using-images.
- Taggart, T. (2023). *DALL-E 2 Tutorial: How to Get Image Consistency.* https://www.youtube.com/watch?v=MU_yXYsfBR0.

From knowledge to wisdom: Adding next steps

9

This chapter covers

- Introducing wisdom and next steps
- Reviewing all the case studies
- Strategic placement of next steps

If you are reading this chapter, you have nearly climbed to the top of the DIKW pyramid and have built a fantastic chart. You should be delighted with your work and have probably already shown it to colleagues and perhaps even friends and relatives. It pains me, however, to tell you that your work is incomplete. The last piece is missing, the one that will transform your graph into a real story: wisdom. Wisdom is something very profound, which cannot be extracted simply from data. Wisdom also comes from one's cultural background, experiences, and the ability to have good judgment. In this chapter, we will not make philosophical disquisitions on wisdom. Still, we will consider wisdom as the ability to propose the next steps for the audience to follow after interacting with our story. We will also use generative AI as a cultural background and shared experience to help us define the next steps.

In this chapter, we will start with the basic concepts behind wisdom and next steps (at the top of the DIKW pyramid in the context of data storytelling). Next, we will review all the case studies described in this book in the light of the defined concept of wisdom. Finally, we will describe how to strategically place the next steps in the story.

9.1 Introducing wisdom

The *Collins Dictionary* defines *wisdom* as "the ability to use experience and knowledge to make sensible decisions or judgments." Applied to data storytelling, wisdom enables us to make better decisions based on our data. Wisdom helps us to continue the story after it has ended.

Wisdom is not simply the result of transforming knowledge. It also requires other aspects that depend on our baggage in life. Wisdom is the result of a few different components:

- *Knowledge*—This provides the foundation upon which wisdom is built. We cannot make informed decisions without a broad understanding of our data.
- *Experience*—This provides the context through which our knowledge is applied. It deepens our understanding of the world and helps us to see the nuances and complexities that can't be learned from books or lectures. Experience encompasses all the moments we've lived through, the challenges we've faced, and the lessons we've learned.
- *Good judgment*—This is the ability to weigh the pros and cons of a situation, analyze the risks and benefits, and come to a well-informed and wise decision. It requires the ability to evaluate a situation from multiple perspectives and understand the potential outcomes of each choice.

Figure 9.1 shows how the DIKW pyramid should be modified to include wisdom properly.

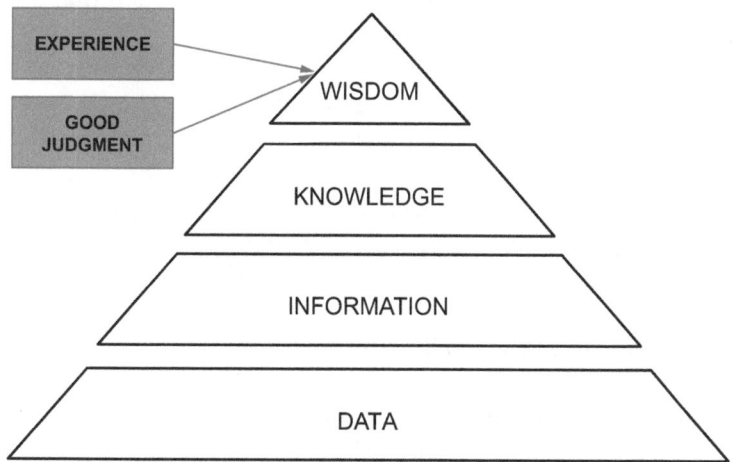

Figure 9.1 The DIKW pyramid enriched with the components of wisdom

Summarizing, we can say that wisdom comprises three main elements: knowledge synthesis, experience, and good judgment. Let's investigate the three elements separately, starting with knowledge synthesis.

9.1.1 *Transforming knowledge into wisdom: Next steps*

Every self-respecting story carries a *message*, something that remains in the hearts of the audience after reading the story. And this is the addition of wisdom to the story. Wisdom allows the audience to be engaged by the story, to be touched, and to be ready to do something after reading the story. However, the message alone is not enough. The audience must do something after reading the story.

Transforming knowledge into wisdom involves adding the *next steps* to the data story. Next steps are the resolution of our data story and defining what the audience should do after reading the story.

Every story ends with an *ending*, which can be good or bad, but there must be one. Even in the case of data storytelling, the addition of an ending to the story is envisaged. Unlike a novel, where the ending concludes the story, data storytelling should be an open story, in the sense that it should not end with the story itself but should expect the audience to do something after reading the story.

To better understand the problem, let's consider an example. Imagine you have started reading a beautiful novel. While you are reading about the protagonist's adventures, suddenly, the novel ends without revealing how or whether the protagonist has overcome the challenges or solved the problems. Well, if you limit your chart only to the knowledge phase, it's like a novel without an end. You shouldn't just present your data findings; you should tell the viewer how you think they should act upon those findings. There's a well-known saying that "knowledge is not power; it is potential power when used for action." Here, then, in your chart, you must add what we'd call *next steps*. These are implemented as a call to action that invites your audience to do something.

The next steps are not something dropped from the sky or, even worse, something theoretical. They are the results of your studies and analysis. They propose to your audience what to do after seeing your chart—for example, regarding possible alternatives or concrete steps to move forward.

The next steps can never be just the fruit of the knowledge acquired through data. They must be anchored to something broader, commonly called *wisdom*. Wisdom is a cultural treasure trove derived from experience (from the data analyzed) and knowledge shared with others.

Based on the context in which you live, the type of audience, and your values, the next steps must be anchored to an ethical framework. You cannot think of proposing a generic next step based only on your own beliefs or experience. It is necessary to anchor the next steps to the values of the society in which you live if they are to have broad application. These values vary from community to community, but fundamental principles must be respected, and the next steps must be anchored to them.

Table 9.1 shows the most popular categories of next steps. We provide each with a short description, purpose, and example.

Table 9.1 The most popular categories of next steps

Category	Description	Purpose	Example
Ask for support	Ask the audience to support the story in some way.	Take advantage of the audience's competencies to solve the problems highlighted in the story.	A text inviting the audience to participate in a survey
Provide different options	Provide the audience with potential alternatives to proceed.	Aid in the audience's decision-making process.	A list of possible alternative next steps: A, B, C
Free interaction	Allow the audience the possibility to freely interact with the story.	Let the audience analyze the data and draw conclusions.	An interactive chart
Learn more	Encourage the audience to delve deeper into the topic or insights presented in the data story.	Direct the audience to additional resources, articles, studies, or references for a more comprehensive understanding.	A link to an in-depth analysis report
Propose a plan	Propose a plan outlining the sequence of actions to be taken.	Let the audience continue working on the story after its end.	A list of possible sequential next steps
Sharing	Encourage the audience to share the data story, for example, on their social networks.	Leverage the audience's networks, foster discussions, and increase visibility to amplify the reach of the data story.	Using social media buttons to share the story

Now, you have learned the different categories of next steps. Next, let's see the second element contributing to defining wisdom, experience, and how we can use ChatGPT to help us generate it.

9.1.2 *Using ChatGPT as a source of experience*

Traditionally, experience derives from our personal baggage, which has some limitations, especially if we don't have much experience in a specific field. In addition to our personal experience, we can use generative AI knowledge to build "collective baggage," to generate wisdom. So far, we have used generative AI only to generate content. Now, it's time to use it as an assistant, a teammate, which assists us in thinking and elaborating concepts. While using generative AI in such a way, be careful about the ethical problems and other issues this may raise, which we will discuss in the next chapter. In practice, we will use ChatGPT as a source of experience. In addition to using our singular experience to propose next steps, we will use ChatGPT as another source of experience.

To be more specific, in this chapter, we will use generative AI in two ways:

- As a content generator, as we have done so far in the book
- As an assistant, helping us in thinking to transform knowledge into wisdom

In this section, we will focus only on the second point, since you already learned how to generate content with generative AI in the previous chapter. In particular, we will focus on ChatGPT.

Imagine ChatGPT as your personal assistant for thinking. We will use ChatGPT as an assistant and, in particular, its ability to interact through an interactive chat, where it remembers the previous conversation. We can use ChatGPT as a natural interlocutor who helps us think and develop strategies. In this chapter, we will use ChatGPT only to propose the next steps, but in general, you can use all the cultural backgrounds of ChatGPT to generate new ideas. For ChatGPT to help us solve our problems, we need to formulate a prompt so that ChatGPT can understand it.

Formulate a prompt consisting of three parts:

1 A definition of the scenario
2 A question
3 A number of possible answers to generate

We can actually combine points 2 and 3 into a single point. After this initial prompt, we can discuss with ChatGPT and ask it to deepen one or more generated answers.

We will see many applications of this prompt structure later in this chapter. For now, it is sufficient to remember this: scenario, question, number of answers.

Once we have extracted an idea from ChatGPT, we can further investigate the topic by looking at additional external material. ChatGPT is only the first point. It acts as a source of ideas. Next, we need to verify the feasibility of ChatGPT's proposals by consulting external materials and sources. Now that you have learned conceptually how to use ChatGPT's cultural baggage as a source of experience, let's move on to the following step: good judgment.

9.1.3 Good judgment: Anchoring the action to an ethical framework

Good judgment is assessing situations accurately and making decisions that align with one's values and goals. In this chapter, we will implement good judgment by anchoring our message to an ethical framework. An ethical framework serves as a guideline for decision making based on principles that guide moral behavior. It provides a framework for evaluating the potential consequences of our actions and helps us identify the best course of action based on what is right and just.

Actions not grounded in ethical considerations can lead to unintended consequences and may even cause harm. Some of the most popular ethical frameworks include the following (Berengueres, 2019):

- *Utilitarianism*—This framework bases ethical decisions on maximizing utility or happiness. According to this view, an action is right if it leads to the greatest happiness for the greatest number of people.
- *Deontology*—This framework focuses on following moral rules or duties. It argues that certain actions are inherently good or bad, regardless of their consequences.
- *Virtue ethics*—This framework emphasizes the importance of developing virtues such as courage, honesty, and compassion. It argues that good people are more important than following rules or maximizing happiness.
- *Care ethics*—This framework emphasizes the importance of relationships and empathy. It bases ethical decisions on caring for others and responding to their needs.

Regardless of the ethical framework chosen, the next steps must be anchored within one if you wish to safeguard your call to action against misinterpretation or misuse. This means considering the potential impact of our actions on others and ensuring that they align with our values and principles. By doing so, we can move beyond simply making decisions based on what is expedient and toward creating positive change rooted in ethical considerations.

To adapt next steps to a specific ethical framework, we can implement the following two phases:

- Using ChatGPT to generate a preliminary text
- Revising the produced content

In the next sections, we will see how to implement these two phases. Now that you have learned the main concepts related to the three components of wisdom—knowledge synthesis, experience, and good judgment—let's apply them to the case studies implemented throughout this book.

9.2 Case studies

In this section, we will review all the case studies implemented throughout the book in the light of more focused next steps. As a quick reminder, figure 9.2 summarizes the case studies implemented in the previous chapters. Let's start with the first case study, the dogs and cats campaign.

9.2.1 Chapter 1: The dogs and cats campaign

Consider the case study we analyzed in chapter 1, the dogs and cats campaign. As a refresher, the case study focused on an event dedicated to pets (dogs and cats). For each pet category, the organizers strategically advertised the event on specialized websites dedicated to that particular pet category. As a result, 9 out of 10 pets participating in the event were cats. The next steps included improving the campaign related to the

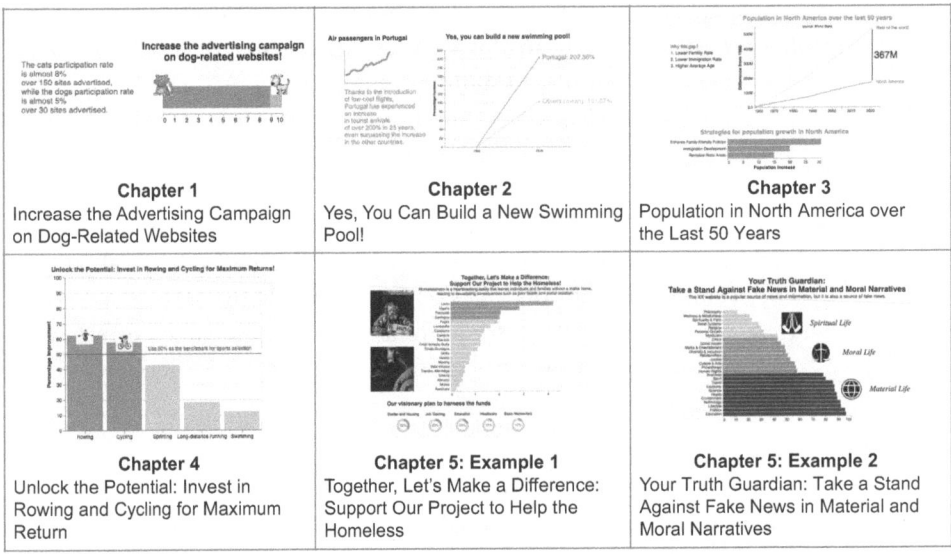

Figure 9.2   A summary of all the examples described in the previous chapters

dog websites to increase the dog participants. Figure 9.3 shows the final data story related to this case study.

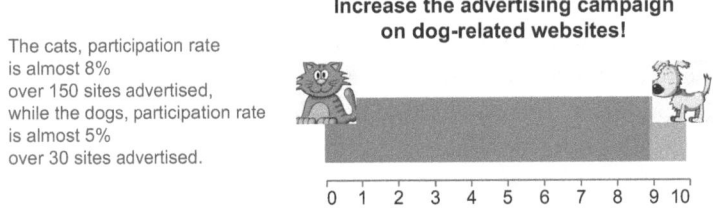

Figure 9.3   The final data story related to the dogs and cats case study

The case study described in chapter 1 contained a generic next step, which invited the audience to increase the dog-related website campaign. Let's implement a more focused next step, using the sharing category. To get some idea of practical implementation, let's ask ChatGPT how we can practically improve these sharing opportunities. Let's formulate the following prompt: *Consider the following scenario: a case study focuses on an event dedicated to pets (dogs and cats). For each pet category, the organizers strategically advertised the event on specialized websites dedicated to that particular pet category. As a result, 9 out of 10 pets participating in the event were cats. Propose 5 ways to invite the audience to share information about the campaign related to the dog websites to increase the number of dog participants.* ChatGPT will write a list of five sharing options. Here is a possible list of options generated by ChatGPT:

- Interactive contests or challenges
- Engaging social media campaigns
- Blogger outreach and partnerships
- Tailored email campaigns
- Community engagement and forums

Here, we have reported only the list of options. You can find the complete answer on ChatGPT's site (https://chatgpt.com/share/75475a1d-7dab-4487-858d-8288f7408737). We can elaborate on the previous options and, for example, reformulate them as three options: online campaign, influencer engagement, and social media promotion.

Let's implement the previous steps. We will implement the sharing options as three rectangles connected by a line, as shown in figure 9.4.

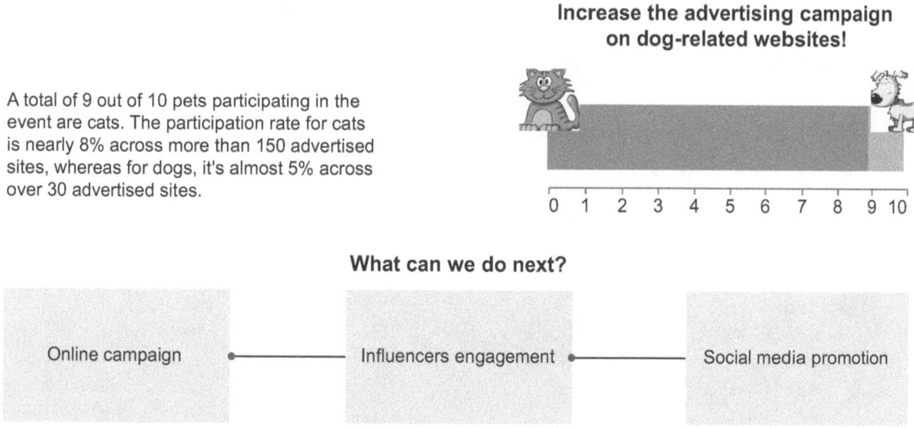

Figure 9.4   The dogs and cats case study enriched with a visualized roadmap

The implemented code is available in the GitHub repository for the book under 09/awareness-and-campaigns/pets-visualized-roadmap.py. We will focus only on the next step (in the bottom part of the chart). Start by defining the rectangle size. We use parametric values to make the code as general as possible. The following listing shows the code to generate the three rectangles at positions 0, 15, and 30.

Listing 9.1   Setting the rectangles' positions

```
import pandas as pd

width = 10
space = 5
N = 3
```

```
x = [i*(width+space) for i in range(N)]
y = [0 for i in range(N)]
x2 = [(i+1)*width+i*space for i in range(N)]
y2 = [10 for i in range(N)]
text = ['Online Campaign', 'Influencers Engagement', 'Social Media Promotion']

df_rect = pd.DataFrame(
    {   'x': x,
        'y': y,
        'x2': x2,
        'y2': y2,
        'text' : text
    }
)
```

NOTE We suppose that we want to generate N (3 in the example) rectangles and that each rectangle has a large `width` (10) and the space between two rectangles is `space` (5). Then, we generate the rectangle coordinates using x, x2, y, and y2 as vertices. We also include the label text for each rectangle. Finally, we create a DataFrame with the defined variables.

Next, draw the rectangles in Altair.

Listing 9.2   Drawing the rectangles

```
import altair as alt

rect = alt.Chart(df_rect).mark_rect(
    color='#80C11E',
    opacity=0.2
).encode(
    x=alt.X('x:Q', axis=None),
    y=alt.Y('y:Q', axis=None),
    x2='x2:Q',
    y2='y2:Q'
).properties(
    width=700,
    height=100,
    title=alt.TitleParams(
        text=['What can we do next?'],
        fontSize=20,
        offset=10
    )
)
```

NOTE Use `mark_rect()` mark to draw rectangles. Use the x, y, x2, and y2 channels and remove the axes (`axis=None`). Finally, set the chart size and the title. Use `alt.TitleParams()` to specify the title font size (`fontSize`) and the title offset from the chart.

Now, we are ready to draw the labels in the rectangles, as shown in the following listing.

Listing 9.3   Drawing the labels

```
text = alt.Chart(df_rect).mark_text(
    fontSize=14,
    align='left',
    dx=10,
).encode(
    text='text:N',
    x=alt.X('x:Q', axis=None),
    y=alt.Y('y_half:Q', axis=None),
).transform_calculate(
    y_half='datum.y2/2'
)
```

> **NOTE**   Use the `mark_text()` mark to draw the labels. Use the `transform_calculate()` method to set the y coordinate for each label.

The following step involves defining the line coordinates. We must draw N-1 lines, with the first starting at position 10 and ending at position 15 and the second starting at position 25 and ending at position 30. Use a formula to calculate these positions.

Listing 9.4   Defining the line coordinates

```
x = [width*i+space*(i-1) for i in range(1,N)]
y = [5 for i in range(N-1)]
y2 = [5 for i in range(N-1)]
x2 = [(width+space)*i for i in range(1,N)]

df_line = pd.DataFrame(
    {   'x': x,
        'y': y,
        'x2': x2,
        'y2': y2
    }
)
```

> **NOTE**   Use a formula to calculate the x coordinates and set the y coordinates to a constant value.

Draw the lines.

Listing 9.5   Drawing the lines

```
line = alt.Chart(df_line).mark_line(
    point=True,
    strokeWidth=2
).encode(
    x=alt.X('x:Q', axis=None),
    y=alt.Y('y:Q', axis=None),
    x2='x2:Q',
    y2='y2:Q'
)
```

> **NOTE** Use the `mark_line()` mark to draw the line. Set `point=True` to show the points.

If we have drawn the previous elements of the dogs and cats chart, we can build the final chart.

Listing 9.6   Building the final chart

```
chart = ((context | (chart + annotation + img)) & (rect + line + text)
).configure_view(
    strokeWidth=0
).resolve_scale(
    color='independent',
    x='independent',
    y='independent'
).configure_axis(
    grid=False
)

chart.save('pets-visualized-roadmap.html')
```

> **NOTE** The final chart includes the `context`, on the left; the main bar chart (`chart`); the `annotation` (texts dog and cat) and images (`img`), on the right; and the next steps (`rect`, `line`, and `text`), below. Set `color`, `x`, and `y` to `independent` in the `resolve_scale()` method to render the chart scales independently.

Now that we have completed this example, let's try out an exercise on the same topic.

EXERCISE 1

Consider again the dogs and cats case study and implement the next steps to Facebook and Instagram, using the social media *sharing* option. Use the `href` channel to include the link to an external website. Figure 9.5 shows a possible result. You can find the solution to this exercise in the GitHub repository for the book under 09/awareness-and-campaigns/pets-social-sharing.py.

A total of 9 out of 10 pets particpating to the event are cats. The participation rate for cats is nearly 8% across more than 150 advertised sites, whereas for dogs, it's almost 5% across over 30 advertised sites.

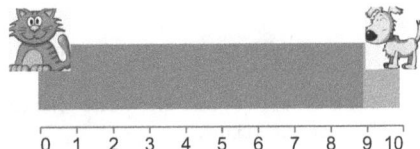

Increase the advertising campaign on dog-related websites!

0 1 2 3 4 5 6 7 8 9 10

Share on social media

Figure 9.5   The dogs and cats case study, enriched with a social sharing call to action

9.2.2   *Chapter 2: The tourist arrivals*

Consider the swimming pool case study we analyzed in chapter 2. As a quick reminder, the case study focused on the possibility of building a new swimming pool in a Portuguese hotel. The data story showed an increasing number of tourists in Portugal in recent years, as shown in figure 9.6.

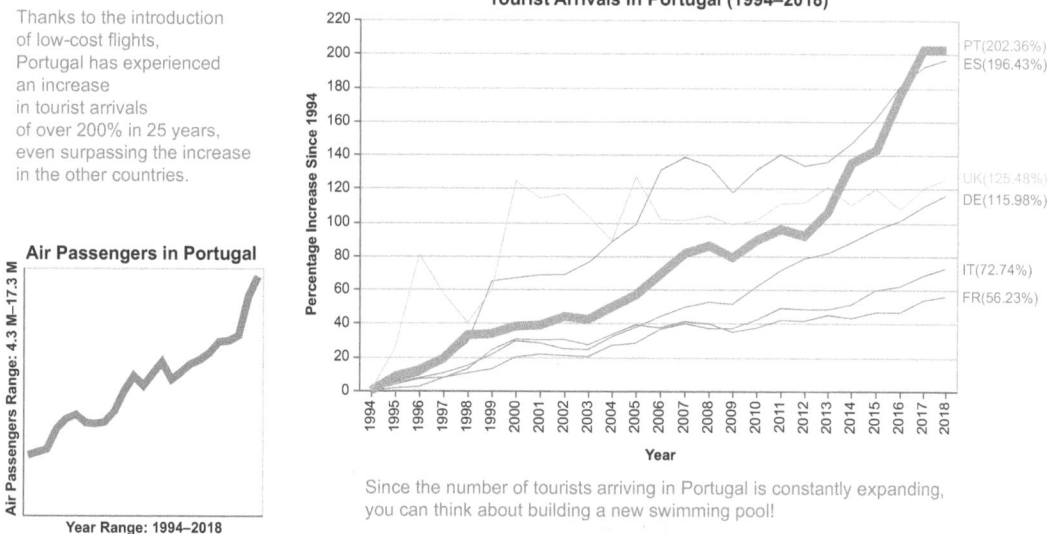

Figure 9.6   The final data story related to the swimming pool case study

The case study described in chapter 2 contained a generic next step, which invited the audience to build a swimming pool. Let's ask ChatGPT to suggest possible next steps in terms of different options to follow. Let's use the following prompt: *Consider a case study focusing on the possibility of building a new swimming pool in a Portuguese hotel. The data story shows an increasing number of tourists in Portugal in recent years. Provide 5 options to further investigate the feasibility of the swimming pool.* Here is a list of a possible ChatGPT answers:

1 Market research and demand analysis
2 Competitive analysis
3 Financial feasibility study
4 Regulatory and environmental assessment
5 Seasonality and utilization analysis

You can find the complete conversation at on ChatGPT's site: https://chatgpt.com/share/28f05921-b548-4edc-938d-ccfcb2d5bba6. For example, we can elaborate on the ChatGPT suggestions and propose the following next steps:

- Ask other hoteliers for their experience.
- Invite the administrative team to calculate the costs.

Let's implement the previous next steps, as shown in figure 9.7.

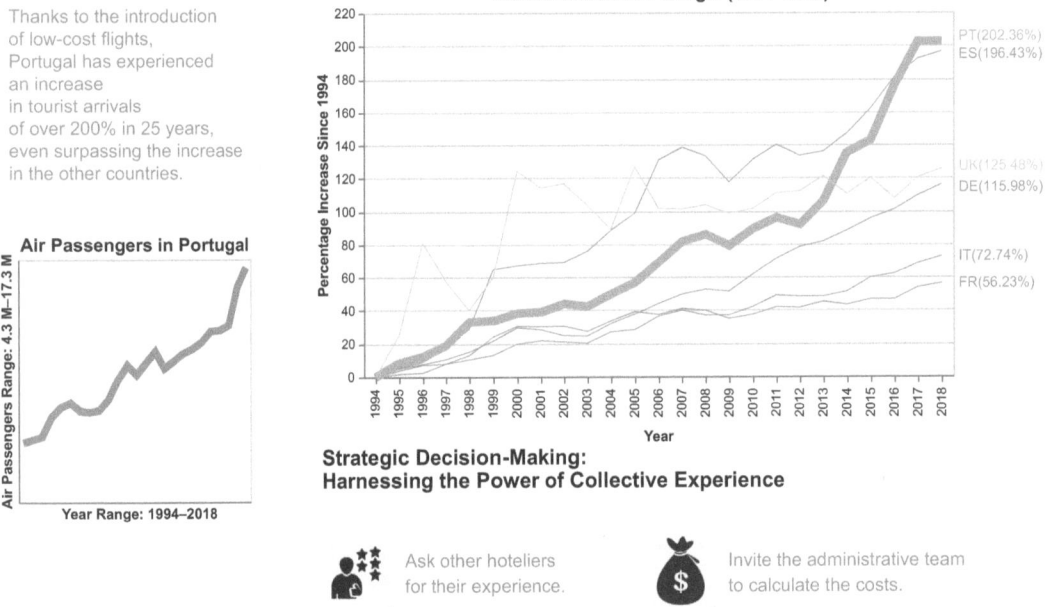

Figure 9.7   The dogs and cats case study enriched with a social sharing call to action

Start by generating the two black icons at the bottom of the data story in the next step section using DALL-E. For the first icon, use the following prompt: *customer experience, black and white icon.* Figure 9.8 shows a possible result produced by DALL-E. Choose one image and download it.

Figure 9.8   A possible output produced by DALL-E using the following prompt: *customer experience, black and white icon*

For the second, use the following prompt: *dollar bag, black and white icon.* Figure 9.9 shows a possible result produced by DALL-E. Choose one image and download it.

Figure 9.9 **A possible output produced by DALL-E using the following prompt:** *dollar bag, black and white icon*

Now, we are ready to write the code to generate the next step section. You can find the code in the GitHub repository for the book under 09/collaborative-efforts/actionable-recommendations.py. Let's focus on the call to action part of the code. Start by creating the DataFrame containing the texts and the images to include in the CTA.

Listing 9.7   The CTA DataFrame

```
df_cta = pd.DataFrame(
    {'text' : [['Ask other hoteliers', 'for their experience'], ['Invite the
      administrative team', 'to calculate the costs']],
     'x' : [0,9.8],
     'img' : ['img/experience.png', 'img/costs.png']})
```

NOTE Create a new DataFrame containing the CTA texts, their position on the x-axis, and the path to the images.

Next, let's use ChatGPT to generate the title for the CTA. Give ChatGPT for the following prompt: *Act as an inviter to action for executives. Generate four titles from the following text: Next steps: Ask other hoteliers for their experience. Invite the administrative team to calculate the costs.* Figure 9.10 shows a possible output.

 ChatGPT
1. "Driving Success: Collaborating with Peers in the Hotel Industry"
2. "Empowering Insights: Engaging Hoteliers for Strategic Growth"
3. "Strategic Decision-Making: Harnessing the Power of Collective Experience"
4. "Cost Evaluation Initiative: Uniting the Administrative Team for Progress"

Figure 9.10   A possible output produced by ChatGPT to generate four titles for the CTA

Let's choose the third title, *Strategic Decision-Making: Harnessing the Power of Collective Experience*, as the title for our next step part of the story. Visually speaking, we implement the next steps as the combination of two charts: a text and an image chart. Since the two charts share some common parts, let's draw a base chart first.

Listing 9.8   Building a base chart for the next steps

```
base_cta = alt.Chart(df_cta
).encode(
    x=alt.X('x:Q', axis=None)
).properties(
    title=alt.TitleParams(
        text=['Strategic Decision-Making:', 'Harnessing the Power of
    Collective Experience'],
        fontSize=18,
        anchor='start'
    )
)
```

NOTE   Build a basic structure of the chart without specifying any mark. Only specify the input DataFrame (`df_cta`), the shared encode channels (`x`), and properties (`title`). Use `TitleParams()` to specify the title details.

Once we have defined the basic chart, we can define each chart detail separately. The following listing describes how to implement the text.

Listing 9.9   Drawing the text chart

```
text_cta = base_cta.mark_text(
    lineBreak='\n',
    align='left',
    fontSize=20,
    y=0,
    dx = 40,
    color='#81c01e'
).encode(
    text='text:N'
)
```

NOTE   The text chart details the `base_cta` chart by adding the text mark property (`mark_text`) and the specific encode for the text (the `text` channel). Within the `mark_text()` property, specify some text properties, such as the line break character (`lineBreak`), how to align the text (`align`), the font size (`fontSize`), the y position (`y`), the x-axis shift from the x position (`dx`), and the text color (`color`).

Next, draw the images, as shown in the following listing.

Listing 9.10   Drawing the image chart

```
img_cta = base_cta.mark_image(
    width=50,
    height=50,
    y=10
).encode(
    url='img'
)
```

> **NOTE**   The image chart details the `base_cta` chart by adding the image mark property (`mark_image`) and the specific encode for the images (the `url` channel). Within the `mark_image()` property, specify some image properties, such as the image width (`width`), height (`height`), and y position (`y`).

Now, we can build the final chart. We assume that we have already implemented the previous parts of the data story (chart, annotation, commentary, and airport chart). The following listing shows how to layer the charts to render the final data story.

Listing 9.11   Building the final data story

```
chart = ((commentary & airports) | ((chart + annotation) & img_cta +
    text_cta)
).resolve_scale(
    x='independent',
).configure_view(
    strokeWidth=0
)

chart.save('actionable-recommendations.html')
```

> **NOTE**   The final chart includes the context (`commentary` and `airport`), the main chart (`chart` and `annotation`), and the next steps (`image_cta` and `text_cta`).

EXERCISE 2

Consider again the swimming pool case study, and implement the *learn more* next step. Use the `href` channel to include the link to an external website. Figure 9.11 shows a possible result. You can find the solution to this exercise in the book's GitHub repository under 09/collaborative-efforts/learn-more.py.

9.2.3   *Chapter 3: Population in North America*

Consider the case study we analyzed in chapter 3, Population in North America Over the Last 50 Years, which aims to study the population growth in North America. The case study discovered that there was a gap between the population growth in North America and the rest of the world. The story, as a next step, proposed some strategies to increase the population growth in the Unite States, as shown in figure 9.12.

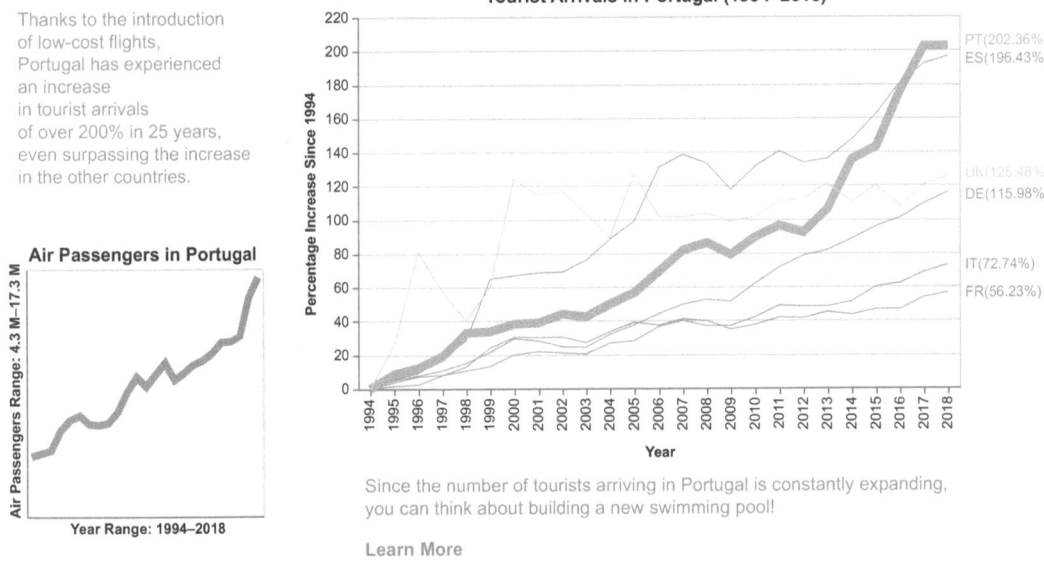

Figure 9.11   **The final data story related to the swimming pool case study**

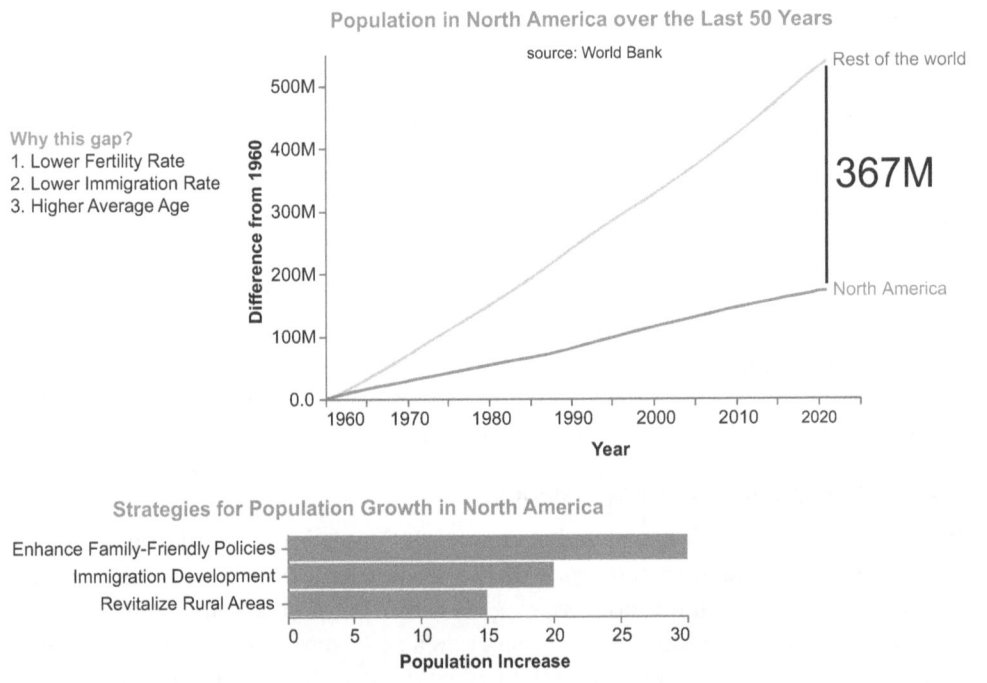

Figure 9.12   **The final data story related to the North American population case study**

This case study already contained some next steps as possible options. We can modify them, for example, by proposing a plan to follow sequentially. Let's start a conversation with ChatGPT with the following prompt: *Consider a case study that focuses on population growth in North America. The study discovered a gap between North American population growth and the rest of the world. Here are three possible options to increase the population growth in North America: enhance family-friend policies, develop immigration, and revitalize rural areas. Order the three options sequentially, based on which you would apply before and which after.* ChatGPT answers by ordering the three options and explains why it chose that order. In my case, it provided the following order:

1 Enhance family-friendly policies.
2 Develop immigration.
3 Revitalize rural areas.

We can also ask to prioritize one specific option, such as revitalizing rural areas, by providing the following prompt: *What if you first revitalized rural areas?* In my case, the answer was not interesting. We could further refine the topic by asking ChatGPT the following: *What is the easiest option to perform before?* According to ChatGPT, the easiest option is to enhance family-friendly policies. The conversation can continue until we find what we are looking for. You can find the complete conversation with ChatGPT used in this example on ChatGPT's site (https://chat.openai.com/share/78c6c6a1-f468-4271-aa60-ac0d410ce117). Once we have extracted this information from ChatGPT, we can further investigate the topic by looking at additional external material. ChatGPT is only the first point. It acts as an experience. Next, we need to verify the feasibility of the ChatGPT proposals.

This case study provided an example of engaging in a conversation with ChatGPT and using it as a teammate. Don't be afraid to ask ChatGPT. Whatever question you ask, even the most uncomfortable, it will always answer you.

9.2.4 *Chapter 4: Sport disciplines*

Consider the case study we analyzed in chapter 4, which is related to selecting the best sports disciplines to train in to achieve good results in the upcoming competitions (figure 9.13). As the next steps for our story, we proposed investing in cycling and rowing. This next step was too generic, so we can improve it.

In this example, we have embedded the next steps in the title. Let's improve them by detailing the investment we could make to improve rowing and cycling. Let's ask ChatGPT the technical skills required to improve results in the top two disciplines. Use the following prompt: *Consider a scenario related to sports disciplines to train in and achieve good results in upcoming competitions. As the next steps for our story, we proposed investing in cycling and rowing. Propose the required technical skills to improve these two disciplines.* ChatGPT generates a detailed list of skills for each of the two disciplines, including, for example, bike handling and positioning for cycling, stroke techniques, and balance and stability for rowing. You can find the complete answer generated by ChatGPT on their

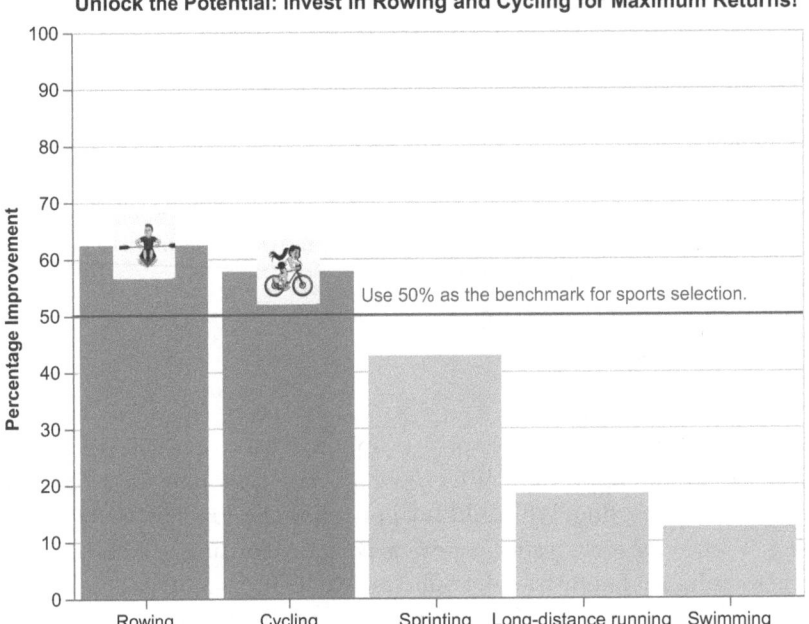

Figure 9.13   The case study in chapter 4

site (https://chat.openai.com/share/9a28714f-dca3-4bbd-9c11-72aa9e371959). Now, we can continue the conversation by asking ChatGPT to group the skills in more general categories. Use the following prompt: *Group the proposed skills in four general categories.* A possible generated output includes the following categories:

- Technical skills
- Physical skills
- Tactical skills
- Nutritional and mental skills

We can use the generated skills as possible next steps. We only modify nutritional and mental skills into psychological skills. Figure 9.14 shows the modified story. Once we have found this information, we can further deepen the topic by searching for additional material, such as the estimated contribution of each skill to the final result.

We suppose that the performance in each sport is given by four factors: physical, psychological, tactical, and technical. Each factor contributes to the performance differently, based on the sports discipline. Figure 9.14 shows some hypothetical values. In addition, the data story imagines that we want to focus on technical factors to improve the athletes' performance. Table 9.2 shows the structure of the dataset containing the contribution of each factor to sports performance.

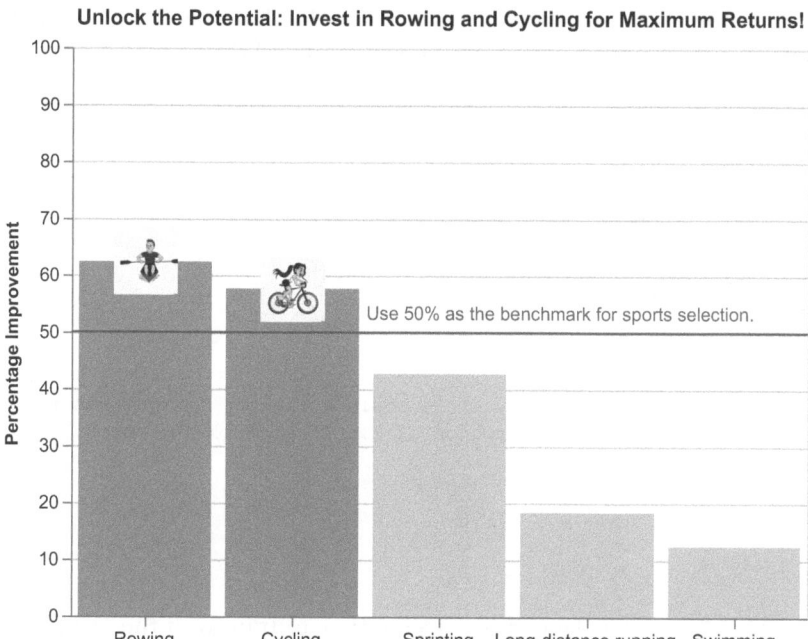

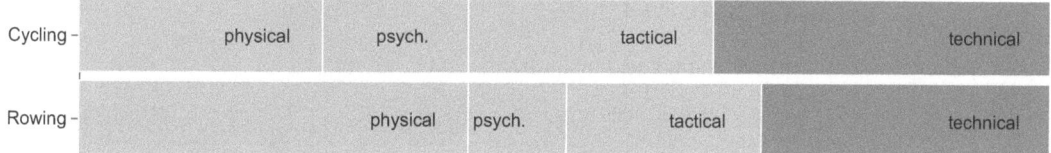

Figure 9.14   The athletes case study with an actionable recommendation as a next step

Table 9.2   The factor dataset

Rowing	Cycling	Factor
40	25	Physical
30	35	Technical
20	25	Tactical
10	15	Psychological

We will implement the next step as a stacked bar chart with a layered text. You can find the code of this example in the GitHub repository for the book under 09/training-

and-development/support-and-assistance.py. Let's start by defining the basic chart, with the encodings and properties shared by the bar chart and the text.

Listing 9.12   Building the basic chart

```
df_cta = pd.read_csv('source/factors.csv')

df_cta = df_cta.melt('Factor', var_name='Sport', value_name='Value')

base_cta = alt.Chart(df_cta).encode(
    y=alt.Y('Sport', title=''),
    x=alt.X('Value', axis=None, title='',  stack=True),
).properties(
    title=alt.TitleParams(
        text='Investing in technical skills could bring the maximum benefit',
        subtitle='What is the contribution of each factor to the sports
     performance?'
    ),
    width=600,
    height=100
)
```

NOTE   First, read the factors dataset, and use `melt()` to transform the Data-Frame from wide to long format. Next, create the base chart (`base_cta`) by defining the input dataset (`alt.Chart(df_cta)`), the encodings (x and y), and the properties (`title`, `width`, and `height`). For the x channel, specify `stack=True` for the stacked chart.

Now, we can build the stacked bar chart.

Listing 9.13   Building the stacked bar chart

```
cta = base_cta.mark_bar(
    strokeWidth=3,
    stroke='white'
).encode(
    color=alt.Color('Factor',
            scale=alt.Scale(
                    range=['lightgrey', '#80C11E',
                        'lightgrey', 'lightgrey'],
                    domain=['physical', 'technical',
                        'taktical', 'psycol.']
            ),
            legend=None),
)
```

NOTE   Use the base chart to build the stacked bar chart. Set the mark property (`mark_bar`) and the color channel.

Finally, draw the text and combine it with the other parts of the data story, as shown in the following listing.

Listing 9.14   Drawing the text

```
text_cta = base_cta.mark_text(
    xOffset=-35,
    fontSize=14,
    color='black'
).encode(
    text = 'Factor:N',
)
chart = ((chart + annotation) & (cta + text_cta)
).configure_view(
    strokeWidth=0
)
```

> **NOTE**   Use the base chart to build the text chart. Add the mark property (`mark_text`) and the text channel. Next, draw the final chart, which includes the main chart (`chart`), the `annotation`, and the next steps (`cta` and `text_cta`) below.

Challenge: Improving the next steps

Improve the next steps in figure 9.14 by replacing the y labels (Cycling and Rowing) with the same icons shown in the main chart.

Now that you have learned how to implement training and development next steps, let's move on to the next category: strategic actions.

9.2.5   *Chapter 5: Homelessness*

Consider the case study on homelessness by Italian region we analyzed in chapter 5, shown in figure 9.15. The case study focused on searching for funds to finance a project about homeless people. In the next steps section, it proposed implementing some strategic actions to mitigate the homeless problem.

Challenge: Choose another next step

Choose one among the following next steps for the homelessness case study and explain why you chose it: ask for support, free interaction, learn more, propose a plan, sharing.

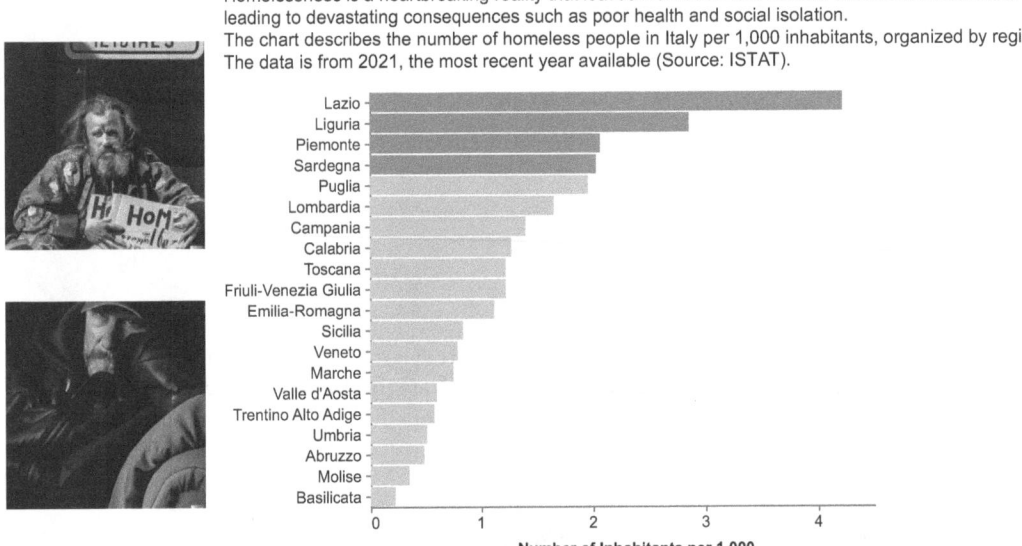

Together, Let's Make a Difference: Support Our Project to Help the Homeless!
Homelessness is a heartbreaking reality that leaves individuals and families without a stable home, leading to devastating consequences such as poor health and social isolation.
The chart describes the number of homeless people in Italy per 1,000 inhabitants, organized by region. The data is from 2021, the most recent year available (Source: ISTAT).

Our Visionary Plan to Harness Funds

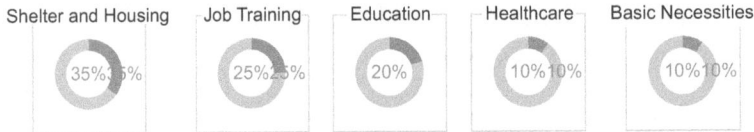

Figure 9.15   The case study on homelessness described in Chapter 5

9.2.6   *Chapter 5: Fake news*

Consider the case study on fake news we analyzed in chapter 5, as shown in figure 9.16. This case study focused on identifying the news categories most likely to contain fake stories. We included the next steps in the title.

Your Truth Guardian:
Take a Stand Against Fake News in Material and Moral Narratives
The LatestNews website is a popular source of news and information, but it is also a source of fake news.

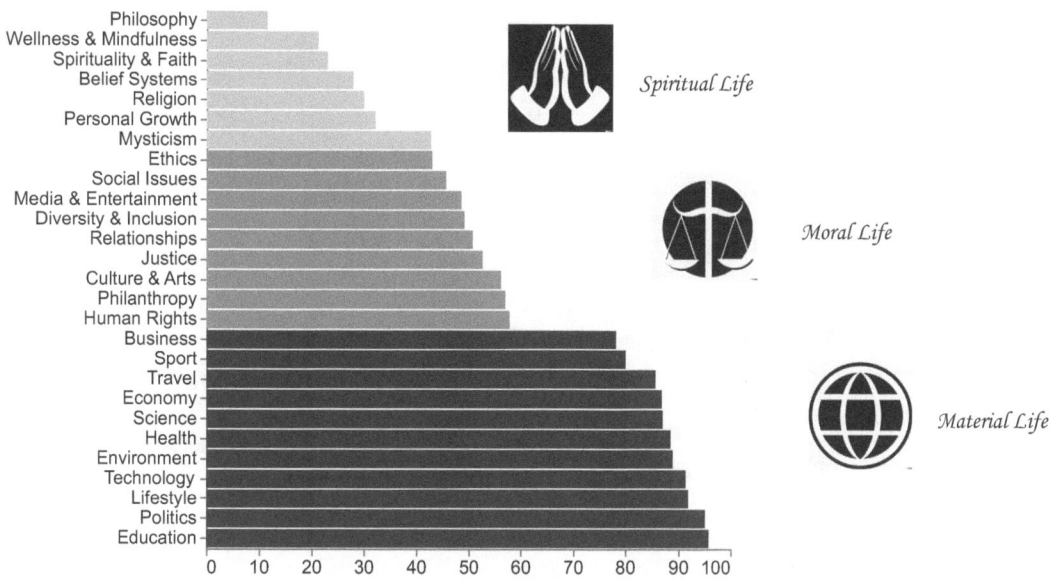

Figure 9.16   A summary of all the examples described in the previous chapters

We can improve our next steps by adding an interactive box showing a sample article title and headline whenever the audience clicks a specific bar, as shown in figure 9.17. In this example, we will not use ChatGPT. Specifically, we use the free interaction call to action to implement the next step of this example.

We implement the text containing the title and the headline as a text chart that changes based on the selected bar. Initially, no text is shown. You can find the code related to this example in the book's GitHub repository under 09/engagement-and-communication/engagement-and-interaction.py.

We will build a selection point and we will connect it to the Category field of the bar chart. Next, we will use the selection point to filter the text to show in the next step part. We will assume a dataset containing an article title and a headline for each category, as shown in table 9.3.

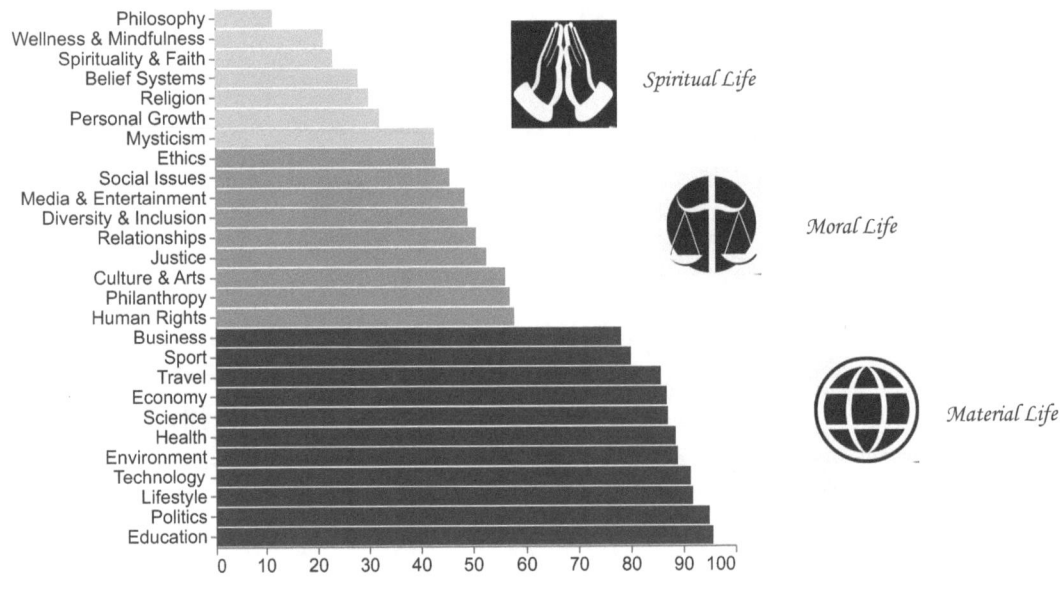

Your Truth Guardian:
Take a Stand Against Fake News in Material and Moral Narratives
The LatestNews website is a popular source of news and information, but it is also a source of fake news.

Click on a category bar to read a sample title
and headline of a fake article for that category.

Title: The Rise of Cryptocurrency: Disrupting Traditional Financial Systems

Headline: Exploring the Impact of Digital Currency on Global Markets

Figure 9.17   A summary of all the examples described in the previous chapters

Table 9.3   The article's dataset

Category	Title	Headline
Politics	The Future of Global Diplomacy: Navigating 21st Century Challenges	How International Relations Are Shaping Our World Today
Economy	The Rise of Cryptocurrency: Disrupting Traditional Financial Systems	Exploring the Impact of Digital Currency on Global Markets
Justice	Reforming Criminal Justice: Toward a Fair and Equitable System	Addressing Inequality within Legal Systems Across Nations
Religion	Interfaith Dialogue in Modern Society: Building Bridges Amidst Differences	Understanding Diversity in Religious Beliefs and Practices

Start by creating a selection point for your chart. Add the Category field of your dataset as a working field for your interactive element, as shown in the following listing.

Listing 9.15   Defining the selection point

```
click = alt.selection_point(name='Select',
                            fields=['Category'], empty=False)
```

NOTE Define a selection point named `Select` that allows users to select data points based on their `Category` values. The `empty` parameter set to `False` ensures that at least one data point must be selected.

Now, use the defined selection point to set the color of a selected bar, as shown in the following listing. The code from the original version implemented in chapter 5 is set in bold.

Listing 9.16   Building the interactive bar chart

```
color=alt.Color('Macro Category:N',
        scale=alt.Scale(
            range=['#991111', '#f38f8f','gray'],
            domain=['Material Life', 'Moral Life', 'Spiritual Life']
        ),
        legend=None
    )

chart = alt.Chart(df).mark_bar(
).encode(
    y=alt.Y('Category:N',
            sort='x',
            title=None,
            axis=alt.Axis(labelFontSize=14)
            ),
    x=alt.X('Percentage of Fake Articles:Q',
            title=None,
            axis=alt.Axis(labelFontSize=14,
                          titleFontSize=14),
    ),
    color=alt.condition(click | ~click, color, alt.value('lightgray')
    )
).properties(
    width=400,
    height=400
).transform_calculate(
    'Percentage of Fake Articles', alt.datum['Number of Fake
    Articles']/alt.datum['Number of Articles']*100
).add_params(
    click
)
```

NOTE First, define the color channel as a variable (we will use it for the text chart as well, so we define it as a variable to avoid writing it twice). Next, build the bar chart, by defining the y and x channels. Also, define the color channel as dependent on the `click` selection point. Use the expression `click |  ~click` to set the color when click is `True` (this happens with a selected bar) and when it is `False` (this happens when no bar is selected because we have set `empty` to `False`). Then, set the properties and add the click interactivity through the `add_params` method.

Now, we can draw the text chart as the combination of two charts: the title and the headline. First, build the base chart, with the details shared between the two charts.

Listing 9.17   Building the basic chart

```
df_cta = pd.read_csv('source/articles.csv')
df_cta['Macro Category'] = df_cta['Category'].apply(lambda x: 'Material Life'
    if x in material_life else ('Moral Life' if x in moral_life else
    'Spiritual Life'))

base_cta = alt.Chart(df_cta).mark_text(
    fontSize=20,
    align='left',
).encode(
    color=color
).transform_filter(
    click
)
```

NOTE   Start by reading the article dataset. Then, add the macro category, used to color the text. Finally, define the basic chart, by specifying the mark property (`mark_text`), the color channel, and the filter used to select the text dynamically.

Now, let's proceed with the article title.

Listing 9.18   Building the article title

```
title_cta = base_cta.encode(
    text='Label:N',
).properties(
    title=alt.TitleParams(
        text=['Click on a category bar to read a sample title', 'and headline
    of a fake article for that category'],
        fontSize=25,
        offset=20,
        anchor='start'
    )
).transform_calculate(
    Label= 'Title: ' + alt.datum.Title
)
```

NOTE   Use the base chart to draw the title article text. Specify the `text` channel, using the `Label` column, generated through the `transform_calculate` method. Also, set the chart properties.

Finally, build the article headline, using a procedure similar to that for the article title.

Listing 9.19   Building the article headline

```
headline_cta = base_cta.encode(
    text='Label:N'
).transform_calculate(
    Label= 'Headline: ' + alt.datum.Headline
)

chart = (chart & (title_cta & headline_cta)).configure_axis(
    grid=False
).configure_view(
    strokeWidth=0
)
chart.save('engagement-and-interaction.html')
```

> **NOTE**   Use the base chart to draw the headline article text. Next, combine the charts to obtain the final data story (suppose that chart contains the main chart). Use `configure_axis(grid=False)` to remove the grid from the charts.

So far, you have learned how to implement a variety of next steps. In the last case study, you'll learn how to anchor your story to an ethical framework. Let's apply this strategy to the salmon aquaculture case study implemented in the previous chapters.

9.2.7   Chapters 6–8: The salmon aquaculture case study

Consider the salmon aquaculture case study. As a quick recap, the salmon aquaculture case study aims to study the problem of safety in salmon aquaculture in the United States. So far, we have represented the dollars earned in the U.S. from salmon exports. We considered two scenarios: a positive one, where we focused on an increase in earnings, and a negative one, where we focused on the disease period. In this chapter, we only focus on the positive scenario, but you can easily adapt the described concepts to the negative one as well. Figure 9.18 shows the data story produced at the end of the previous chapter: turning information into knowledge.

Let's suppose that as the next step for our data story, we want to propose some strategic actions to continue the positive increase in the amount of dollars trend line. We propose the following next steps:

- Emphasize salmon safety measures.
- Enhance salmon aquaculture practices.
- Promote sustainable salmon consumption.

Let's use ChatGPT to adapt the proposed next steps to the different ethical frameworks. Use the following prompt for ChatGPT to generate the utilitarianism-based texts: *Consider the following ethical framework: utilitarianism.* This framework bases ethical decisions on maximizing utility or happiness. According to this view, an action is right if it leads to the greatest happiness for the greatest number of people.

Aquaculture Exports of Salmon in the U.S.
Aquaculture seafood in the U.S. is regulated by the FDA to ensure safety. Strict standards are in place to monitor water quality, feed, and disease control. Regular inspections and testing are conducted to minimize risks and protect consumers. (Source: U.S. Food and Drug Administration)

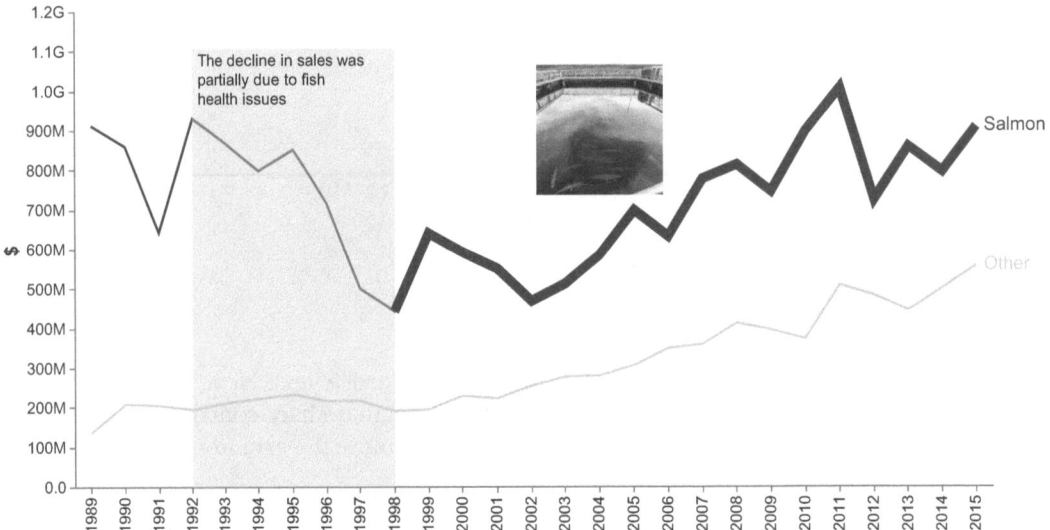

Figure 9.18   The story produced after turning information into knowledge

Reformulate the following next steps to adapt to the utilitarianism framework:

- Emphasize salmon safety measures.
- Enhance salmon aquaculture practices.
- Promote sustainable salmon consumption.

Figure 9.19 shows a possible output produced by ChatGPT.

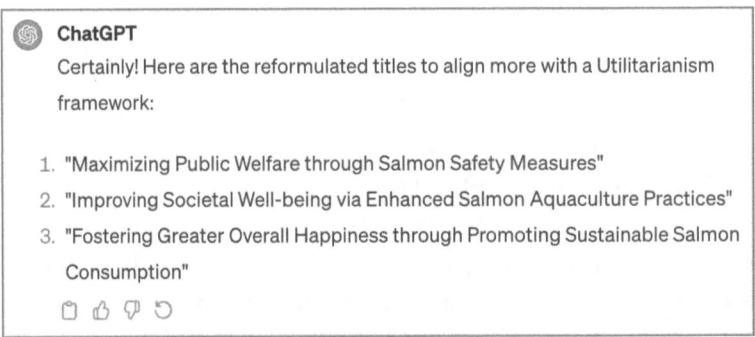

Figure 9.19   A possible output generated by ChatGPT

Now, let's review by asking the following question: Is this output really anchored to a utilitarian framework? If the answer is yes, we can incorporate the output in our story; otherwise, we must correct it before adding it to our story. In the next chapter, we will see more of the details regarding the ethical problems related to ChatGPT and generative AI in general. For now, it's sufficient to control the produced output.

In our case, the produced output contains words such as *welfare, societal well-being,* and *overall happiness,* which is in line with the utilitarianism framework.

Apply the same procedure to the other ethical frameworks. Table 9.4 shows some possible outputs by ChatGPT.

Table 9.4   The outputs generated by ChatGPT for each ethical framework

Ethical Framework	Emphasize Salmon Safety Measures	Enhance Salmon Aquaculture Practices	Promote Sustainable Salmon Consumption
Utilitarianism	Maximizing Public Welfare through Salmon Safety Measures	Improving Societal Well-Being via Enhanced Salmon Aquaculture Practices	Fostering Greater Overall Happiness through Promoting Sustainable Salmon Consumption
Deontology	Uphold Salmon Safety Measures	Adhere to Enhanced Salmon Aquaculture Practices	Advocate for Ethical and Sustainable Salmon Consumption
Virtue ethics	Cultivating Ethical Stewardship for Salmon Well-Being	Fostering Virtuous Aquaculture Practices for Salmon Health	Advancing Virtues through Sustainable Salmon Consumption
Care ethics	Prioritizing Salmon Well-Being: Implementing Safety Measures	Nurturing Salmon: Improving Aquaculture Practices with Empathy	Caring Consumption: Fostering Sustainable Salmon Choices

Once you have chosen your specific ethical framework, you can use it to implement the next steps for your case study.

EXERCISE 3

Consider the output produced by ChatGPT for a utilitarian ethical framework, and implement it as actionable recommendations. For example, you can generate the final data story described in figure 9.20. You can find the solution to this exercise in the GitHub repository for the book under CaseStudies/aquaculture/from-knowledge-to-wisdom/chart.py.

Now that you have learned how to anchor your next steps to an ethical framework, let's move on to the last part of this chapter: how to place your next steps strategically in your data story.

9.3   *Strategic placement of next steps*

Next steps must follow the main point of your data story. For this reason, I suggest you place your next steps in one of the following three main positions in your data

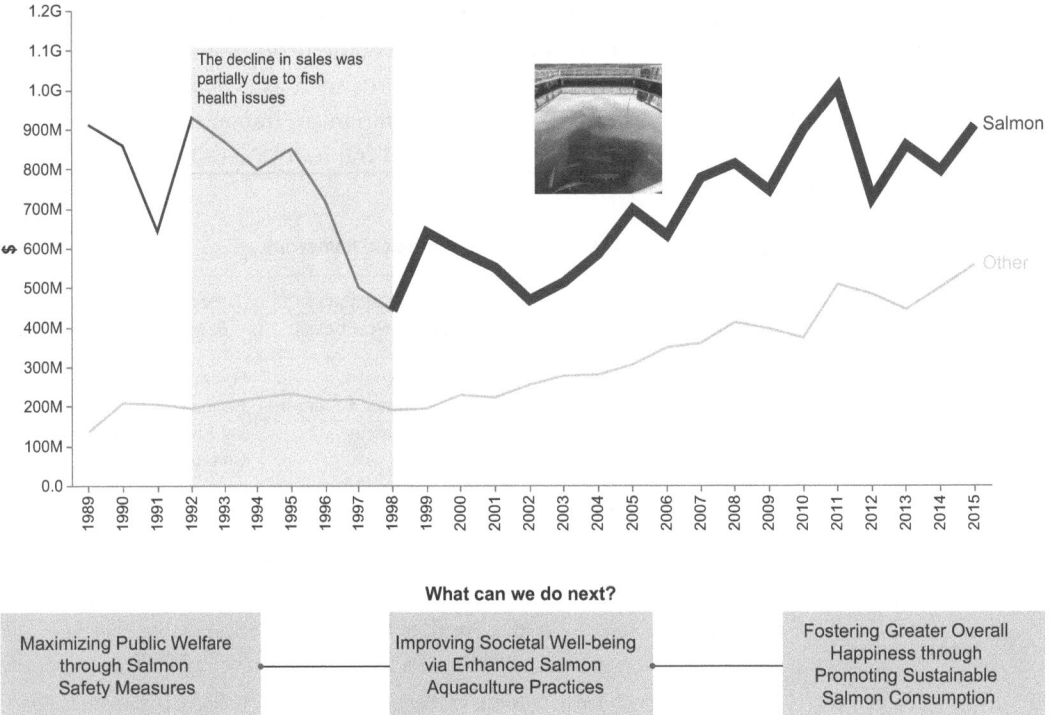

Aquaculture Exports of Salmon in the U.S.
Aquaculture seafood in the U.S. is regulated by the FDA to ensure safety. Strict standards are in place to monitor water quality, feed, and disease control. Regular inspections and testing are conducted to minimize risks and protect consumers. (Source: U.S. Food and Drug Administration)

Figure 9.20   The final data story for the aquaculture case study

visualization chart: in the title, on the right, or below the chart. In the remainder of this section, we will investigate each next step position separately.

9.3.1 Title placement

Adding next steps in the title means adding it to the top part of the chart, as shown in figure 9.21. Use the title to state what the audience should do with the data story.

> **Challenge: Next steps in the title**
> Consider the case studies described in chapters 1–5 (figure 9.2). Which ones added the next steps in the title?

9.3.2 Right placement

Adding next steps on the right of a chart involves adding it immediately after the chart, as shown in figure 9.22.

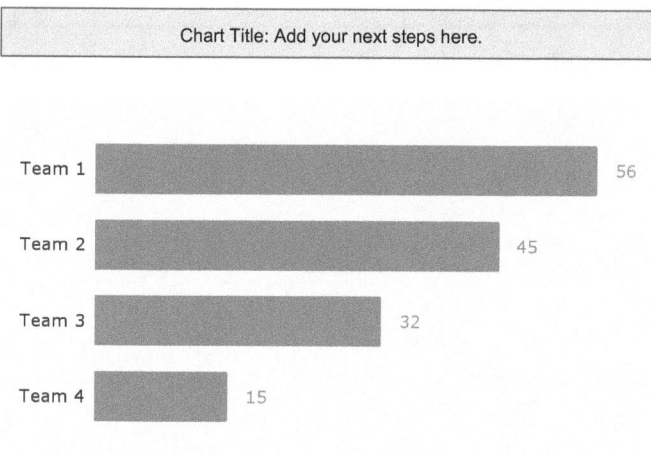

Figure 9.21 **The next step position at the top of the chart**

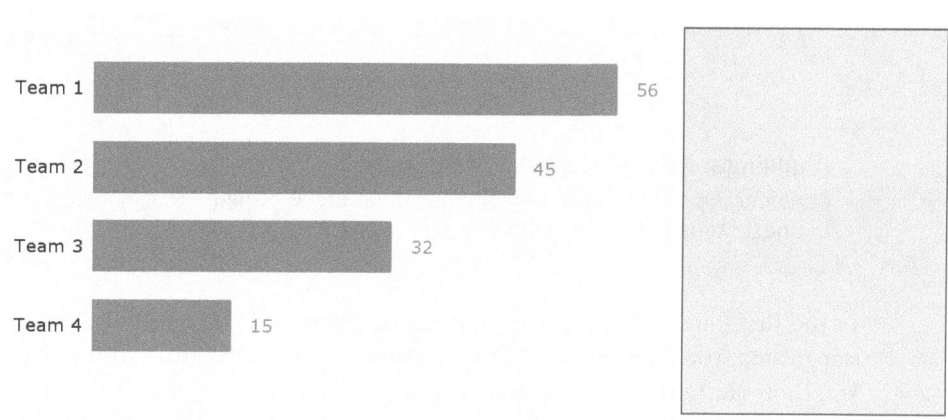

Figure 9.22 **The next step position on the right of the chart**

Use the right part of the chart to add deep details, such as textual descriptions. Use other charts to explain to the audience what to do after reading the previous parts.

Challenge: Next steps on the right
Consider the case studies described in chapters 1–5 (figure 9.2). Which ones added the next steps on the right?

9.3.3   *Below placement*

Next steps below a chart differ from the previous case only in their position, as shown in figure 9.23.

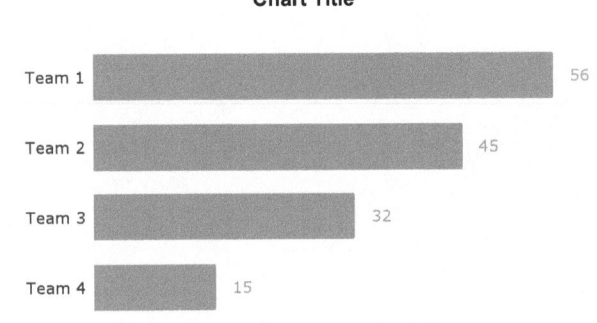

Figure 9.23   The next step position below the chart

Challenge: Next steps below the chart
Consider the case studies described in chapters 1–5 (figure 9.2). Which ones added the next steps below the chart?

In the first part of this chapter, you learned about the concept of wisdom and next steps. Next, you learned how to use ChatGPT as an alternative source of experience. You then saw how to anchor your next steps to an ethical framework. Finally, you saw how to place the next steps strategically in your story. In the next chapter, you will see how to publish your data story and what ethical implications the use of generative AI may have in data storytelling.

Summary

- Wisdom is the ability to use your experience and knowledge to make sensible decisions or judgments.
- Transforming knowledge into wisdom involves adding the next steps to the data story. Next steps are the resolution of our data story and define what the audience should do after reading the story.

- Classify the next steps into different categories based on the objectives they aim to achieve: ask for support, provide different options, free interaction, learn more, propose a plan, or sharing.
- Use ChatGPT's knowledge to help you generate possible next steps to include in your story. However, you must always check and refine the suggested next steps manually.
- An ethical framework serves as a guideline for decision making based on principles that guide moral behavior. Next steps must be anchored to an ethical framework.

References

- Berengueres, J. and Sandell, M. (2019). *Introduction to Data Visualization & Storytelling: A Guide for The Data Scientist (Visual Thinking)*. Self-published.
- McDowell, K. (2021). Storytelling Wisdom: Story, Information, and DIKW. *Journal of the Association for Information Science and Technology*, 72 (10), 1223–1233.

Part 3

Delivering the data story

The data, information, knowledge, wisdom (DIKW) pyramid and the use of generative AI have helped you to build a data story tailored to the audience. First, you have turned data into information by extracting interesting insights using different techniques, such as connections, coincidences, curiosity, and contradictions. You have built a basic chart representing the extracted insight using Altair and GitHub Copilot. Next, you have turned information into knowledge by adding context to your chart, which can be textual or visual. You have used ChatGPT to generate the textual context and DALL-E for the visual context. Finally, you have turned knowledge into wisdom, adding the next steps to your story. You have learned that there are different types of next steps, including awareness and campaigns, collaborative efforts, advocacy and policy, training and development, strategic actions, and engagement and communications. Each of them depends on the specific objective you want to achieve.

Your data story is now ready, but your work is not complete yet. You have the responsibility to tell your story ethically, meaning that you should consider the culture, the traditions, and all the frameworks your audience trusts. Next, you are ready to publish your story.

In chapter 10, you'll consider the common issues while using generative AI, including hallucination, bias, and copyright. You'll also learn the UNESCO guidelines to monitor the generative AI output to ensure it does not discriminate against people.

In chapter 11, you'll learn popular strategies for publishing your data story using Streamlit, Tableau, Power BI, and Comet. You will use Streamlit as a standalone framework for building web applications. You will also learn how to integrate your data story into two popular tools for data storytelling: Tableau and Power BI. Finally, you'll use Comet, an experimentation platform for machine learning that also offers a system for report building. Finally, you'll see how to split a single-chart data story into multiple slides for a live presentation.

Common issues while using generative AI

Your data story is now ready. However, before disseminating it to your audience, you should reflect on the possible issues associated with using generative AI. We will not discuss the technical details of these issues, but the concepts described will help you complete your overall understanding of generative AI. For more details, you are encouraged to read some of the detailed books on the topic—some of which are listed in the references of this chapter.

In the first part of this chapter, we'll focus on hallucinations, bias, and copyright. Next, we will focus on the guidelines for using generative AI. Finally, we'll see how to correctly credit the sources, including data sources and generative AI. Before disseminating the story, let's start with the first point: hallucination, bias, and copyright.

10.1 Hallucination, bias, and copyright

Generative AI is not an omniscient oracle but is the result of human effort, and as such, it reflects the characteristics of humanity, including intelligence, scientific progress, and so on but also discrimination, inequalities, and injustices. Generative AI, while a product of human ingenuity and scientific advancement, inherently carries the biases and limitations of the data it is trained on, reflecting not only our knowledge but also our societal and historical prejudices. It's crucial to remember that these systems are limited by their training data and algorithms and do not possess any innate understanding of consciousness.

All this, everything that humanity is, is therefore reflected in AI, both what is good and what is less good in humanity. For this reason, when generative AI is used as a work tool, particularly in data storytelling, it must always be handled carefully, with special attention paid to the benefits and possible damage it could cause to the most vulnerable people. Generative AI products have the potential to generate outputs biased against minority populations and can even be used to build manipulated stories with a plausible appearance.

To get a practical idea of discrimination that may occur inadvertently while using Generative AI, consider the following prompt for DALL-E: *a photo of a woman wearing a red dress.* Figure 10.1 shows a possible output generated by DALL-E.

Figure 10.1   A photo of a woman wearing a red dress

Now, consider this prompt for DALL-E: *a woman wearing an orange sweater drinking a coffee.* Figure 10.2 shows a possible output generated by DALL-E.

Figure 10.2   A woman wearing an orange sweater drinking a coffee

Finally, consider the following prompt: *a man sitting on a yellow chair.* Figure 10.3 shows a possible output generated by DALL-E.

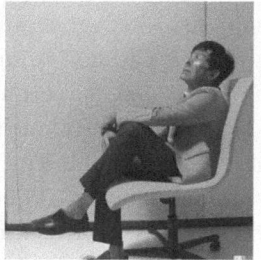

Figure 10.3   A man sitting on a yellow chair

We may continue generating more and more images, but we will probably always obtain similar results. What do figures 10.1, 10.2, and 10.3 have in common? The problem is that almost all the represented persons have dark hair (with the possible exception of the woman in image number 4 in figure 10.2). This may be a minor problem if we generate a small number of images. However, if we use DALL-E to generate images at a large scale, using only dark-haired people may generate a sort of bias against people with lighter hair.

This example shows a simple case of generative AI being partial. The problem is probably in the data used to train the generative model, which may contain more dark-haired than blonde-haired people. This is a relatively benign example, but it clearly illustrates that you may need to be mindful of bias and partiality in AI outputs.

Challenge 1: Other types of discrimination

Can you find other kinds of discrimination while generating images with people in DALL-E? For example, try to generate a woman with blonde hair. What kind of stereotype would you obtain?

Challenge 2: Color and mood

DALL-E associates color and mood. For example, try the following prompts: (1) a woman wearing a yellow sweater, (2) a woman wearing a blue dress. Do you obtain different moods, although you didn't specify any?

These issues caused by generative AI depend on different causes, such as AI hallucinations, bias, and copyright. We will not delve into the technical details related to these issues; you can read the resources described in this chapter's references for more information. In the remainder of this section, we will briefly describe the potential issues generative AI may introduce. Let's start with hallucinations.

10.1.1 AI hallucinations

Hallucination happens when generative AI generates content that does not correspond to reality. Hallucinations within AI can create misleading or entirely fabricated data. Hallucination is a problem brought about by generative AI's very design, in the sense that the large language models (LLMs) behind generative AI are generated without any communication of intent. It simply generates a statistically more probable text given the training dataset. Additionally, for instance, if a generative AI model is primarily trained on datasets containing English-language texts, it might struggle with accurately understanding or generating content in less-represented languages, reflecting a statistical bias toward English. Hallucinations can lead to ethical problems due to their potential misuse for generating content for user manipulation.

Hallucination in generative AI can manifest in various ways. For example, a data story completely generated by AI could describe a nonexistent political scandal, potentially misleading the public and impacting decision making based on false information. To mitigate the hallucination problem, we always recommend having a *human-in-the-loop validation*. Before incorporating any AI-generated content in your story, please judge and review it.

Challenge 3: Joking with generative AI

Just for fun, consider the swimming pool case study in chapter 2. As a quick reminder, the case study focused on the possibility of building a new swimming pool in a Portuguese hotel. Use the following prompt to generate the next steps: *Consider the following scenario: the case study focused on the possibility of building a new swimming pool in a Portuguese hotel. The data story showed an increasing number of tourists in Portugal in recent years. Write some hallucinated next steps.* What do you obtain?

To mitigate hallucinations, you can try to set the following parameters in your prompt:

- *Temperature*—The temperature controls the degree of randomness applied during the output generation process. It allows users to tailor the generated content's level of creativity and unpredictability. It ranges from 0 for more structured and predictable outputs to 2 for more creative and unexpected results. The default value rests at 1. To introduce temperature in your prompt, simply add the text *set temperature = N* (e.g., *use temperature = 2*). We can use a lower temperature value to reduce the probability that the model hallucinates.

- *Top P*—Top P is also known as nucleus sampling or penalty-free sampling. It helps to control the diversity of the generated text. Use this technique to generate responses that don't completely deviate from the topic. The range is between 0 and 1. A higher top P makes the output more diverse, while a lower value makes the model more deterministic. The default value is 1. To introduce top P in your prompt, add the text *set top P = N* (e.g., *use top P = 1*).

Usually, you set one parameter per prompt.

Challenge 4: Setting temperature and P value

Set the temperature or the top P value to 0 in the prompt of challenge 3, and then compare the new result with the original output of challenge 3. Do you notice some differences?

You can find a detailed description of how to set the temperature and the top P parameters in my blog post, "How to Improve Your ChatGPT Outputs Using Configuration Parameters" (https://mng.bz/pp4G). Now that we've covered hallucinations, let's move to the next problem: bias.

10.1.2 *Bias*

Bias is a systematic and often unconscious inclination, prejudice, or tendency that influences decision making, actions, perceptions, or judgments in favor of or against a particular person, group, object, or idea. Bias relies on human beliefs, stereotypes, or discrimination. Since LLMs are trained on datasets mostly created by humans, inevitably, these datasets contain bias. Bias is intrinsic to humans, so we can't remove it from our datasets.

Even in a hypothetical scenario where people are equal and free, and there is no discrimination and war, bias may occur. In fact, *bias is multifaceted.* Bias in AI is not limited to negative topics, like discrimination or war. Bias can also be in the form of cultural preferences, idiomatic expressions, or even what is considered "normal." An LLM trained on data from even an ideal world might still develop biases based on what is prevalent or dominant in that data.

In addition, even with ideal data, LLMs might still develop biases, due to their design, the algorithms they use, or the inherent limitations in understanding and processing human language. In other words, our LLM could still exhibit bias even in this hypothetical scenario.

There are different perspectives to classify bias, such as those proposed by Baer in his book, *Understand, Manage, and Prevent Algorithmic Bias* (Apress, 2019). One possible approach classifies bias into the following types:

- *Data bias*—This refers to the presence of bias in the training set. It derives from different causes, such as the underrepresentation of some groups in the training set. LLMs are trained with data extracted from the internet. However, the text on the web is written by a small percentage of humans. Tons of books representing the knowledge of all humankind from their beginning to today lie in libraries and are not available on the internet. This means that although big data is used to train LLMs, size does not guarantee diversity.

- *Algorithm bias*—This type of bias derives from the assumptions and decisions made during the design, coding, or implementation of machine learning (ML) algorithms. This bias can arise due to feature selection, model complexity, and other technical problems related to the algorithms.
- *Measurement bias*—This occurs when the methods or tools used to collect data systematically misrepresent or distort the information being gathered. This bias can originate from different factors, such as instruments, human observers, and so on.

At the time of writing this book, there is no definitive solution to remove bias from generative AI tools. However, some possible techniques to mitigate bias could include data cleaning and balancing, inserting a human in the middle, model evaluation, and so on. Removing bias from generative AI would also mean first removing it from humanity, which would be significant progress for the world but is unlikely to happen anytime soon. Anyway, we could mitigate bias by always paying attention to the generated output and anchoring our data story to an ethical framework, as explained in the previous chapter. Having considered the problem of bias, let's move to the next problem: copyright.

10.1.3 Copyright

Generative AI models have been trained on huge quantities of data, derived especially from content shared freely in the public space. However, the creators of the original datasets used to train the models do not allow access to them, so we don't know whether proprietary data has also been used to train the models. For this reason, AI-generated content might raise questions about intellectual property rights and ownership. Copyright questions may be connected to the fact that generative AI models are black boxes and the data used to train them are unavailable.

For example, consider an AI system trained on an extensive database of music that generates a piece similar to an existing copyrighted song. Determining the original creator becomes complex, raising questions about the ownership of AI-generated content and the rightful attribution of intellectual property, potentially leading to legal disputes between creators and AI systems. Before using generative AI, you should always understand copyright law, use clear licensing agreements for data sources, create original datasets, emphasize attribution and acknowledgment, implement copyright filters, and seek regular legal consultation.

All the issues described in this book remain unsolved at the time of writing. For this reason, I recommend consistently controlling the output produced by generative AI when using it for data storytelling and, in general, for all the application fields. It's your responsibility to use generative AI ethically, ensuring that everyone is treated equally, minority populations are not underrepresented, and so on. To help you control the generative AI output and modify it if needed, you can apply the common guidelines for ethically using generative AI. So, let's move on and learn them!

10.2   Guidelines for using generative AI

Many initiatives exist to regularize the use of AI and generative AI in all domains, such as the European Union AI Act (European Commission, 2021) and the White House's Executive Order on the Safe, Secure, and Trustworthy Development and Use of Artificial Intelligence (White House, 2023). You can find the links to these documents at the end of the chapter, in the references section.

Referring to the field of data storytelling, following these guidelines means respecting human values. The UNESCO AI ethics guidelines emphasize four core values:

- *Human rights and dignity*—Every data story, including the pieces generated through generative AI, must respect human rights and dignity. The UNESCO guidelines say explicitly that "no human being or human community should be harmed or subordinated, whether physically, economically, socially, politically, culturally or mentally, during any phase of the life cycle of AI systems."
- *Peaceful and just societies*—AI should be used to foster harmony and equity within communities, promoting fairness, transparency, and accountability in decision-making processes.
- *Diversity and inclusiveness*—AI should respect the richness of human diversity in all its forms, including but not limited to race, gender, ethnicity, culture, and more.
- *Environmental flourishing*—AI should prioritize sustainability and contribute positively to environmental preservation.

We can apply the previous guidelines to check if AI-generated content is ethically correct. For all AI-generated content, we should answer the following questions:

- Does the AI-generated content respect human rights?
- Does the AI-generated content respect society?
- Does the AI-generated content respect diversity and inclusiveness?
- Does the AI-generated content respect the environment?

To understand how to apply these guidelines to a practical data story, consider the case study we analyzed in chapter 4, related to selecting the optimal sports disciplines to train in to achieve good results in upcoming competitions. As next steps, we proposed to invest in cycling and rowing. Figure 10.4 shows the final data story.

Let's look at each part generated by AI:

- *General title*—Unlock the Potential: Invest in Rowing and Cycling for Maximum Returns!
- *Image 1*—The white man practicing rowing
- *Image 2*—The white woman practicing cycling

The only problem with the AI-generated content for this scenario regards diversity and inclusiveness. The two images, in fact, describe two white-skinned individuals. To

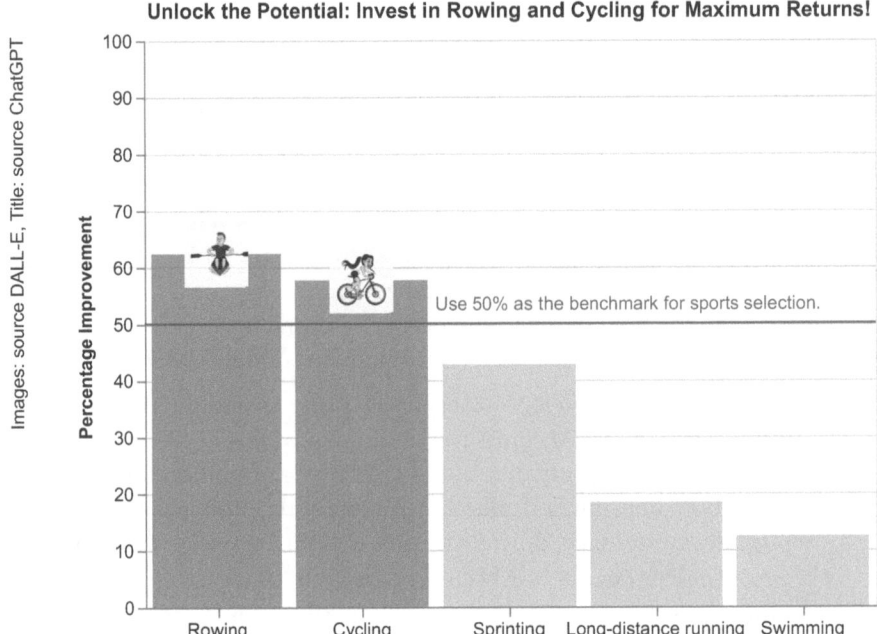

Figure 10.4   **The sports disciplines data story described in chapter 4**

make the data story fit the ethical guidelines, we could replace one of the two images with an individual of another ethnicity.

EXERCISE 1

Consider the case study described in chapter 5 on homelessness, shown in figure 10.5. The case study focused on searching for funds to finance a project on reducing homelessness in Italy. Now, follow these steps:

1 Identify the AI-generated content.
2 For each element of AI-generated content, answer the following questions:
   a Does the AI-generated content respect human rights?
   b Does the AI-generated content respect society?
   c Does the AI-generated content respect diversity and inclusiveness?
   d Does the AI-generated content respect the environment?

Now that you have learned the guidelines for ethically using generative AI, let's move on to the following step: determining your role while using it.

Together, Let's Make a Difference:
Support Our Project to Help the Homeless!
Homelessness is a heartbreaking reality that leaves individuals and families without a stable home,
leading to devastating consequences such as poor health and social isolation.

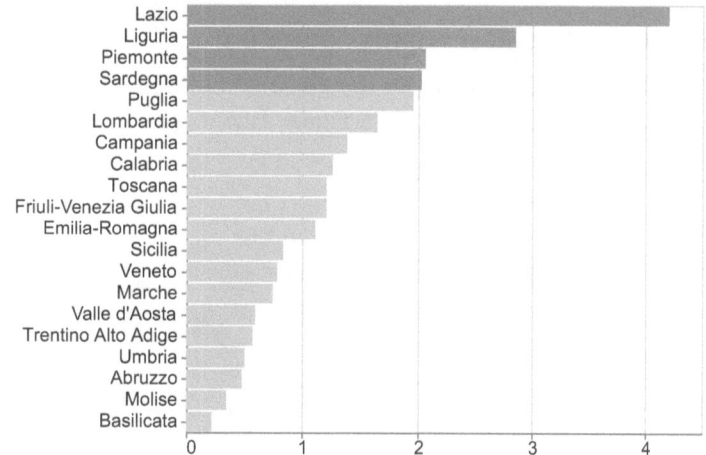

Our Visionary Plan to Harness the Funds

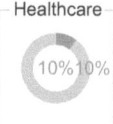

Figure 10.5   The homelessness case study described in chapter 5

10.3   Crediting the sources

Crediting the sources means referencing the sources used in a data story. This is particularly important because it allows you to recognize the work done by others. It also adds credibility to the story, as the audience can personally check the sources used in the story. What types of sources should be credited? In general, any source used to make the story should be credited, but particularly the following sources:

- The data source
- The fact that generative AI was used
- Any documents used for fine-tuning or retrieval augmented generation (RAG)

While you can use your creativity to place credits wherever you want, traditionally, we add credits to a data story in one of four places:

- Under the title/subtitle
- Under the main chart

- Under the next steps
- Sideways

Let's investigate each of these separately.

10.3.1 *Under the title or subtitle*

Placing the credits under the title or subtitle generates a sense of trust in the audience from the story's beginning. Figure 10.6 shows an example of credits placed under the title or subtitle.

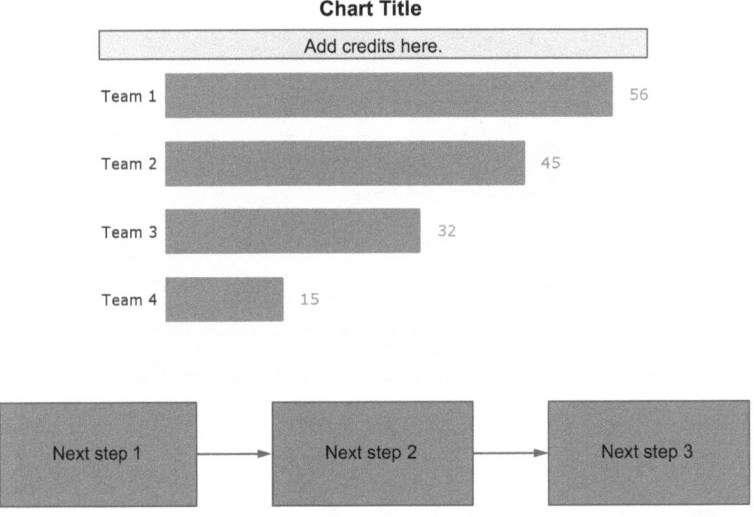

Figure 10.6   An example of a data story with credits under the title or subtitle

Use this placement if you want your audience to know the sources from the story's beginning. Although this placement may generate trust, it could also be distracting, since the audience may leave your story to search for the sources.

10.3.2 *Under the main chart*

Placing the credits under the main chart involves adding a detail to the main point of the story. This helps reinforce the essential points of the story. Figure 10.7 shows an example of credits placed under the main chart. Use this placement if you want to reinforce the main message of your chart.

10.3.3 *Under the next steps*

In this case, credit the sources at the end of your story, as an appendix to the next steps, as shown in figure 10.8. Use this placement if you want to reinforce the next steps of your story.

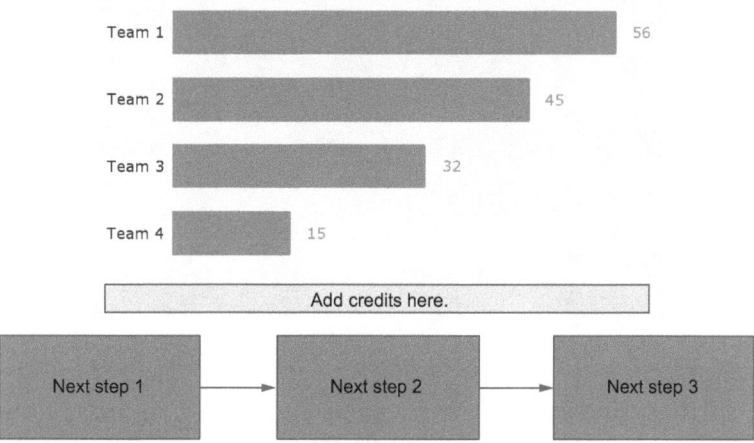

Figure 10.7   An example of a data story with credits under the main chart

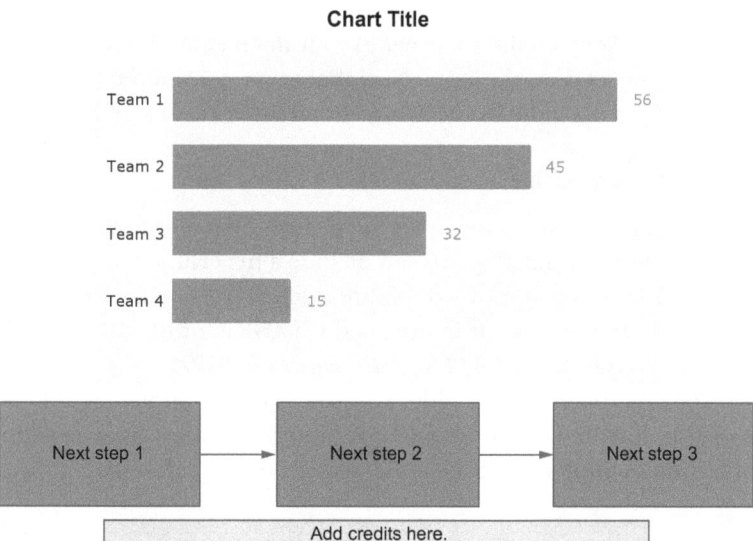

Figure 10.8   An example of a data story with credits under the next steps

10.3.4 *Sideways*

Placing credits sideways means considering them as external to the main data story workflow. You can place credits either on the left or on the right, as shown in figure 10.9.

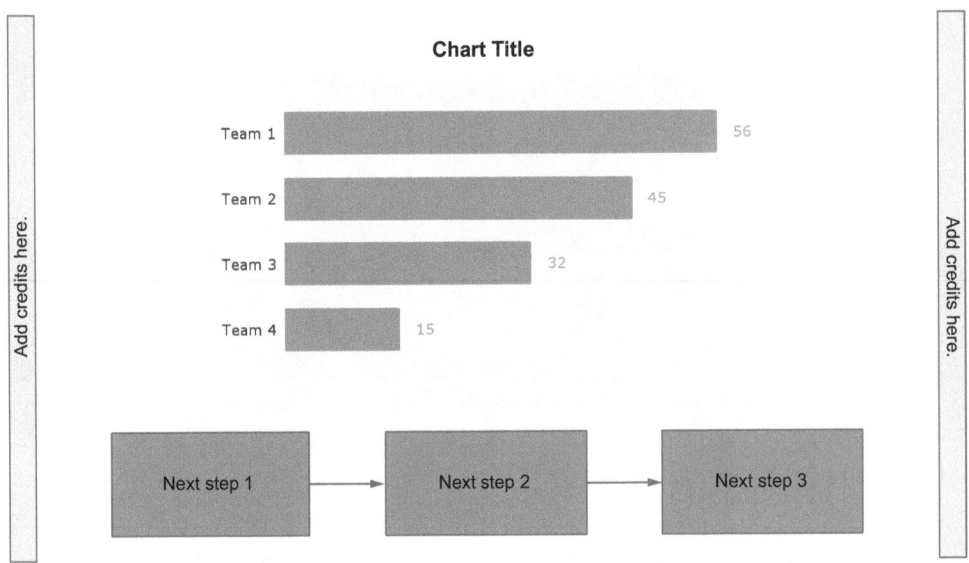

Figure 10.9   An example of a data story with credits on the left and the right

Use this placement to keep credits external to your main data story workflow and keep the audience concentrated on the story. Now that we've considered the various places you can place credits, let's move on to how to implement credits practically.

10.3.5 *Implementing credits in Altair*

To implement credits in Altair, use `mark_text()` with a smaller font than the one you used for the main story. Optionally, you can include a hyperlink to the original source.

Consider the case study on sports disciplines described in chapter 4 and shown in figure 10.4. Let's credit DALL-E for images and ChatGPT for the title. Add the following text as credits: *Images: source DALL-E. Title: source ChatGPT.*

We will place credits on the left side. You can find the implemented code in the GitHub repository for the book under 10/crediting/left-sideways. The following listing shows the code to implement the credits section on the left.

Listing 10.1   Adding the credits on the left

```
credits_df = pd.DataFrame({'text': ['Images: source DALL-E, Title: source
    ChatGPT']})
credits = alt.Chart(credits_df
        ).mark_text(
            size=10,
            align='left',
            color='black',
            angle=270,
```

```
        x=-70,
        y=200

    ) .encode(
        text='text'
    )

chart = (credits + chart + annotation)
```

NOTE Use `mark_text()` to add credits. Use the `angle` attribute to rotate the text 270 degrees. Adapt `x` and `y` to the chart. Try different values to obtain the best result.

As usual, you can ask Copilot to generate the code for you. Figure 10.10 shows the resulting chart.

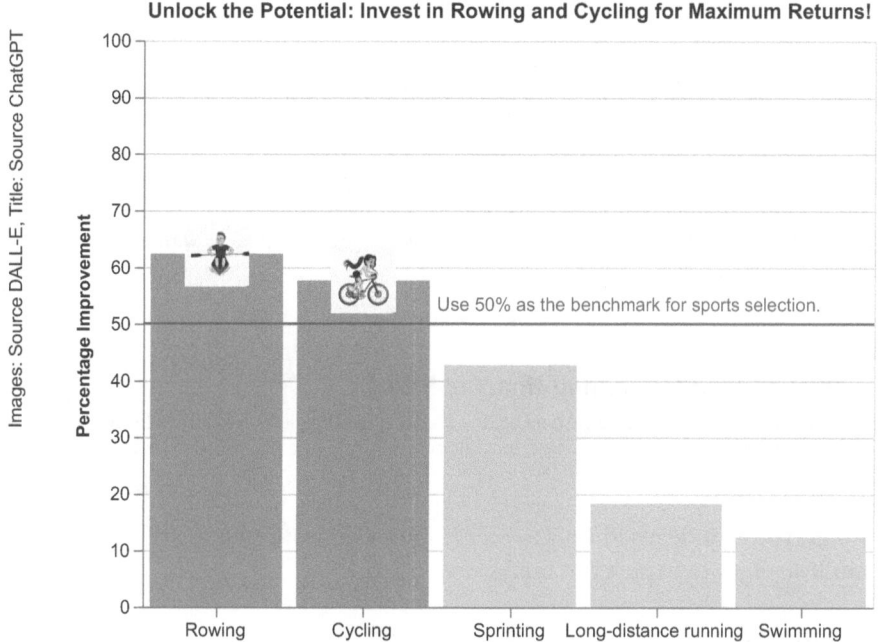

Figure 10.10 The sports disciplines case study described in chapter 4, with credits on the left

EXERCISE 2

Modify the previous example by implementing credits under the title, as shown in figure 10.11.

You can find the solution to this exercise in the GitHub repository for the book under 10/crediting/under-the-title.

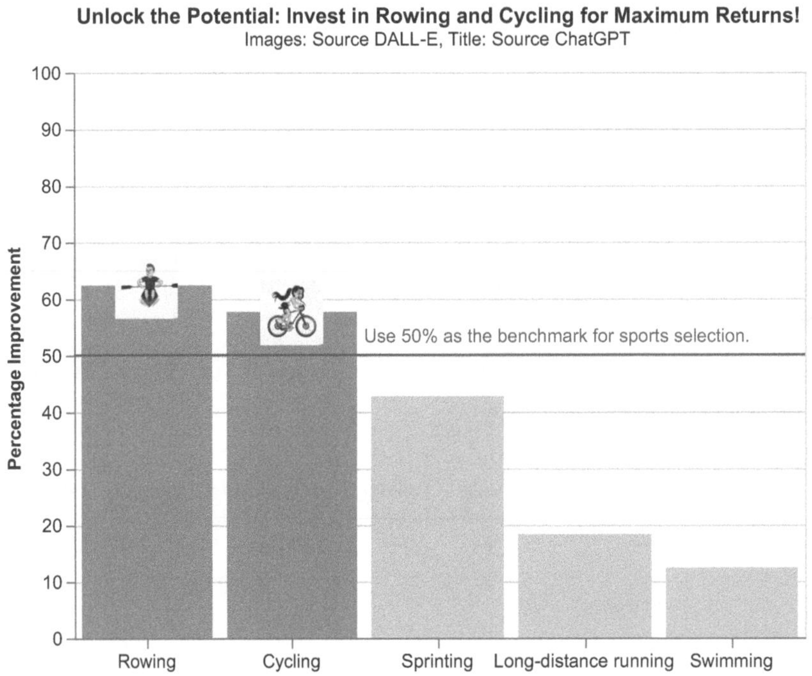

Figure 10.11   The sports disciplines case study described in chapter 4, with credits under the title

Challenge 5: Comparing Reading Flows
Compare figures 10.10 and 10.11. Can you distinguish any difference in terms of the reading flow?

So far, you have learned how to credit sources in your data story. In the next chapter, you'll see how to export the final story.

Summary

- Using generative AI in any application may surface different issues, such as bias and discrimination. Thus, it is very important to review the content provided by generative AI.
- Using generative AI ethically means respecting people, society, and the environment, according to UNESCO principles.
- Before publishing your data story, make sure to credit sources. It's a way to recognize the work done by others and to generate a greater sense of trust in your audience

References

Generative AI issues

- Baer, T. (2019). *Understand, Manage, and Prevent Algorithmic Bias.* Apress.
- Tomczak, J. M. (2022). *Deep Generative Modeling.* Springer.

Ethics and AI

- EU AI Act. (2021). https://eur-lex.europa.eu/legal-content/EN/TXT/HTML/?uri=CELEX:52021PC0206.
- White House Executive Order on the Safe, Secure, and Trustworthy Development and Use of Artificial Intelligence. (2023). https://www.whitehouse.gov/briefing-room/presidential-actions/2023/10/30/executive-order-on-the-safe-secure-and-trustworthy-development-and-use-of-artificial-intelligence/.
- UNESCO Recommendation on the Ethics of Artificial Intelligence. (2022). https://unesdoc.unesco.org/ark:/48223/pf0000381137 .

Publishing the data story

Throughout this book, we have built our data story. Now, in this final chapter of the book, it is time to show it to our audience. This chapter is all about taking your data story and getting it to your audience—and doing so in an ethical manner. In the first part of this chapter, we'll focus on the different techniques to export a data story. Next, we'll describe Streamlit, a Python library fully integrated with Altair. Streamlit helps you build a complete standalone website hosting your Python code. Then, we'll see some alternative ways to publish your story, including techniques to integrate it into some popular tools for data analytics and visualization: Tableau and Power BI. Afterward, we'll describe how to integrate your data story in Comet, an experimentation platform for machine learning (ML). Finally, we'll see how to present your data story through slides. Let's start with the first point: exporting the story.

11.1 Exporting the story

Altair provides different formats to export your data story. Throughout this book, we have used `chart.save('chart.html')`, but Altair also supports other formats, including the following:

- JPEG
- PNG
- SVG
- JSON
- PDF

To save the chart into a specific format, simply add the format extension to your file, such as `chart.save('chart.png')` for PNG files. For PNG files, you can also specify the pixels per inch (ppi) through an additional parameter: `chart.save('chart.png',ppi=300)`.

You may wonder why, throughout this book, we have used the HTML format and not PNG or SVG. The main reason is that HTML is the only supported format that shows interactivity. In addition, once you have created the HTML file, Altair also provides a default action menu on the top-right part of the chart, enabling you to download the story into another format, as shown in figure 11.1.

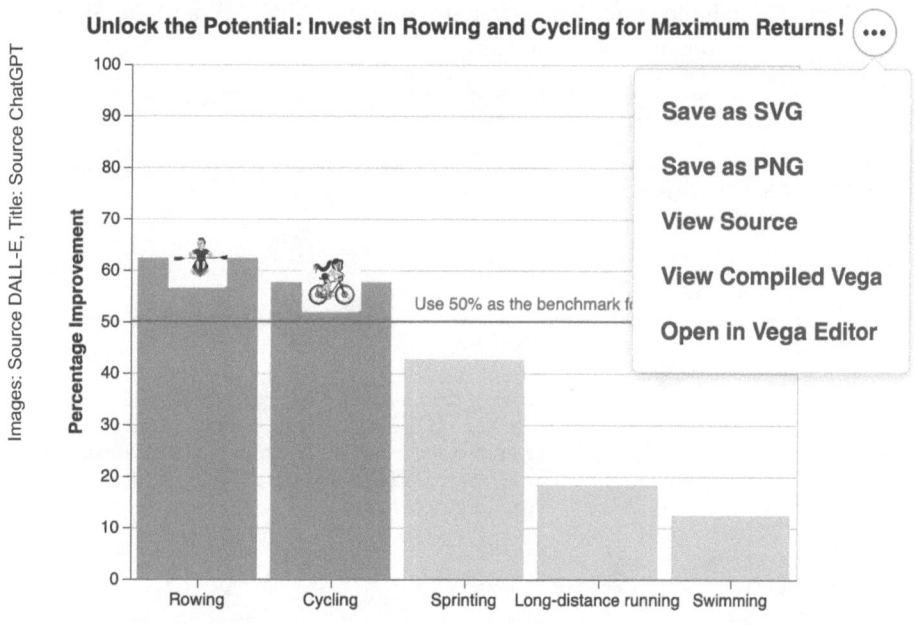

Figure 11.1 The top-right menu in the HTML file enables you to export the story into another format.

If you prefer not to show the action menu, you can disable it, as shown in the following listing.

Listing 11.1 Disabling the action menu

```
chart.save('chart.html', embed_options={'actions': False})
```

Your journey from transforming raw data into a data story through the DIKW pyramid, generative AI, and Altair is complete. Your story is finally ready to be disseminated to your audience!

In the remaining two sections, we will see some alternative techniques to publish your story. Feel free to skip them if you are not interested. First, we will see how to publish the story using Streamlit, and next, we will see how to embed our story into some popular tools. Let's start with Streamlit.

11.2 *Publishing the story over the web: Streamlit*

Streamlit (https://streamlit.io/) is an open source framework enabling you to build web applications quickly using Python. To start with Streamlit, refer to appendix A. The main advantage of using Streamlit compared to other options is that it is fully integrated with Altair.

In this section, we'll see how to publish a data story using Streamlit. Streamlit uses specific functions to show content on a web page, such as st.Title() to set the page title.

Consider the example shown in listing 11.2. Start by importing the Streamlit library. Then, build your chart as you usually do in Altair. To add the Altair chart wherever on the web page, use st.altair_chart().

Listing 11.2 Building the app in Streamlit

```
import streamlit as st
import pandas as pd
import altair as alt

# Read the data
df = pd.read_csv('../data/population.csv')
df = df.melt(id_vars='Country Name', var_name='Year', value_name='Population')
df['Year'] = df['Year'].astype('int')

continents = ['Africa Eastern and Southern', 'Africa Western and Central',
              'Middle East & North Africa', 'Sub-Saharan Africa',
              'Europe & Central Asia', 'Latin America & Caribbean',
              'North America', 'East Asia & Pacific']
df = df[df['Country Name'].isin(continents)]

# Create Streamlit app
st.title('Population of Continents')

# Add a slider for year selection
selected_year = st.slider('Select a year', min_value=1960, max_value=2021,
    value=2021, step=1)

# Filter data based on selected year
filtered_df = df[df['Year'] == selected_year]
```

```
# Create Altair chart
chart = alt.Chart(filtered_df).mark_bar(color='#80C11E').encode(
    y=alt.Y('Country Name:O', sort='-x', title=''),
    x=alt.X('Population:Q', title='Population')
).properties(
    width=600,
    height=300
).configure_view(
    strokeWidth=0
).configure_axis(
    grid=False
)

# Display chart in Streamlit app
st.altair_chart(chart)
```

> **NOTE** Start by importing the required libraries. Then, load the data as a pandas DataFrame. Next, build the chart in Altair as you usually do. Use `st.slider()` to build a slider in Streamlit. Finally, display the chart in the Streamlit app using `st.altair_chart()`.

To launch the app, run the following command in a terminal: `streamlit app.py`, with app.py being the name of the script containing the code shown in listing 11.2. Figure 11.2 shows the resulting app.

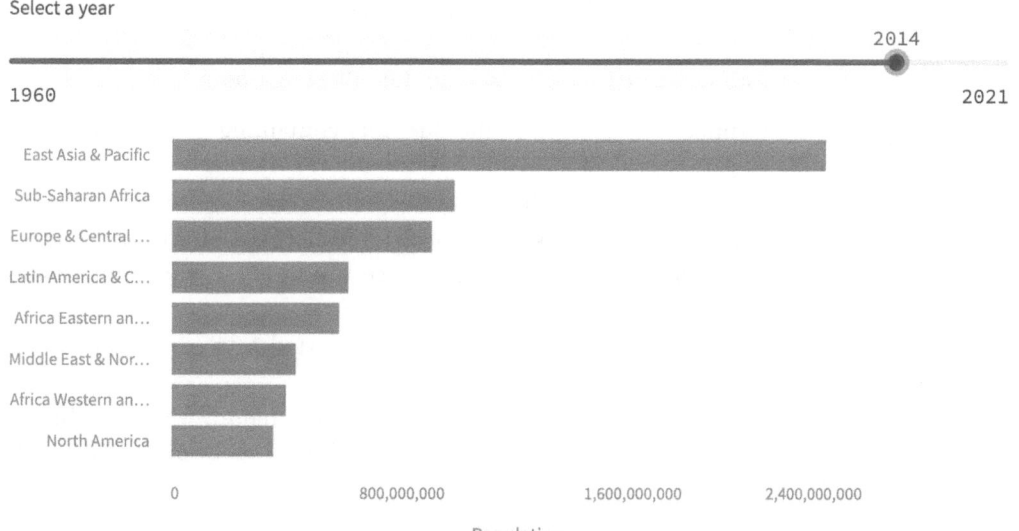

Figure 11.2 The resulting app in Streamlit

You can download the code described in this section from the GitHub repository for the book under 11/streamlit/app.py. Now that you have learned how to publish your data story in Streamlit, let's move on to some alternative ways to publish the story. If you are not interested in them, feel free to skip directly to the conclusions.

There are different ways to publish your data story, including reports, presentations, HTML pages, and so on. In this section, we will see three tools:

- Tableau
- Power BI
- Comet

Let's start with the first one: Tableau.

11.3 *Tableau*

Tableau (https://www.tableau.com/) is a business intelligence tool that helps users to explore and analyze data. Unlike Streamlit, Tableau does not require any programming skills, since its interface is completely visual. Combining Tableau and Altair may be useful if you already know Tableau and use it to implement your dashboards. You can integrate your data story produced in Altair into Tableau as an additional item of your dashboard.

Before using Tableau, you must download it on your local computer and pay for a license. Tableau also offers a trial version and a completely free license for students and teachers. In the bibliography of this chapter, you can find some interesting resources to get started with Tableau. In this section, we describe how to import your data story in Tableau. If you are not interested, feel free to skip this section and move to the next one. Consider the example in chapter 08/slider.py and shown in figure 11.3.

To import your chart in Tableau, execute the following operations:

1 Open a terminal and point it to the directory containing 08/slider.py. Run the command `python3 slider.py` to generate the HTML file.
2 In the terminal, launch a Python server as follows: `python3 -m http.server`.
3 Point to localhost:8000 in your browser, and access slider.html; you should see the chart shown in figure 10.14. Move the slider to check if the code is working correctly.
4 Start your Tableau Desktop application, import the 08/data/population.csv file, and create a new dashboard.
5 In Tableau, double-click Web Page from the Object menu, and enter the following URL in the dialog box, as shown in figure 11.4: http://localhost:8000/slider.html.

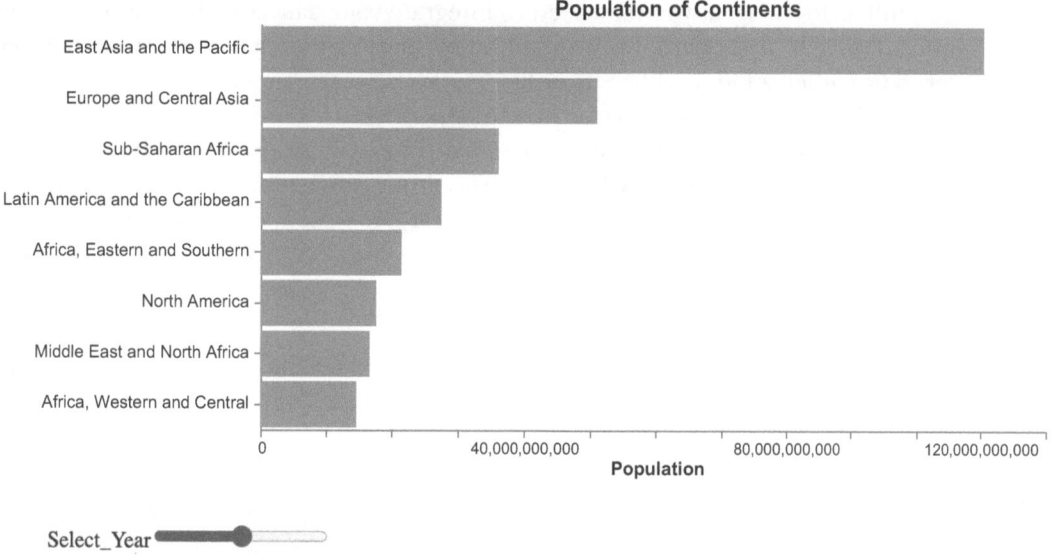

Figure 11.3   A chart with a slider

Figure 11.4   The Edit URL dialog box in Tableau

Now, you should see your Altair chart in your Tableau Dashboard. Now that you have learned how to include your data story in a Tableau dashboard, let's move on to the next alternative: including your data story in a Power BI report.

> **NOTE**   In some cases, the web page importer fails because it does not support the JavaScript wrapper generated by Altair to include the chart. In this case, export the chart as a .png file and use the Image object to import it into Tableau.

11.4   Power BI

Power BI (https://mng.bz/o0Pd) is another business intelligence tool released by Microsoft. Power BI offers both a desktop and an online version. Like Tableau, you must buy a license to use it; however, you can test using a trial version. In the bibliography of this chapter, you can find some resources to start with Power BI.

Similar to Tableau, you may need to integrate your data story produced in Altair into a Power BI dashboard if you already know and use Power BI. Otherwise, it is better to use other solutions, like Streamlit.

In this section, we only describe how to import a chart generated in Altair to Power BI. We will use the online version. Specifically, we will use the Deneb plugin, which enables you to import data written in Vega. From Altair, you can always export the Vega or Vega-Lite JSON file directly from the HTML version, as shown in figure 10.12.

For simplicity, in the following example, we will directly use the Vega code, available in the GitHub repository for the book under 03/vega/json/spec.json. Execute the following steps:

1   Log in to your Power BI account, and click New Report → Paste or Manually Enter Your Data → Create a Blank Report (figure 11.5).

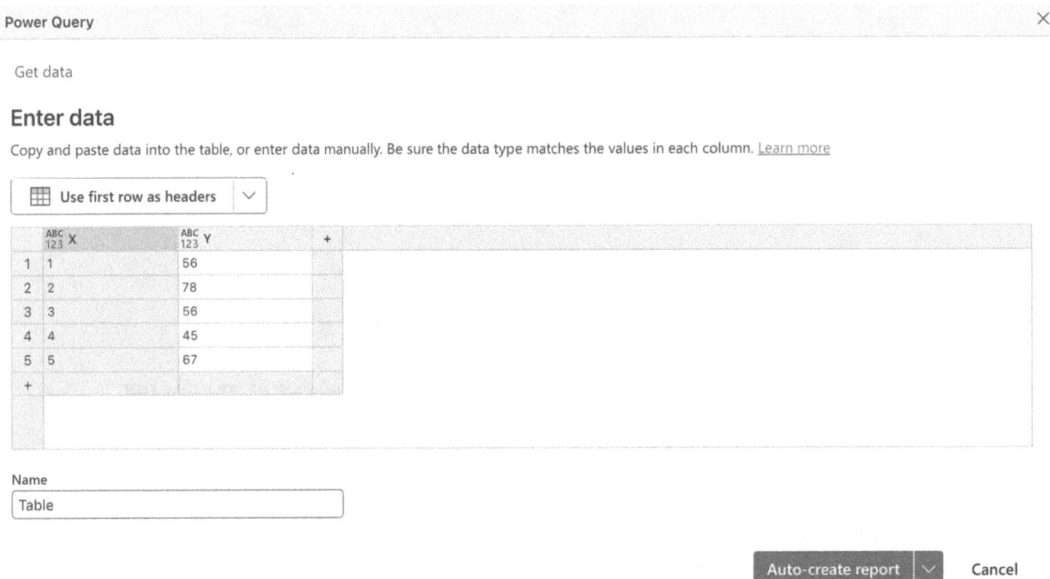

Figure 11.5   A snapshot of the Power BI interface to create a new report

2   Click Visualizations → (three dots) Get More Visuals → Vega → Deneb: Declarative Visualization in Power BI → Add (figure 11.6).
3   In the visualization menu, select Deneb.
4   In the Data menu, add X and Y to Deneb. Click on the arrow near X and Y, and then click Don't Summarize (figure 11.7).

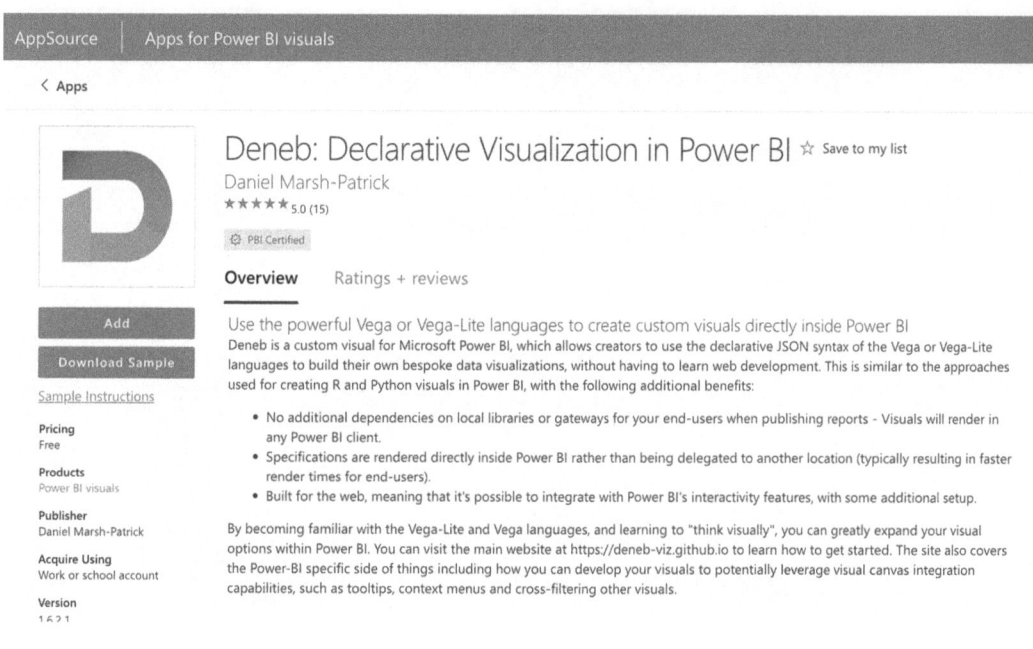

Figure 11.6   The Deneb plugin

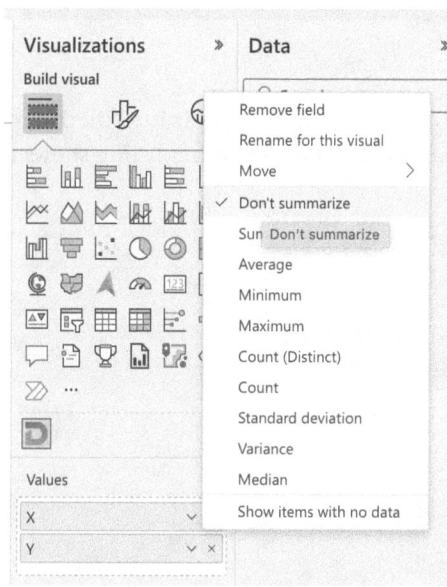

Figure 11.7   The selection menu for each variable

5   On the left side of the app, you should see the Deneb box. Click the three dots in the Deneb box in the report, and then click Edit → Vega - Empty (figure 11.8).

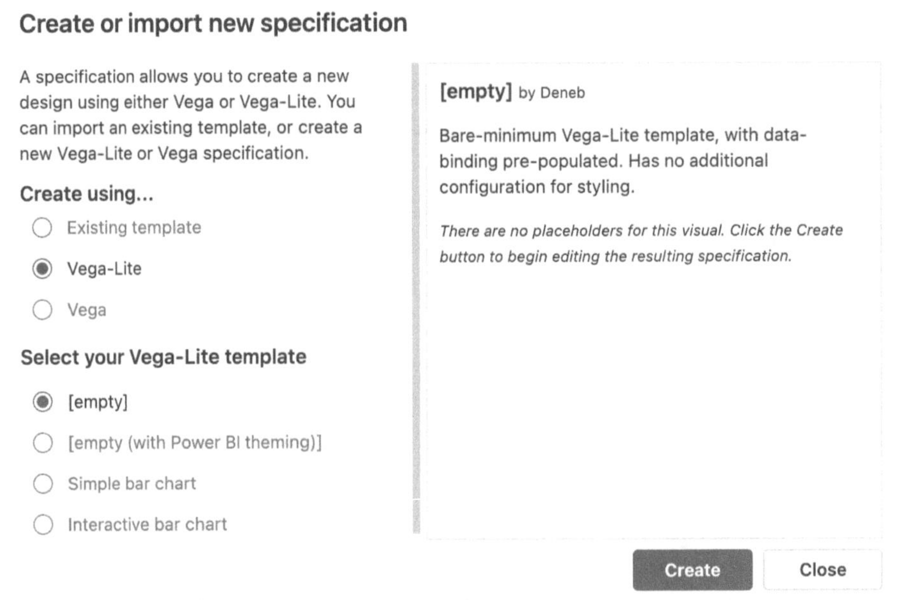

Create or import new specification

A specification allows you to create a new design using either Vega or Vega-Lite. You can import an existing template, or create a new Vega-Lite or Vega specification.

Create using...

○ Existing template

◉ Vega-Lite

○ Vega

Select your Vega-Lite template

◉ [empty]

○ [empty (with Power BI theming)]

○ Simple bar chart

○ Interactive bar chart

[empty] by Deneb

Bare-minimum Vega-Lite template, with data-binding pre-populated. Has no additional configuration for styling.

There are no placeholders for this visual. Click the Create button to begin editing the resulting specification.

Create   Close

Figure 11.8   The selection menu for each variable

6   Copy the code in 03/vega/json/spec.json, and paste it into the Deneb editor. You need to remove data to make it work. The following listing shows the modified version. Next, click Create.

Listing 11.3   The Vega representation of the chart

```
{
  "description": "A basic line chart",
  "width": 400,
  "height": 200,
  "padding": 5,
  "data": [
    {
      "name": "dataset"
    }
  ],

  "scales": [
    {
      "name": "xscale",
```

```
      "domain": {"data": "dataset", "field": "X"},
      "range": "width"
    },
    {
      "name": "yscale",
      "domain": {"data": "dataset", "field": "Y"},
      "range": "height"
    }
  ],

  "axes": [
    { "orient": "bottom", "scale": "xscale" },
    { "orient": "left", "scale": "yscale" }
  ],

  "marks": [
    {
      "type": "line",
      "from": {"data":"dataset"},
      "encode": {
        "enter": {
          "x": {"scale": "xscale", "field": "X"},
          "y": {"scale": "yscale", "field": "Y"},
          "stroke": { "value": "#636466"}

        }
      }
    },
    {
      "type": "symbol",
      "from": {"data":"dataset"},
      "encode": {
        "enter": {
          "x": {"scale": "xscale", "field": "X"},
          "y": {"scale": "yscale", "field": "Y"},
          "shape": {"value": "circle"},
          "size" : {"value": 100},
          "fill": { "value": "#636466"}
        }
      }
    }
  ]
}
```

NOTE Replace the data section with a link to the dataset imported in Power BI.

Figure 11.9 shows the resulting chart. Now that you have learned how to import a Vega chart in Power BI, let's move on to the final tool: Comet.

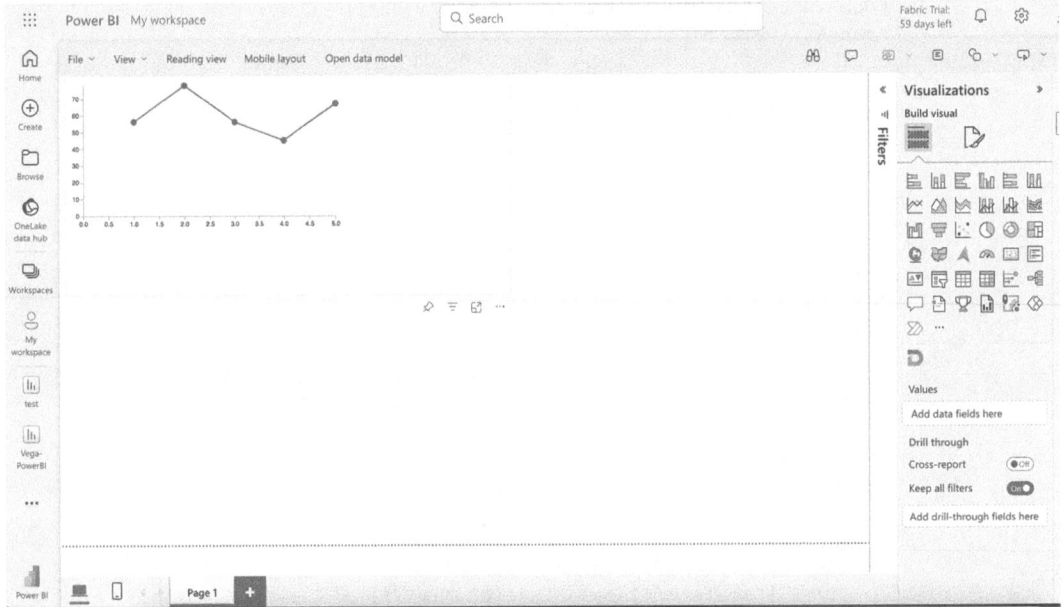

Figure 11.9   The resulting chart in Power BI

11.5 *Comet*

Comet (https://www.comet.com/site/) is an experimentation platform for ML model tracking and testing. In my previous book, *Comet for Data Science*, I thoroughly described how to start with Comet and how to track ML, deep learning (DL), time series, and natural language processing tasks. So, for more details about Comet, please refer to this book.

You may decide to integrate your data story produced in Altair with Comet if your data storytelling story is the result of a complete data science project, which includes model experimentations and evaluations. Since Comet supports Python, using it to export your data story does not require any particular effort.

In Comet, you can also create reports. A *report* is a collection of text and panels showing the results of your experiments. This is a very helpful feature because you can use just one tool for experiments and reporting. Unlike Tableau and Power BI, Comet is free for personal use, so you don't need to pay to access the service. In addition, Comet supports many programming languages, including Python, so you can continue using Python for your reports.

In this section, we will describe how to include a simple interactive chart in Comet. We will use the HTML code generated by 08/slider.py. We will write the code in HTML/JavaScript, but Comet also supports Python. If you don't have the slider.html file, open a terminal, move to the 08 directory, and run the command `python slider.py`.

Execute the following steps:

1 Log in to Comet, and start a new project by following these steps: New Project → Project Name, Description, Project Type: Experiment Tracking.

2 Create a new report by clicking Report → New Report → Add Panel → New.

3 The Comet SDK should open. By default, it is configured to work in JavaScript. If you want to work with Python, on the top-right part, click JavaScript and select Python.

4 Under the Resources tab, add the following three JavaScript scripts (figure 11.10):

   a https://cdn.jsdelivr.net/npm/vega@5

   b https://cdn.jsdelivr.net/npm/vega-lite@5.8.0

   c https://cdn.jsdelivr.net/npm/vega-embed@6

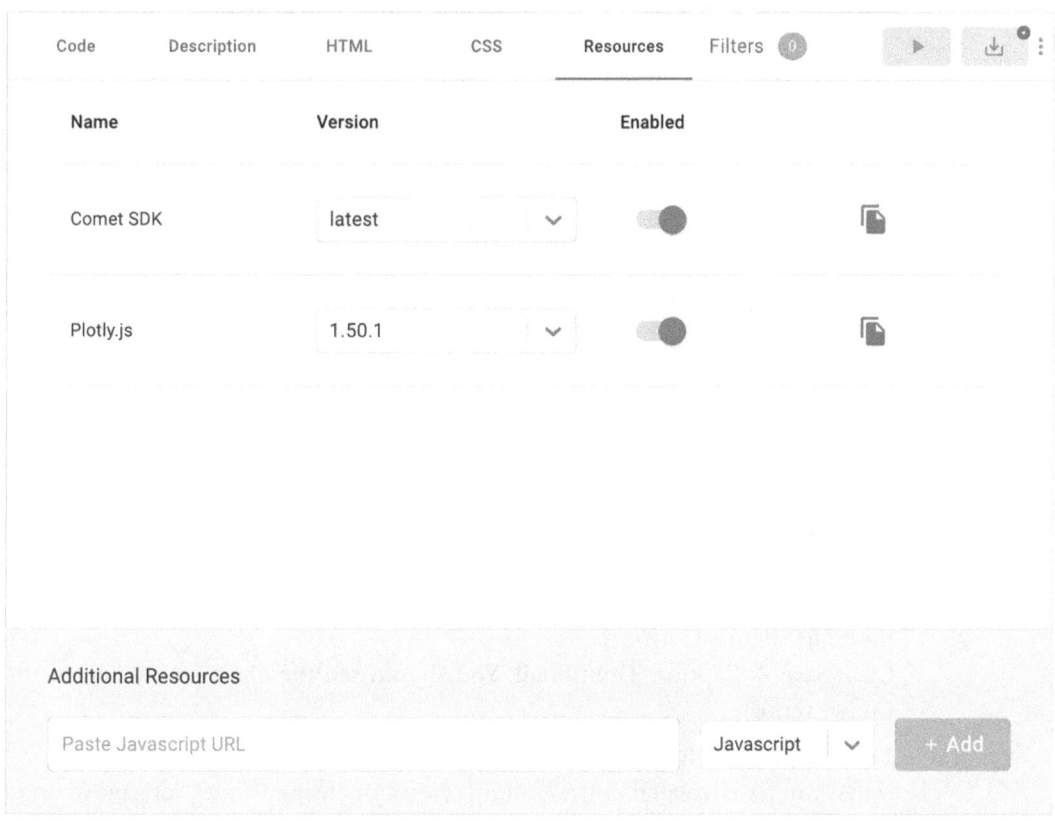

Figure 11.10   The Resources tab in the Comet SDK

5 Switch to the HTML tab, and change the `id` name to `vis`, as shown in figure 11.11.

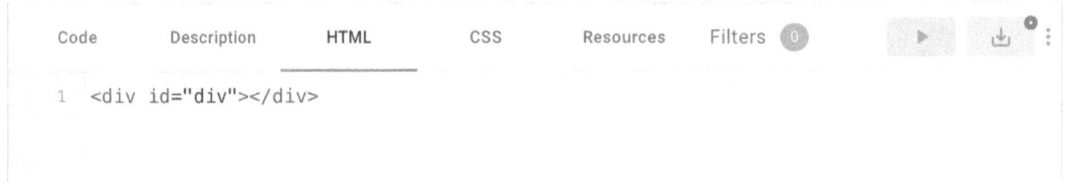

Figure 11.11   The HTML tab in the Comet SDK

6 Open the slider.html file in Edit mode, and copy the code between the <script> and </scripts> tags (as shown in the following listing). Don't worry if this code is very long; it contains all the Vega specs, including the original dataset. Paste the copied code in the Code tab of the Comet SDK.

Listing 11.4   The JavaScript code generating the chart

```
(function(vegaEmbed) {
    var spec = {"config": {"view": ...}};
    var embedOpt = {"mode": "vega-lite"};

    function showError(el, error){
        el.innerHTML = ('<div style="color:red;">'
                        + '<p>JavaScript Error: ' + error.message + '</p>'
                        + "<p>This usually means there's a typo in your
chart specification. "
                        + "See the javascript console for the full
traceback.</p>"
                        + '</div>');
        throw error;
    }
    const el = document.getElementById('vis');
    vegaEmbed("#vis", spec, embedOpt)
      .catch(error => showError(el, error));
}) (vegaEmbed);
```

7 Click the Run button (a green triangle). You should see the rendered chart on the right (figure 11.12).

8 Click Save & Capture Thumbnail. You should see the chart embedded in the Comet report.

Table 11.1 summarizes the pros and the cons of each analyzed tool.

Now that you have learned how to publish your story using Comet, let's move on to the next way: presenting your data story through slides.

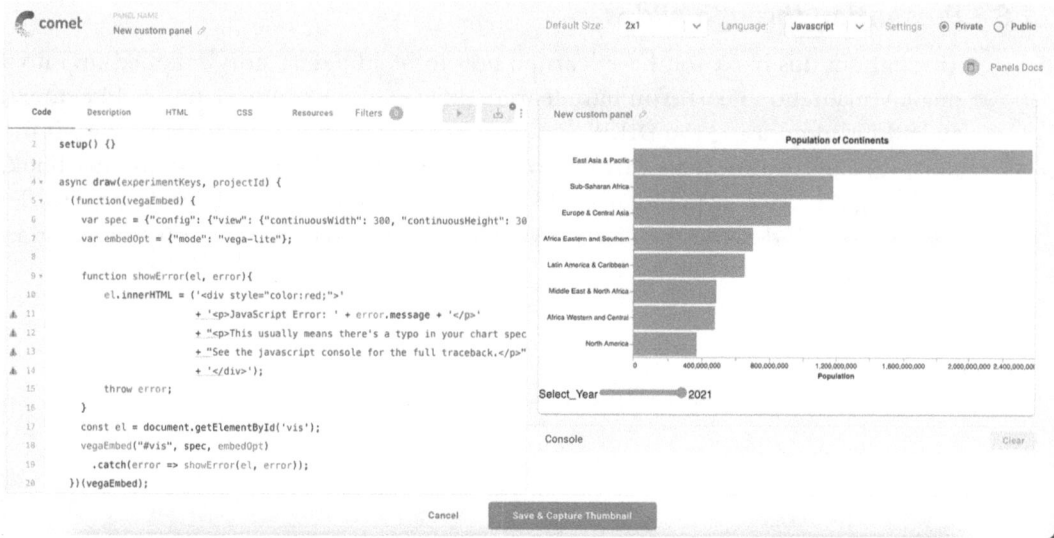

Figure 11.12   The rendered chart in the Comet SDK

Table 11.1   Pros and cons of the analyzed tools

Tool Name	Description	Pros	Cons
Streamlit	A Python library to implement a web server	▪ Simple to use ▪ Fully integrated with Altair ▪ Use Python to configure the interface ▪ Free	▪ You must write your dashboard from scratch. ▪ You must have a machine to host the web server.
Tableau	A software for data exploration and data analysis	▪ Does not require any programming skills ▪ Fast dashboard building	▪ Not free ▪ You must learn the basics to use it.
Power BI	A software for data exploration and data analysis	▪ Does not require any programming skills ▪ Fast dashboard building	▪ Not free ▪ You must learn the basics to use it. ▪ It does not support Altair directly. You must use Vega instead.
Comet	An experimentation platform for data science	▪ Free for personal usage ▪ Enables you to track all your data science workflow, up to data storytelling ▪ Fully integrated with Altair	▪ You must learn the basics to connect the data to a data visualization chart.

11.6   *Presenting through slides*

Throughout this book you have learned how to build a data story wrapped up into a single visualization chart. You have learned how to extract information, add context, and add wisdom using the DIKW pyramid. But what happens if you have to split the story through slides? You can exploit all the principles described in this book, including the DIKW pyramid, to build a story organized in more slides or sections. To organize your slides, use the data storytelling arc discussed in chapter 5 and shown in figure 11.13.

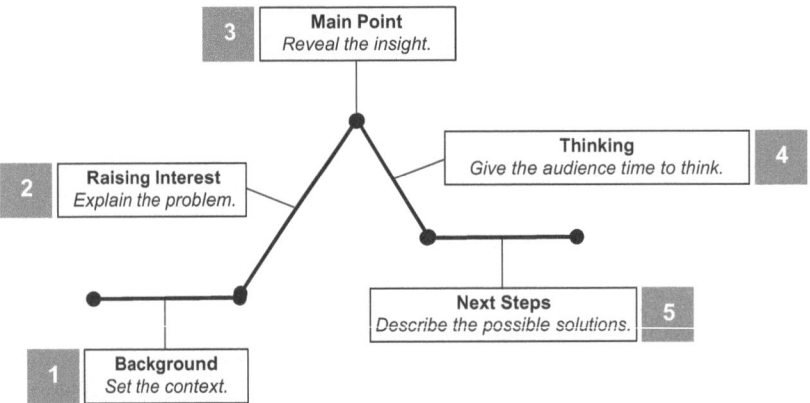

Figure 11.13   The data storytelling arc

Start by showing the background, with the context of your story. Next, raise interest, by explaining the problem. Then, define the main point, which reveals the insight and leaves the audience the time to think. Finally, propose the resolution.

Let's implement a practical example to better understand how to split a single-story chart into multiple slides. We will focus on the homelessness case study, shown in figure 11.14.

Organize the slides into five parts, one for each point of the data storytelling arc. Implement each part using a single slide or multiple slides. Let's see a possible implementation:

1 *Background*—Use two slides. Split the first paragraph of the commentary under the title into two parts, and put each part in a separate slide. Add the pictures in the left part of the original data story to the slides, one for each slide (figure 11.15). To enrich the background part, you may add other slides with additional details, such as interviews with involved people or other documents that engage the audience with the problem.

Together, Let's Make a Difference: Support Our Project to Help the Homeless!
Homelessness is a heartbreaking reality that leaves individuals and families without a stable home,
leading to devastating consequences such as poor health and social isolation.
The chart describes the number of homeless people in Italy per 1,000 inhabitants, organized by region.
The data is from 2021, the most recent year available (Source: ISTAT).

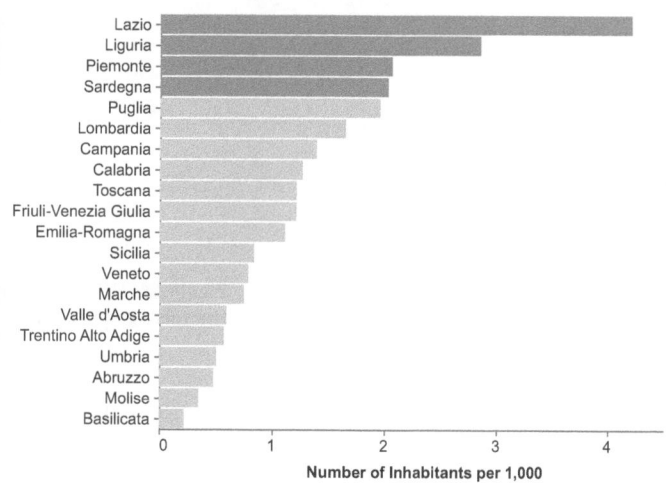

Our Visionary Plan to Harness the Funds

Figure 11.14   The homelessness case study implemented as a single story

Figure 11.15   The slides in the background part

2   *Raising interest*—Add a single slide, which explains the problem (figure 11.16). In the single-chart data story, this text was not present because of the space limitation and the presence of the chart, which already highlighted the problem.

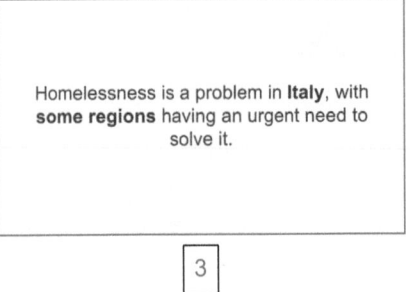

Figure 11.16   **The slide in the raising interest part**

3   *Main point*—Implement a single slide with the main chart (figure 11.17).

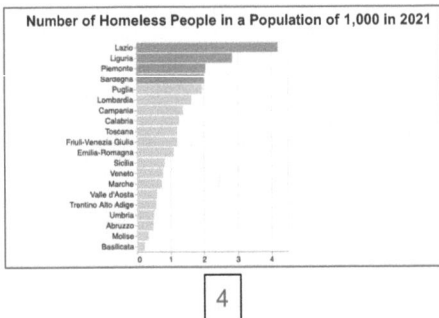

Figure 11.17   **The slide in the main point part**

4   *Thinking*—Add a slide with a question that leaves the audience the time to think (figure 11.18).

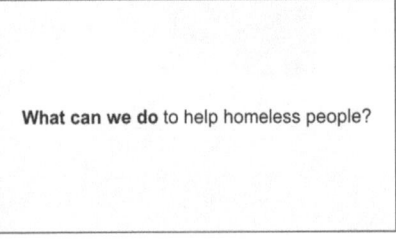

Figure 11.18   **The slide in the thinking part**

5   *Next steps*—Add a single slide for each donut chart (figure 11.19).

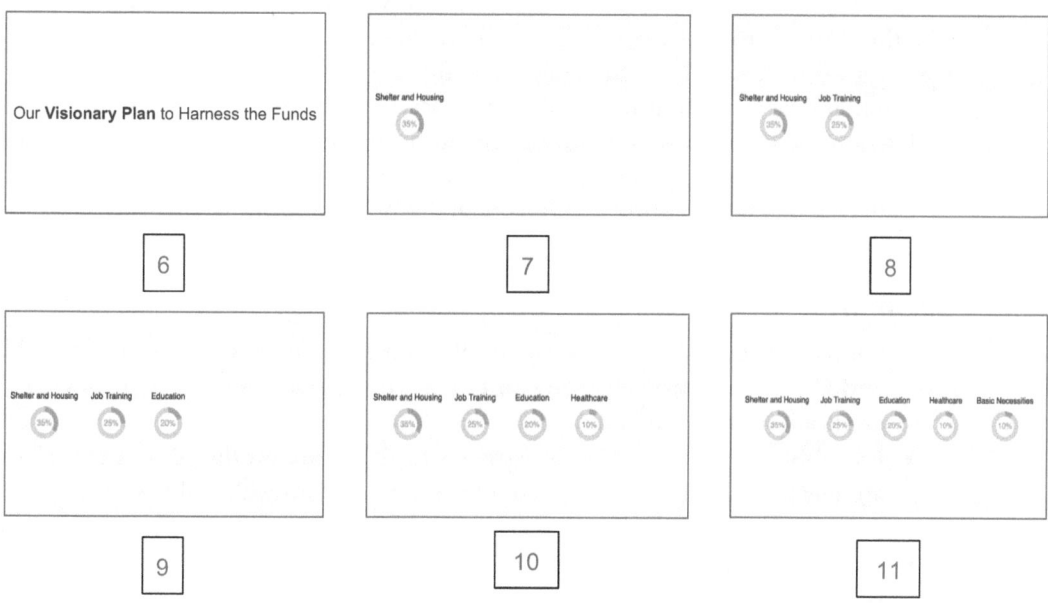

Figure 11.19   The slides in the next steps part

As a final step, add the title to your presentation as the first slide. Use the title of the original data story.

To implement each chart in the slides, you can still use Altair, as you did throughout this book, and your preferred tool for slide editing, such as PowerPoint, Google Slides, and so on.

Realize your slides using the charts implemented in Altair. You can find the complete implementation of the slides described in this section in the GitHub repository for the book under 11/slides/homeless-people.pdf.

In this final chapter of the book, you have learned how to publish the final data story using different tools: Streamlit, Tableau, Power BI, and Comet. Finally, you have seen how to transform a single data story into multiple slides for presentation.

11.7   Final thoughts

Our journey together has come to an end. Dear reader, I thank you for sharing a stretch of the road with me, for trusting me, and for getting to the end of the reading. I hope that through this book, you have learned to love data storytelling as an effective tool for communicating your data. I also hope that although the book is finished, your journey into the world of data storytelling can continue, reading other books, perhaps more interesting than this one, updating yourself with articles, courses, and anything else good you can find in the world. In the book, here and there, I have indicated interesting resources that you can draw on, but do not limit your personal research.

Throughout the book, I have tried to equip you with the basic tools to undertake the difficult role of the data storyteller. The task of using them responsibly is now up to you. Always have the courage and strength to look for the truth behind the data, and after finding it, maintain the desire to delve deeper into it and look for it again. I recommend one more thing: never forget to handle the data that you gradually tell with care because, behind the data, there are always people waiting for someone to tell their story. I wish you all the best and hope to meet you again in the future!

Summary

- You can embed your data story in different tools, such as Tableau, Power BI, and Comet. Alternatively, you can publish your data story as a standalone web application using Streamlit.
- Use Tableau or Power BI to integrate your data story produced in Altair, if you already have some knowledge about these tools. This will enable you to enrich your dashboards quickly.
- Use Comet to integrate your data story produced in Altair, if the story is the result of a complete data science project that included some experimentations, such as model tracking and evaluation. Use Streamlit to export your data story produced in Altair if you need a web application to share it with your audience.
- In all cases, after completing your data story, remember to share it with your audience. And after having told the first story, begin to work on the second. And then work on the third, and so on, until you become an effective data storyteller. Happy work! Happy data storytelling!

References

- Deckler, G. and Powell, B. (2022). *Mastering Microsoft Power BI: Expert Techniques to Create Interactive Insights for Effective Data Analytics and Business Intelligence* (2nd ed.). Packt Publishing.
- Milligan, J. N. (2022). *Learning Tableau 2022: Create Effective Data Visualizations, Build Interactive Visual Analytics, and Improve Your Data Storytelling Capabilities* (5th ed.). Packt Publishing.
- Lo Duca, A. (2022). *Comet for Data Science.* Packt Publishing.

appendix A
Technical requirements

This appendix describes how to install the software used in this book.

A.1 Cloning the GitHub repository

To clone the GitHub repository for the book on your local machine, you can adopt one of the following two strategies:

- Using the terminal
- Using GitHub Desktop

A.1.1 Using the terminal

To clone the GitHub repository for the book from the command line, follow these steps:

1 Install the Git suite (https://git-scm.com/downloads), if you do not have it yet.
2 Open the terminal on your computer.
3 Navigate to the directory where you want to clone the repository.
4 Run the command `git clone` `https://github.com/alod83/Data-Story-telling-with-Altair-and-AI/tree/main`.
5 Wait for the repository to be cloned to your local machine.

A.1.2 Using GitHub Desktop

To clone the GitHub repository for the book from GitHub Desktop, follow these steps:

1 Download and install GitHub Desktop from their official website: https://desktop.github.com/.
2 Launch GitHub Desktop.

3 Sign in to your GitHub account or create a new account if needed.
4 Click the File menu, and select Clone Repository.
5 In the Clone a Repository window, choose the URL tab.
6 Enter the repository URL (https://github.com/alod83/Data-Storytelling-with-Altair-and-AI/tree/main) in the Repository URL field.
7 Choose the local path where you want to clone the repository.
8 Click the Clone button.
9 Wait for GitHub Desktop to clone the repository to your local machine.

A.2   *Installing the Python packages*

The examples illustrated in this book use Python 3.8. You can download it from the official website: https://www.python.org/downloads/release/python-3810/.

The examples described in this book use the following Python packages:

- langchain==0.1.12
- langchain-community==0.0.28
- langchain-core==0.1.32
- langchain-openai==0.0.8
- langchain-text-splitters==0.0.1
- altair==5.3.0
- chromadb==0.4.22
- jupyterlab==3.5.1
- ydata-profiling==4.6.0
- matplotlib==3.5.0
- numpy==1.24.4
- pandas==1.3.4
- unstructured==0.10.19

You can install the latest versions of these packages simply by running the command `pip install <package_name>`. However, since technology evolves rapidly, these packages may have already been updated by the time you read this book. To ensure that you can still run the code used in this book as it is, create a virtual environment on your computer and install the specific package versions used in this book. To create a virtual environment on your computer, open a terminal and install the `virtualenv` package using `pip install virtualenv`. Then, run the commands described in the following listing.

Listing A.1   Creating and running a virtual environment

```
python -m venv env
source env/bin/activate
```

To deactivate the virtual environment, simply run `deactivate`. Within the virtual environment, install the preceding packages by running the command `pip install`

`<package_name>==<version>` for each package. Alternatively, use the requirements.txt file placed in the root of the GitHub repository for the book and run the following command: `pip install -r requirements.txt`.

A.3 Installing GitHub Copilot

To use GitHub Copilot, you must set up a free trial or subscription for your personal GitHub account. If you are a teacher or a student, you can set up a free subscription plan at the following link: https://education.github.com/discount_requests/pack_application.

Once your account is set up to use GitHub Copilot, configure it as an extension of Visual Studio Code (VSC), a free, open source code editor designed for developers to write and debug code. Download VSC from its official website: https://visualstudio.microsoft.com/it/downloads/.

To start GitHub Copilot, open Visual Studio and navigate to the Extensions tab. Download and install the GitHub Copilot extension, and then select Connect to Your Account from the dashboard. Enter your GitHub credentials. Once logged in, the extension will detect existing repositories and provide options for configuring new projects.

A.4 Configuring ChatGPT

To use ChatGPT, you must set up an account on the Open AI website (https://openai.com/). At the time of writing this book, ChatGPT version GPT-4 is free.

To access ChatGPT, visit https://chat.openai.com/, log in to your account, and start writing your prompts in the input text box like a live chat. Whenever you want to start a new topic, create a new chat session by clicking on the top-left New Chat button.

ChatGPT retains the entire history of prompts within a single chat session. This means you can write step-by-step instructions, build on previous prompts, and maintain a coherent conversation.

The web interface also provides a paid account that gives some additional features, such as the ability to use advanced models. In this book, we use the free version of the web interface.

A.5 Installing Open AI API

Point to https://openai.com/, log in to your account, or create a new one. Once logged in, click API. To use the Open AI API, you must upgrade your existing account to a paid one. Click your personal logo in the top-right of the screen, next to Manage Account | Billing. Next, add credits by adding a payment method. The OpenAI official pricing page (https://openai.com/pricing) lists pricing details. A credit of $5 should be sufficient for the experiments run in this book.

Next, you can configure an API key by performing the following steps:

1 Click the Personal button on the top right of your dashboard.
2 Select View API Keys in your dropdown menu.

3 Click Create a New Secret Key.

4 Insert the key name, and create a secret key.

5 Click the Copy symbol to copy the key value, and paste it in a secure place.

6 Click Done.

Once you have created the API key, you can use the Open AI API in your Python scripts.

The following listing shows an example of usage of the Open AI API, asking ChatGPT to produce some output given an input prompt. You can also find the following code in the GitHub repository for the book under AppendixA/openai-test.py.

Listing A.2 Calling the ChatGPT API

```
import openai
openai.api_key = 'MY_API_KEY'
prompt = 'What is the capital of Italy?'

messages = [{"role": "user", "content": prompt}]
response = openai.chat.completions.create(
    model="gpt-3.5-turbo",
    messages=messages,
    temperature=0,          Temperature [0,1] defines the
)                           grade of randomness in the output.

print(response.choices[0].message.content.strip())
```

NOTE Use the `chat.completions.create()` method to call the ChatGPT API.

A.6 *Installing LangChain*

You can use LangChain with your preferred large language model (LLM). In this book, we use OpenAI, which requires an API key. For more details, refer to the section related to OpenAI installation.

The following listing shows an example of using LangChain to interact with the OpenAI API in Python. You can also find the code of this example in the GitHub repository for the book under AppendixA/langchain-test.py

Listing A.3 Using LangChain

```
from langchain_openai import ChatOpenAI
import os

os.environ["OPENAI_API_KEY"] = 'MY_API_KEY'

llm = ChatOpenAI(temperature=0, model='gpt-3.5-turbo')

print(
    llm.invoke("What is the capital of Italy?")
)
```

Windows users may get the error `"ModuleNotFoundError: No module named 'pwd'"` when using LangChain. This is because the `pwd` module is not available on Windows platforms. One workaround to this problem is to add the following line of code before using LangChain.

Listing A.4   Using `pwd` on Windows

```
try:
    import pwd
except ImportError:
    import winpwd as pwd
```

For more details, see the LangChain official documentation (https://www.langchain .com/).

A.7   Installing Chroma

Chroma uses SQLite as a database. The following listing shows how to connect to a Chroma database.

Listing A.5   Importing Chroma

```
import chromadb

client = chromadb.Client()
```

For more details, see the Chroma official documentation (https://docs.trychroma .com/getting-started).

A.8   Configuring DALL-E

To use DALL-E, you must set up an account on the Open AI website (https://openai .com). Then, you can use two ways to interact with DALL-E: the web interface or the Open AI API. In this book, we prefer using the web interface for a couple of advantages. The Open AI API requires a paid account, while the web interface requires buying credits to generate images. In addition, if you created your DALL-E account before April 6, 2023, every month, you have a pool of free credits.

To use the web interface, visit https://labs.openai.com, log in to your account, and write your prompts in the input text box. To use the Open AI API, you must upgrade your existing account to a paid one. Next, you can configure an API as described in section 2.5. Once you have created the API key, you can use the Open AI API in your Python scripts.

The following listing shows an example of using the Open AI API: asking DALL-E to produce an image given an input prompt. The code is also available in the GitHub repository for the book under AppendixA/dalle-test.py.

Listing A.6   Calling the DALL-E API

```
import openai
import requests

openai.api_key = 'MY_API_KEY'
prompt = 'Create a painting of a beautiful sunset over a calm lake.'

n=1
response = openai.images.generate(
  prompt=prompt,
  n=n,                    ⟵————  The number of
  size='1024x1024'               images to generate
)

i = 0
for image_data in response.data:
    print(image_data.url)
    img = requests.get(image_data.url).content
    with open(f"image-{i}.png", 'wb') as handler:
        handler.write(img)
    i += 1

print(output)
```

> **NOTE**   Use the `Image.create()` method to call the DALL-E API. The output contains the URL to the produced image.

Additionally, you can modify an existing image using the `Image.create_edit()` method. Refer to the DALL-E official documentation (https://platform.openai.com/docs/guides/images/usage) for more details.

appendix B
Python pandas
DataFrame

This appendix describes an overview of the pandas DataFrame and the methods used in this book.

B.1 An overview of pandas DataFrame

Python pandas is a data manipulation, analysis, and visualization library. It provides tools to load and allow you to manipulate, analyze, and visualize data. In this book, we use the pandas DataFrame, a two-dimensional structure composed of rows and columns. The DataFrame stores data in a tabular form, enabling you to manipulate, analyze, filter, and aggregate data quickly and easily.

There are different ways to create a pandas DataFrame. In this book, we consider two ways: from a Python dictionary and from a CSV file. You can download the code described in this appendix from the GitHub repository for the book under AppendixB/Pandas DataFrame.ipynb.

B.1.1 Building from a dictionary

Consider the following listing, which creates a pandas DataFrame from a Python dictionary.

Listing B.1 Creating a DataFrame from a dictionary

```
import pandas as pd

data = {
    'Name': ['Alice', 'Bob', 'Charlie'],
    'BirthDate': ['2000-01-30', '2001-02-03', '2001-04-05'],
    'MathsScore': [90, 85, None],
```

```
        'PhysicsScore': [87, 92, 89],
        'ChemistryScore': [92, None, 90],
        'Grade' : ['A', 'B', 'A']              Defines the
}                                              dictionary

df = pd.DataFrame(data)          Creates the DataFrame

df['BirthDate'] = pd.to_datetime(df['BirthDate'], format='%Y-%m-%d')
```

Parses the
BirthDate field
as a date

NOTE Use `DataFrame()` to create a new DataFrame from a dictionary.

B.1.2 Building from a CSV file

Use the `read_csv()` file to load a CSV file into a pandas DataFrame.

Listing B.2 Creating a DataFrame from a CSV file

```
import pandas as pd

df = pd.read_csv('data.csv')
```

NOTE Use `read_csv()` to create a new DataFrame from a CSV file.

Now that you have seen how to create a pandas DataFrame, we'll discuss the main DataFrame functions used in this book.

B.2 dt

The `dt` variable within a pandas Dataframe enables you to access Python's built-in DateTime library. Use it to store and manipulate DateTime values, such as the year, month, day, hour, minute, and second. Consider the following listing, which extracts the year from a DateTime column.

Listing B.3 How to use pandas dt

```
import pandas as pd

data = {
    'Name': ['Alice', 'Bob', 'Charlie'],
    'BirthDate': ['2000-01-30', '2001-02-03', '2001-04-05'],
    'MathsScore': [90, 85, None],
    'PhysicsScore': [87, 92, 89],
    'ChemistryScore': [92, None, 90],
    'Grade' : ['A', 'B', 'A']
}

df = pd.DataFrame(data)
df['BirthDate'] = pd.to_datetime(df['BirthDate'], format='%Y-%m-%d')

year = df['BirthDate'].dt.year
month = df['BirthDate'].dt.month        Extracts the month from
                                        the BirthDate column
```

Extracts the year from
the BirthDate column

```
day = df['BirthDate'].dt.day                    ◄──────        Extracts the
weekOfYear = df['BirthDate'].dt.isocalendar().week  ◄──┐       day from the
                                                       │       BirthDate column
                        Extracts the week from         │
                        the BirthDate column
```

NOTE Use pandas `dt` to access the DateTime functions of Python's DateTime library.

B.3   *groupby()*

The pandas `groupby()` method splits data into groups based on the values of certain columns. This process often involves creating an aggregate statistic for each group, such as a sum or mean.

Listing B.4   How to use pandas `groupby`

```
import pandas as pd

data = {
    'Name': ['Alice', 'Bob', 'Charlie'],
    'BirthDate': ['2000-01-30', '2001-02-03', '2001-04-05'],
    'MathsScore': [90, 85, None],
    'PhysicsScore': [87, 92, 89],
    'ChemistryScore': [92, None, 90],
    'Grade' : ['A', 'B', 'A']
}

df = pd.DataFrame(data)
df_grouped = df.groupby(by='Grade').mean().reset_index()
```

NOTE Use pandas `groupby` to group by instrument and calculate the average score by grade. Use the `reset_index()` method to restore the indexer column (`Grade` in the example).

Table B.1 shows the result.

Table B.1   The result of `groupby()` in listing B.4

Grade	MathsScore	PhysicsScore	ChemistryScore
A	90.0	88.0	91.0
B	85.0	92.0	

B.4   *isnull()*

The pandas DataFrame `isnull()` method returns a new Boolean DataFrame indicating which values in the DataFrame are null (`NaN`). Use this method to detect missing values in a DataFrame.

Listing B.5   How to use pandas `isnull`

```
import pandas as pd

data = {
    'Name': ['Alice', 'Bob', 'Charlie'],
    'BirthDate': ['2000-01-30', '2001-02-03', '2001-04-05'],
    'MathsScore': [90, 85, None],
    'PhysicsScore': [87, 92, 89],
    'ChemistryScore': [92, None, 90],
    'Grade' : ['A', 'B', 'A']
}

df = pd.DataFrame(data)
df_isnull = df.isnull()
```

> **NOTE**   Use pandas `isnull()` to check if a DataFrame contains missing values. You can apply the `isnull()` method to a single column as well (e.g., `df['ChemistryScore'].isnull()`).

Table B.2 shows the result, a Boolean DataFrame indicating which values in the DataFrame are null (`NaN`).

Table B.2   The result of `isnull()` in listing B.5

Name	BirthDate	MathsScore	PhysicsScore	ChemistryScore	Grade
False	False	False	False	False	False
False	False	False	False	True	False
False	False	True	False	False	False

B.5   *melt()*

We use the pandas `melt()` function to reshape data by turning columns into rows. This function unpivots a DataFrame from wide to long format, optionally leaving identifiers set.

Listing B.6   How to use pandas `melt`

```
import pandas as pd

data = {
    'Name': ['Alice', 'Bob', 'Charlie'],
    'BirthDate': ['2000-01-30', '2001-02-03', '2001-04-05'],
    'MathsScore': [90, 85, None],
    'PhysicsScore': [87, 92, 89],
    'ChemistryScore': [92, None, 90],
    'Grade' : ['A', 'B', 'A']
}
```

```
df = pd.DataFrame(data)

df_melted = df.melt(id_vars='Name',
        var_name='Subject',
        value_name='Score',
        value_vars=['MathsScore', 'PhysicsScore', 'ChemistryScore']
)
```

NOTE Use pandas `melt()` to transform the DataFrame from wide to long format. Set the `id_vars` argument to specify which variables to keep as an identifier and the `var_name` and `value_name` arguments to set the column names for the new variables in the resulting melted DataFrame. Use `value_ vars` to select the columns to group.

Table B.3 shows the result of melting the data used in listing B.6.

Table B.3 The result of the melting operation described in listing B.6

Name	Subject	Score
Alice	MathsScore	90.0
Bob	MathsScore	85.0
Charlie	MathsScore	Null
Alice	PhysicsScore	87.0
Bob	PhysicsScore	92.0
Charlie	PhysicsScore	89.0
Alice	ChemistryScore	92.0
Bob	ChemistryScore	Null

B.6 *unique()*

We use the pandas `unique()` method to obtain distinct values from a specific column in a DataFrame. This method returns an array-like object, containing the unique values found in the specified column.

Listing B.7 How to use pandas `unique`

```
import pandas as pd

data = {
    'Name': ['Alice', 'Bob', 'Charlie'],
    'BirthDate': ['2000-01-30', '2001-02-03', '2001-04-05'],
    'MathsScore': [90, 85, None],
    'PhysicsScore': [87, 92, 89],
    'ChemistryScore': [92, None, 90],
    'Grade' : ['A', 'B', 'A']
}
```

```
df = pd.DataFrame(data)
unique_grades = df['Grade'].unique()
```

NOTE   Use pandas `unique()` to get the unique values of a column.

The following listing shows the result of calculating the unique values for the column `Grade` (from listing B.7).

Listing B.8   The results of pandas `unique()`

```
array(['A', 'B'], dtype=object)
```

NOTE   The method returns an array containing the unique values for a column.

appendix C
Other chart types

This appendix describes some of the most popular charts not covered in chapter 6. Use it as a reference if you want to try new charts. The code for the charts is described in the book's GitHub repository, under 06, in addition to other, less popular charts. Let's investigate each chart type separately, starting with the cooking chart family.

C.1 Donut chart

A *donut chart* is a type of circular data visualization that displays data in a ring shape. It is similar to the pie chart but with a hole in the center, creating a visual representation of proportions or percentages of different categories.

To convert a pie chart into a donut chart, simply add the `innerRadius` and `outerRadius` properties to the `mark_arc()` method, as shown in the following listing.

Listing C.1   The code to generate a donut chart

```
import pandas as pd
import altair as alt

data = {
    'percentage': [0.7,0.3],
    'label'     : ['70%','30%'],
    'color'     : ['#81c01e','lightgray']
}

df = pd.DataFrame(data)

chart = alt.Chart(df).mark_arc(
    innerRadius=100,
    outerRadius=150
).encode(
    theta='percentage',
```

```
    color=alt.Color('color', scale=None),
    tooltip='label'
).properties(
    width=300,
    height=300
)
```

NOTE   The code to generate a donut chart in Altair. Use the `innerRadius` and `outerRadius` to transform a pie chart into a donut chart.

If your donut chart compares two values and you want to focus on just one value, add this value at the center of the donut, as shown in figure C.1.

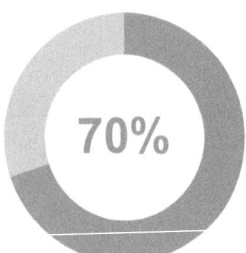

Figure C.1   A donut chart in Altair with a label in the center

Ask Copilot to generate the label within the donut.

Listing C.2   The prompt to generate the label within the donut chart

```
# Add text to the center of the donut chart
# - Use df.head(1) to get the first row of the dataframe
# - Use the `label` column for text channel
# - Use the `color` column for color
```

NOTE   The sequence of instructions for Copilot to generate the label within the donut chart.

The following listing shows the generated code.

Listing C.3   The code generated by Copilot

```
text = alt.Chart(df.head(1)).mark_text(
    align='center',
    baseline='middle',
    fontSize=60,
    fontWeight='bold'
).encode(
    text='label',
    color=alt.Color('color', scale=None)
).properties(
```

```
    width=300,
    height=300
)
```

NOTE   Change the `fontSize` property to increase or decrease the label size.

From a data storytelling perspective, you can concatenate multiple donut charts, as illustrated in figure C.2, to show many percentage values, for example, the output of a questionnaire on which meals the customers of a restaurant like. The total sum is greater than 100% because a customer may like more than one meal.

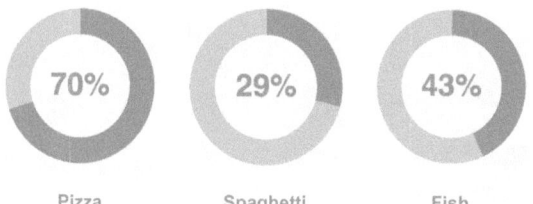

Which meal do our customers like?

Pizza   Spaghetti   Fish

Figure C.2   Multiple donut charts showing the output of a questionnaire

In this scenario, a cumulative total of 100% might not be essential. However, it's worth noting that other business perspectives could potentially warrant such a requirement. Now that you have learned how to draw the cooking charts in Altair and Copilot, let's move on to the next family of charts: the bar charts.

C.2   Bar charts family

The bar charts family includes column charts, columns chart with multiple series, pyramid charts, stacked column charts, 100% stacked column charts, and histograms. Let's take a closer look.

C.2.1   Column chart

A *column chart* is similar to a bar chart but with inverted axes; it shows categories on the x-axis and values on the y-axis. Since the x-axis is also used to represent temporal data, you can use a column chart to describe periods as categories. The following listing shows how to draw a column chart. Also, increase the chart width to 600 to leave more space for each column.

Listing C.4   The code to create a column chart

```
import pandas as pd
import altair as alt

chart = alt.Chart(df).mark_bar(
    color='#81c01e'
```

```
).encode(
    x=alt.X('Meal Type', sort='-y'),
    y='Number of Likes'
).properties(
    width=600,
    height=300
)

chart.save('column-chart.html')
```

NOTE   Invert the x and y channels. Also, change the `sort` property value.

Figure C.3 represents the generated column chart.

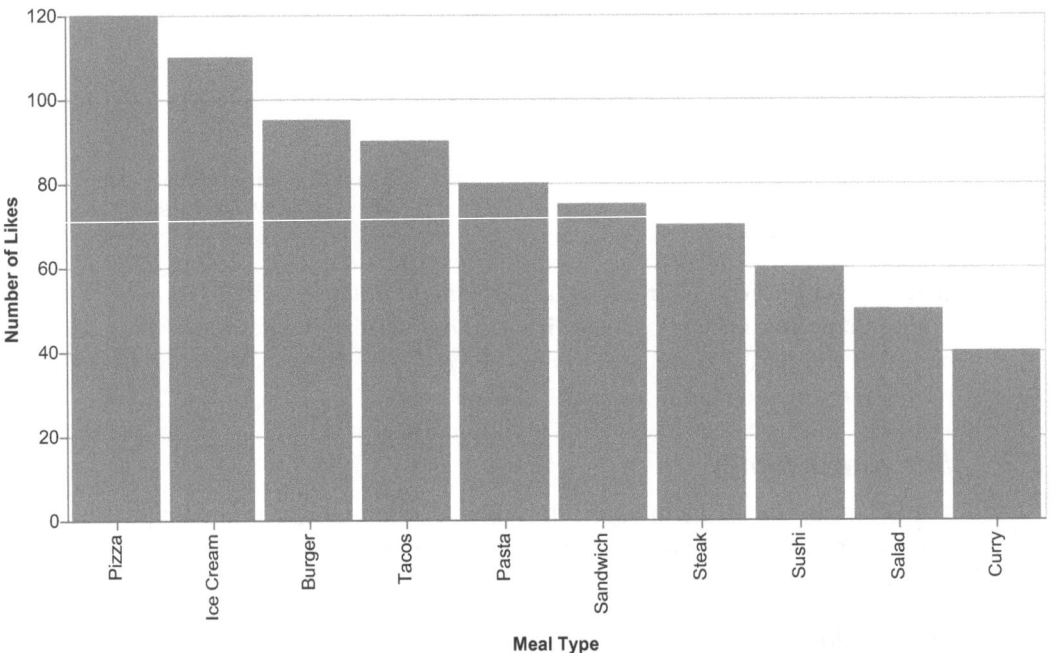

Figure C.3   A column chart

The x-axis of the column chart shows rotated labels. Also, the x-axis title is unnecessary. Let's proceed with decluttering. Start by formatting the code in the Visual Studio Code editor, and then write the comment # `Rotate`, as shown in figure C.4. Copilot will suggest how to complete the sentence. Press Tab, change 45 degrees to 0 degrees, and press Enter, and then Copilot will add the desired code in the row below your comment (`axis=alt.Axis(labelAngle=0),`).

```
14   chart = alt.Chart(df).mark_bar(
15       color='#81c01e'
16   ).encode(
17       x=alt.X('Meal Type',
18               sort='-y',
19               # Rotate the labels by 45 degrees
20           ),
21       y='Number of Likes'
22   ).properties(
23       width=300,
24       height=300
25   )
```

Figure C.4   How to add a comment within the code. Copilot will suggest the next steps.

Next, add a comma after `labelAngle=0`, and start writing the word `title`. Copilot will suggest `=None`. Press Tab to confirm. This Copilot property is fantastic! You can add a comment at any point of your code and ask Copilot to generate new code for you! Since Copilot is a generative AI tool, it may happen that, in your case, the suggested prompts are different. Anyway, you can easily adapt the code suggested by Copilot to your needs.

Now, let's declutter the y-axis. Remove the y-axis by changing the `y='Number of Likes'` channel to `y=alt.Y('Number of Likes',axis=alt.Axis(grid=False))`. Then, add the number of likes at the top of each bar. Let's use Copilot to do the job for us. Start writing the following text: `# Add a text mark to the chart with the following options:`. Copilot will suggest the next steps. Confirm them by pressing Tab and then Enter. At some point, Copilot will stop suggesting. Press Enter, and Copilot will generate the code for you. Test the generated code. If it does not satisfy you, change it manually, or use Copilot again. In my case, I had to slightly change the Copilot comments to make it generate the desired code. The following listing shows the Copilot instructions.

Listing C.5   The prompts generated by Copilot and slightly modified

```
# Add a text mark to the chart with the following options:
# - Use the `Number of Likes` column for y channel
# - Use the `Meal Type` column for x channel and sort by the number of likes
# - Set the color of the text to '#81c01e'
# - Set the text to the `Number of Likes` column
# - Set the font size to 14
# - Set the font weight to 600
# - Set the text baseline to 'center'
# - Set the text align to 'middle'
# - Set the y offset to 10
```

NOTE Ask Copilot to generate the code to draw labels at the top of each column

The following listing describes the produced code.

Listing C.6   The code generated by Copilot to add labels

```
text = alt.Chart(df).mark_text(
    color='#81c01e',
    fontSize=14,
    fontWeight=600,
    baseline='middle',
    align='center',
    dy=-10
).encode(
    x=alt.X('Meal Type',
            sort='-y',
            axis=alt.Axis(labelAngle=0,title=None),
    ),
    y='Number of Likes',
    text='Number of Likes'
)
```

NOTE Use the `mark_text()` method to add a label at the top of each column. Set the text properties, including color, font size, weight, and so on.

Finally, combine the two charts.

Listing C.7   Combining the charts to build a final chart

```
# Combine the bar chart and text mark into a single chart.
chart = chart + text

chart = chart.configure_view(
    strokeWidth=0
)

chart.save('column-chart.html')
```

NOTE Write the prompt asking Copilot to generate the code to merge charts.

Figure C.5 shows the final chart.

Now that you have learned how to declutter a column chart in Altair and Copilot, let's move on to the next chart. In the following section, we'll cover the column chart with multiple series.

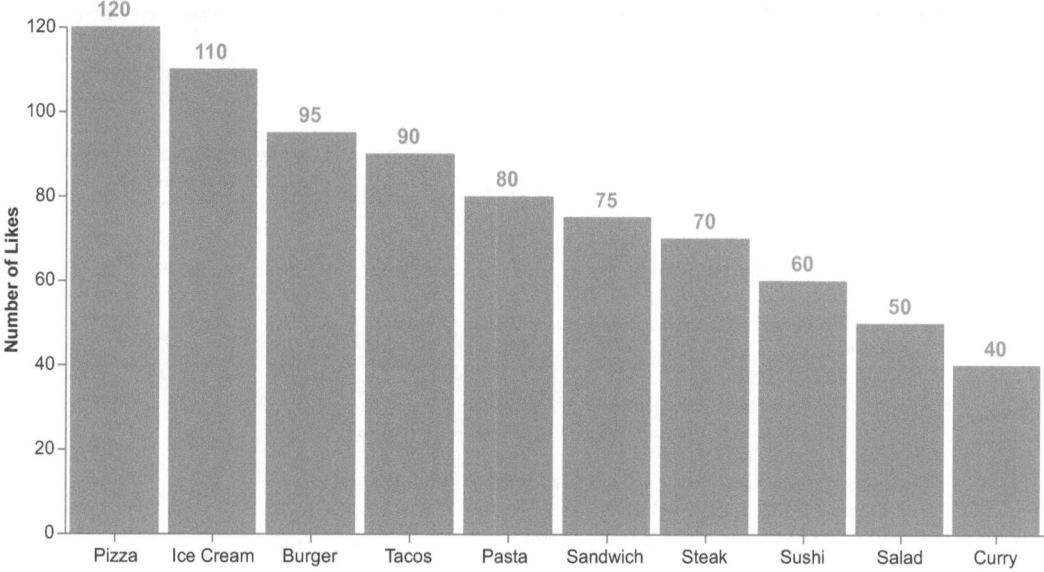

Figure C.5   A decluttered column chart

C.2.2   *Column chart with multiple series*

So far, we have implemented a column chart (as well as a bar chart) with just one series of data. However, in many cases, you want to compare two or more series of data. In this case, you must add the `column` channel to your chart. Consider the dataset shown in table C.1, describing the number of likes for each type of meal for 2022 and 2023.

Table C.1   A sample dataset with two series of data, one for 2022 and the other for 2023

Meal Type	Number of Likes in 2022	Number of Likes in 2023
Pizza	120	145
Burger	95	88
Pasta	80	97
Sushi	60	67
Salad	50	52
Steak	70	66
Tacos	90	78
Ice Cream	110	134
Curry	40	43
Sandwich	75	59

The following listing describes the code to build a multiple-series column chart.

> **Listing C.8   How to create a multiple-series column chart**

```
import pandas as pd
import altair as alt                                    Load data data/
                                                         meals.csv as a pandas
df = pd.read_csv('data/meals-by-year.csv')       ◁——    DataFrame.

df = df.melt(id_vars=['Meal Type'],var_name='Year',value_name='Number of Likes')

chart = alt.Chart(df).mark_bar(
).encode(
    x=alt.X('Year',
            # Rotate the labels by 0 degrees
            axis=alt.Axis(title=None, labels=False)
    ),
    y=alt.Y('Number of Likes',axis=alt.Axis(grid=False)),
    column=alt.Column('Meal Type',
                      header=alt.Header(
                          labelOrient='bottom',
                          title=None
                      )),
    color=alt.Color('Year',scale=alt.Scale(range=['lightgray','#81c01e']))
).properties(
    width=50,
    height=300
).configure_view(
    strokeWidth=0
)

chart.save('multiple-column-chart.html')
```

NOTE   Use the column channel to add multiple series to your chart. By default, the column chart adds the column labels to the top of the chart. To move them to the bottom, set the header attribute in the alt.Column() channel.

Figure C.6 shows the resulting chart. Now that you have learned how to implement a column chart with multiple series, let's move on to the following chart: the pyramid chart.

C.2.3   *Pyramid chart*

A *pyramid chart* consists of two back-to-back bar charts, with the two sides representing contrasting categories, such as males and females. The horizontal axis represents quantities, and the vertical axis shows categories to compare—typically, periods. Consider the dataset shown in table C.2, describing the number of orders of pizza and spaghetti from January to December 2023.

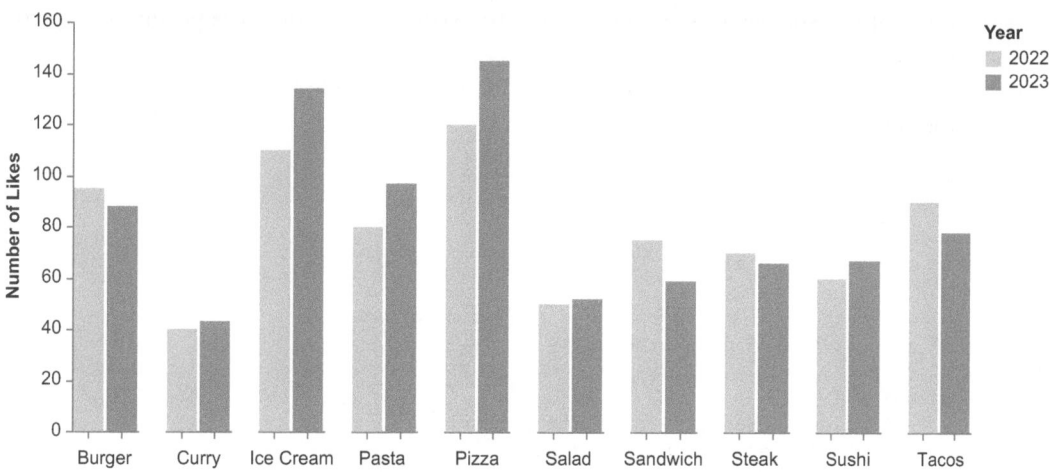

Figure C.6   A column chart with multiple series

Table C.2   A sample dataset with the number of orders of pizza and spaghetti from January to December 2023

Month	Pizza	Spaghetti
January	200	18
February	220	19
March	240	10
April	230	20
May	250	20
June	280	40
July	300	60
August	320	80
September	310	70
October	290	50
November	270	30
December	240	20

A pyramid chart is composed of three main elements: the left part, the middle part, and the right part. The left and right parts are bar charts, with the left part flipped. The middle part, instead, contains the labels of both of the bar charts. We will not write prompts for Copilot to generate code. Instead, we will use the suggestions it proposes and will show directly the produced code, without providing the screenshots of the Copilot suggestions. I suggest you try writing the code to experiment with the Copilot power.

The following listing describes the code to build the left part of the pyramid chart, focusing on pizza.

Listing C.9   How to create the left part of the pyramid chart

```python
import pandas as pd
import altair as alt

# Load data data/orders.csv as a Pandas dataframe.
df = pd.read_csv('data/orders.csv')

months = [
    "January", "February", "March", "April",
    "May", "June", "July", "August",
    "September", "October", "November", "December"
]

left_base = alt.Chart(df).encode(
    y=alt.Y('Month:N', axis=None, sort=months),
    x=alt.X('Pizza:Q', title='',sort=alt.SortOrder('descending'), axis=None),
)

left =  left_base.mark_bar(
    size=20,
    color='#c01e95'
).properties(
    title='Pizza',
    width=300,
    height=300
)

left_text = left_base.encode(
    text=alt.Text('Pizza:N'),
).mark_text(
    color='#c01e95',
    baseline='middle',
    align='right',
    dx=-10,
)
```

NOTE   Build a base chart (`left_chart`), and then use it to build the bar chart and the labels. Set the `sort` attribute in the x channel to `descending` to anchor bars to the right. Also, set the `align` attribute to `right` in the `mark_text()` method.

Now, let's build the middle part of the pyramid chart. This part contains the labels. The following listing shows the code.

Listing C.10   How to create the middle part of the pyramid chart

```
middle = alt.Chart(df
).encode(
    y=alt.Y('Month:N', axis=None, sort=months),
    text=alt.Text('Month:N'),
).mark_text(
    size=20,
).properties(
    width=100,
    height=300,
    title='Number of orders in 2023'
)
```

NOTE   Use `mark_text()` to set the labels. Also, set the `y` and `text` channels.

Finally, let's draw the right part of the pyramid chart. Use the same strategy as the left chart, without sorting the x channel. Also, combine all the charts to build the final chart. The following listing shows the code.

Listing C.11   How to create the right part of the pyramid chart

```
right_base =  alt.Chart(df
).encode(
    y=alt.Y('Month:N', axis=None,sort=months),
    x=alt.X('Spaghetti:Q', title='',axis=None),
)

right = right_base.mark_bar(
    size=20,
    color='#81c01e'
).properties(
    title='Spaghetti',
    width=300,
    height=300
)

right_text = right_base.encode(
    text=alt.Text('Spaghetti:Q')
).mark_text(
    baseline='middle',
    align='left',
    dx=10,
    color='#81c01e'
)

chart = left + left_text | middle | right + right_text

chart = chart.configure_view(
    strokeWidth=0
)
# save chart as 'pyramid-chart.html'
chart.save('pyramid-chart.html')
```

> **NOTE**   Use the + operator to combine each single part and the | operator to combine the parts together.

Figure C.7 shows the resulting chart. Now that you have learned how to build a pyramid chart, let's move on to the following chart, the stacked column chart.

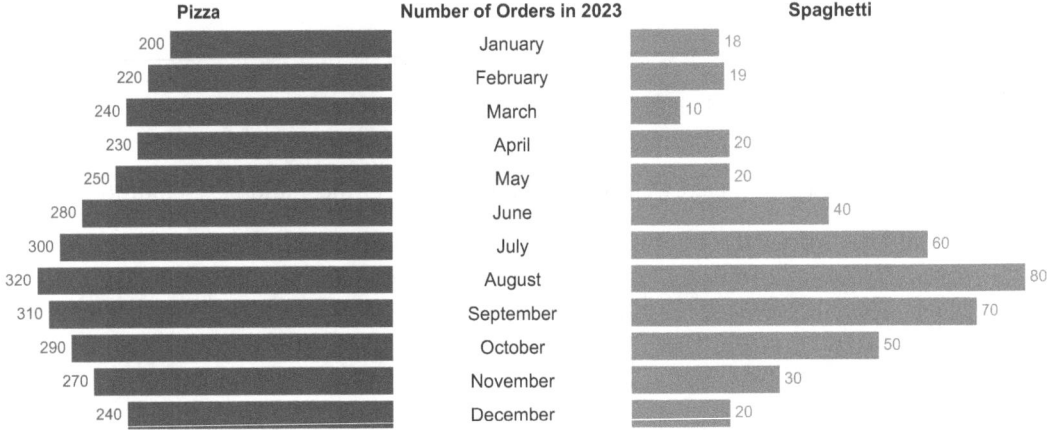

Figure C.7   A pyramid chart

C.2.4   *Stacked column chart*

Stacked column charts are similar to traditional column charts, but they show the contribution of each data series to the total value. Use stacked column charts to show how different factors contribute to a total over time or to compare the relative importance of different data series.

Let's use Copilot to draw a stacked column chart. Surely, you've noticed that as you write code, Copilot suggests more and more code that is close to your programming style. Therefore, we can try a new strategy when using Copilot: writing general instructions. This strategy assumes we've already written a lot of code using Copilot. If we had used generic instructions when we had started using Copilot, we would surely have gotten disappointing results. The following listing shows the generic instructions for Copilot.

Listing C.12   The generic instructions for Copilot

```
# Consider the dataset in the file data/orders.csv.
# The dataset contains information about orders placed by customers in a
    restaurant.
# Each row in the dataset represents the number of orders by month.
# The dataset contains the following columns:
# - `Month`: The month of the year
# - `Pizza`: The number of pizza orders
```

```
# - `Spaghetti`: The number of spaghetti orders
# Build a complete stacked column chart in Altair using the dataset.
```

NOTE   First, specify the dataset structure, and then use Copilot to build a stacked column chart using the described dataset.

The following listing shows the code produced by Copilot. It is quite similar to the codes generated in the previous examples.

Listing C.13   How to create a stacked column chart

```
import pandas as pd
import altair as alt

# Load data data/orders.csv as a pandas DataFrame.
df = pd.read_csv('data/orders.csv')

df = df.melt(id_vars=['Month'],var_name='Meal Type',value_name='Number of
    Orders')
                                                            Use this comment to tell
# Build a list of months                                    Copilot to build the list.
months = ['January','February','March','April','May','June','July','August',
    'September','October','November','December']

chart = alt.Chart(df).mark_bar(
).encode(
    x=alt.X('Month',
            axis=alt.Axis(title=None,
                          labelAngle=0,
            ),
            sort=months
    ),
    y=alt.Y('Number of Orders'),
    color=alt.Color('Meal Type',scale=alt.Scale(range=['#81c01e','gray']))
).properties(
    width=600,
    height=300
).configure_view(
    strokeWidth=0
).configure_axis(
    grid=False
)
chart.save('stacked-column-chart.html')
```

NOTE   Before drawing the chart, transform the DataFrame using the `melt()` method.

Figure C.8 shows the resulting chart. Now that you have learned how to build a stacked column chart in Altair and Copilot, let's see how to transform a stacked column chart into a 100% stacked column chart.

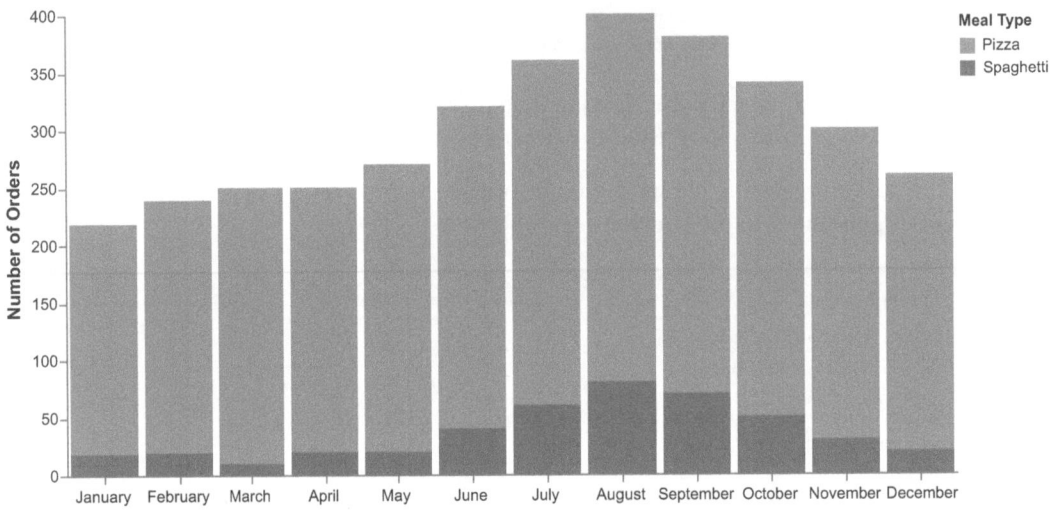

Figure C.8   A stacked column chart

C.2.5   *100% stacked column chart*

A *100% stacked column chart* is a stacked column chart with each column stacked to a height of 100% to show the proportional composition of each category. It is used to compare the contribution of different categories within each column across multiple data series. To transform a stacked column chart into a 100% stacked column chart, set `normalize=True` in the `y` channel of listing 6.4: `y=alt.Y('Number of Orders',` `stack='normalize')`. Figure C.9 shows the resulting chart.

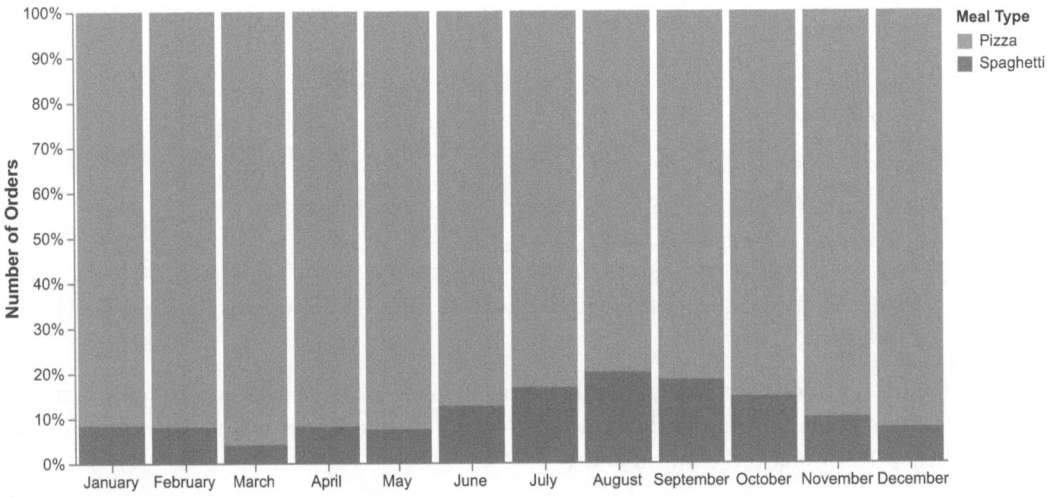

Figure C.9   A 100% stacked column chart

Now we've built a 100% stacked column chart. Next, let's cover the last chart of the bar chart family: histograms.

C.2.6 Histograms

A *histogram* represents the distribution of numerical data, with continuous data divided into intervals called bins and the height of each bar representing the frequency of data points falling within that bin. Use histograms to visualize the spread of a dataset and identify patterns or outliers.

Consider the sample dataset described in table C.3, showing the average rating for different products (ratings range from 1 to 10). The table shows the values for only some products; you can read the full dataset in the GitHub repository of the book, under 06/bar-charts/data/product-ratings.csv.

Table C.3   A sample dataset the with average rating of some products

ProductID	Rating
Product_1	4.8
Product_2	5.7
Product_3	5.3
Product_4	4.8
Product_5	5.9
Product_6	4.0
Product_7	4.7
Product_8	5.9
Product_9	4.0

Let's use Copilot to draw a histogram. The idea is to specify generic prompts to build the chart framework and then to refine the chart manually. The following listing shows the prompts used.

Listing C.14   The generic prompts to build a histogram

```
# Import the required libraries
# Load the 'data/product-ratings.csv' into a dataframe
# Create a histogram of the Rating column using Altair
# Save the chart as a HTML file
```

NOTE   Use these prompts to speed up the chart framework creation.

As an alternative, you could have specified very detailed prompts, as we did for the other charts. Here, we want to illustrate the different strategies to generate prompts for Copilot. In our case, the resulting code is not totally correct. We need to improve it, to produce the code shown in the following listing.

Listing C.15   The code to build a histogram

```
chart = alt.Chart(df).mark_bar(
    color='#81c01e'
).encode(
    x=alt.X('Rating:Q',
            bin=alt.Bin(maxbins=10, extent=[1, 10]),
            title='Rating',
            axis=alt.Axis(
                format='d',
            )
    ),
    y=alt.Y('count()', title='Number of Products')
)
```

> **Add the bin attribute to the original code generated by Copilot.**

> **NOTE**   Use the `bin` attribute to specify the number of bins (`maxbins`) and their extent.

For comparison, we can add a one-dimensional kernel density estimation over our data to the chart by using the `transform_density()` method provided by Altair.

Listing C.16   One-dimensional kernel density estimation

```
line = alt.Chart(df).transform_density(
    'Rating',
    as_=['rating', 'density'],
).mark_line(
    color='red',
).encode(
    x='rating:Q',
    y=alt.Y('density:Q', axis=None)
)

# Combine the bar chart and the density estimator.
chart = chart + line

chart = chart.resolve_scale(y='independent'
).configure_view(
    stroke=None
).configure_axis(
    grid=False
)

chart.save('histogram.html')
```

> **NOTE**   Use the `transform_density()` method to create a kernel density estimation. This method takes the `column` of the DataFrame to use for calculation and the names of the generated columns (`as_` attribute). Also, combine the produced line with the previous chart.

Figure C.10 shows the resulting chart.

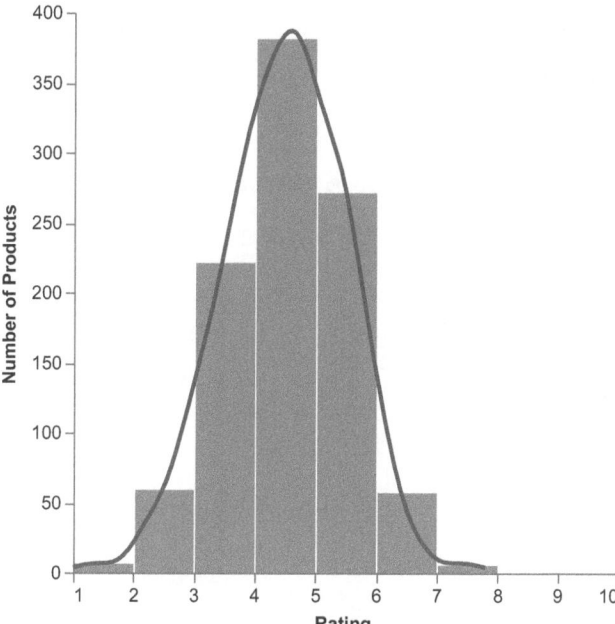

Figure C.10   A histogram

Now, we have completed the bar charts family. Next, let's move on and analyze the line charts family.

C.3   Line charts family

The line charts family includes area charts, slope charts, and dumbbell charts. The following discusses each in turn.

C.3.1   Area chart

An *area chart* shows the cumulative trend over time of a variable. It is similar to a line chart, but the area between the x-axis and the line is filled with a color or pattern to distinguish it from a line chart. The following listing shows how to draw an area chart.

Listing C.17   An area chart

```
import pandas as pd
import altair as alt

# Load data data/orders.csv as a pandas DataFrame.
df = pd.read_csv('data/orders.csv')

df = df.melt(id_vars=['Month'],var_name='Meal Type',value_name='Number of
    Orders')
```

```
# Build a list of months.
months = ['January','February','March','April','May','June','July','August',
        'September','October','November','December']

base = alt.Chart(df).encode(
    x=alt.X('Month:N',
            axis=alt.Axis(title=None,
                            labelAngle=0),
            sort=months
    ),
    y=alt.Y('Number of Orders',axis=alt.Axis(offset=-25)),
    color=alt.Color('Meal
     Type',scale=alt.Scale(range=['#81c01e','gray']),legend=None)
).properties(
    width=600,
    height=300
)

chart = base.mark_area(line=True)

text = base.mark_text(
    fontSize=14,
    baseline='middle',
    align='left',
    dx=10
).encode(
    text=alt.Text('Meal Type:N'),
).transform_filter(
    alt.datum['Month'] == 'December'
)

# Combine the line chart and text mark into a single chart
chart = chart + text

chart = chart.configure_view(
    strokeWidth=0
).configure_axis(
    grid=False
)
chart.save('area-chart.html')
```

NOTE   Compared to a line chart, change `base.mark_line()` to `base.mark_area(line=True)`. Translate the y-axis to align to the x-axis by using the following code: `y=alt.Y('Number of Orders',axis=alt.Axis(offset=-25))`.

Figure C.11 shows the resulting chart. After seeing the area chart, let's investigate the slope chart.

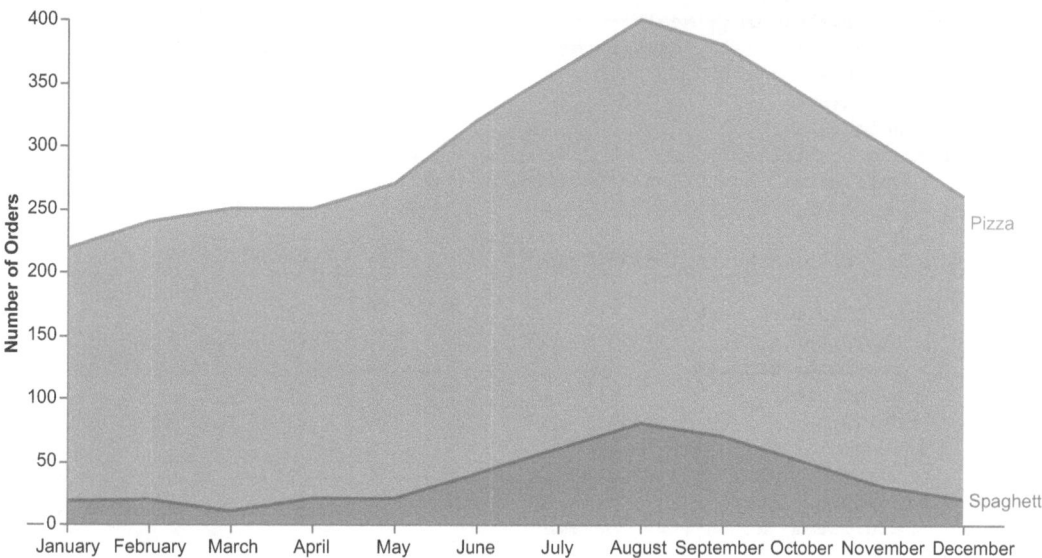

Figure C.11   An area chart

C.3.2   *Slope chart*

Slope charts consist of two sets of data points connected by a line, with the x-axis representing the different categories or periods and the y-axis representing the values of the data. In other words, a slope chart is a line chart with just two points. The following listing shows how to build a slope chart.

Listing C.18   A slope chart

```
import pandas as pd
import altair as alt

# Load data data/orders.csv as a pandas DataFrame.
df = pd.read_csv('data/orders.csv')

df = df.melt(id_vars=['Month'],var_name='Meal Type',value_name='Number of
    Orders')

# Build a list of months.
months = ['January','February','March','April','May','June','July','August',
    'September','October','November','December']

base = alt.Chart(df).encode(
    x=alt.X('Month',
            axis=alt.Axis(title=None,labelAngle=0),
            sort=months
    ),
    y=alt.Y('Number of Orders'),
```

```
    color=alt.Color('Meal
      Type',scale=alt.Scale(range=['#81c01e','gray']),legend=None)
).properties(
    width=600,
    height=300
).transform_filter(
    (alt.datum['Month'] == 'December') | (alt.datum['Month'] == 'January')
)

chart = base.mark_line(point=True)

text = base.mark_text(
    fontSize=14,
    baseline='middle',
    align='left',
    dx=10
).encode(
    text=alt.Text('Meal Type:N'),
).transform_filter(
    alt.datum['Month'] == 'December'
)

# Combine the line chart and text mark into a single chart.
chart = chart + text

chart = chart.configure_view(
    strokeWidth=0
).configure_axis(
    grid=False
)
chart.save('slope-chart.html')
```

> **NOTE**   Add the `transform_filter()` method to select only the first and the last values of your series.

Figure C.12 shows the resulting slope chart. The next chart is the dumbbell chart. Let's move on to analyze it.

C.3.3   *Dumbbell chart*

Dumbbell charts, or *floating bar charts*, consist of two data points connected by a line, one at the beginning and one at the end. Dumbbell charts are similar to slope charts, but usually, we use dumbbell charts to compare the difference between two data points. In contrast, we use slope charts to compare changes in data over time or between different groups.

Figure C.13 shows an example of a dumbbell chart. You can find the associated code in the book's GitHub repository.

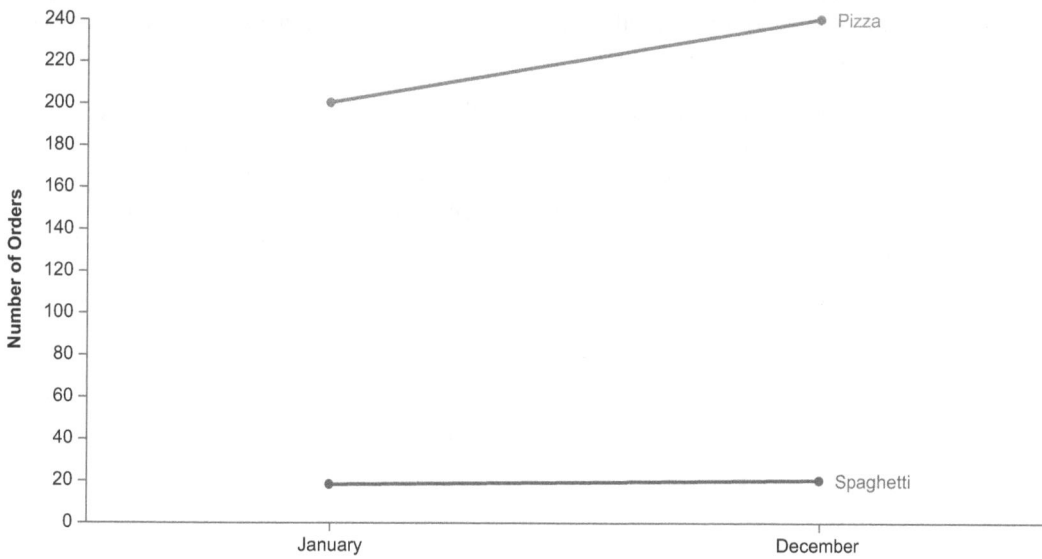

Figure C.12   A slope chart

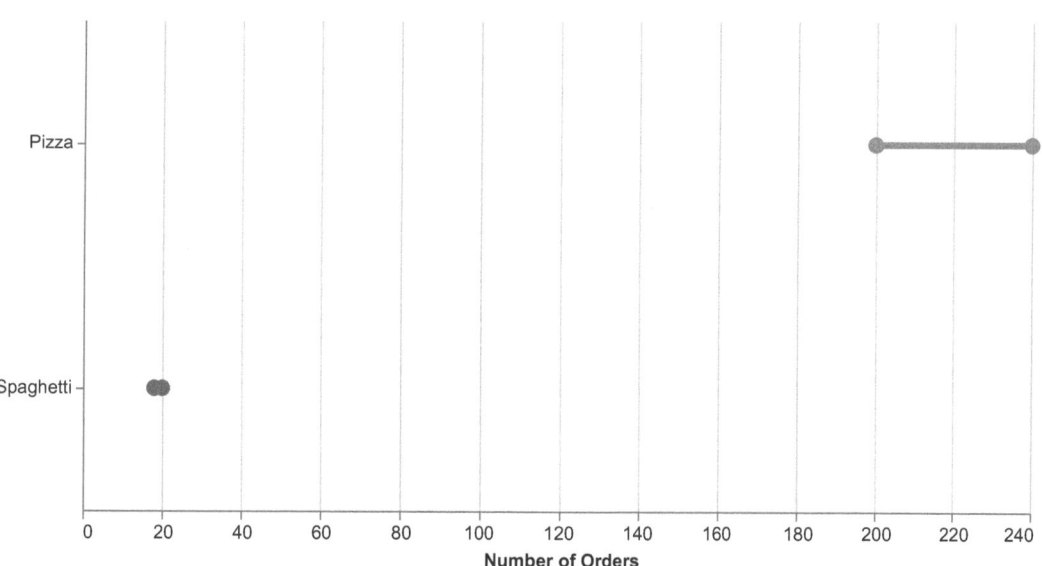

Figure C.13   A dumbbell chart

We have now completed the review of line charts. For some of them, we used Copilot, and for others we didn't directly specify prompts; however, in all cases, Copilot assisted us while writing the code by suggesting new code while writing it.

> **NOTE**   Line charts were first used by William Playfair, a Scottish engineer, in the late eighteenth century. In addition to inventing the line chart, Playfair also introduced other popular data visualizations, such as bar charts and pie charts. His innovative use of graphical representations revolutionized the way we present data today, making him a pioneer in the field of data visualization.

index